ARCTIC

EUROPE

NORTH
AMERICA

MEDITERRANEAN
SEA

ATLANTIC
OCEAN

AFRICA

CARIBBEAN
SEA

RED SEA

SOUTH
AMERICA

ANTARCTICA

Second edition published in Germany 2007
by
IKAN - Unterwasserarchiv
Waldschulstrasse 166
65933 Frankfurt, Germany
e-mail: ikanuw@aol.com

Distribution: ConchBooks   Mainzer Str. 25, 55546 Hackenheim, Germany.
e-mail: conchbooks@conchbooks.de
home-page: http://www.conchbooks.de
ISBN 3-925919-33-3

Conception & Layout: Helmut Debelius
Type-setting & Translation: Ralf Michael Hennemann
Printed in Czech Republic

Ralf M. Hennemann

SHARKS & RAYS
ELASMOBRANCH
GUIDE OF THE WORLD

Pacific Ocean • Indian Ocean • Red Sea
Atlantic Ocean • Caribbean • Arctic Ocean

Over 600 photographs
of elasmobranchs
taken in their natural habitat

TABLE OF CONTENTS

SHARKS

RAYS

CHIMAERAS

PICTURE STORIES

3

Acknowledgements

First of all my thanks go out to Helmut Debelius who not only made the proposal to create this book but who also supplied all the photographs. A major part of the superb photo material comes from the vaults of his renowned IKAN-uw-archive. But whenever we couldn't match a name on our species search list with the photographs available at IKAN, he was the one who unhesitatingly contacted those people who might have them, i.e. underwater photographers and scientists from all over the world. When all the slides and data files were 'in the house' and while I was struggling with the first (English) version of the text, he selected the best and most interesting images and arranged them into what finally became the lay-out of this book.

Next, my sincere thanks go out to all the photographers themselves. No one is able to do all the work necessary for such a pictorial guide alone, not even in a lifetime. Nowadays it is easy to go out and see some sharks as a diving tourist. It is a totally different story, however, to take such magnificent underwater photographs of some hardly known creatures like they have been contributed to this book by enthusiasts from all around the globe. Individual credits for each photo are given in the caption or at a lower corner of an image. Nevertheless, I once again want to say special thanks to the major contributors Clay Bryce, Mark Conlin, Helmut Debelius, Pedro N. Duarte, Howard Hall, Ralf Kiefner, Rudie Kuiter, Kazu Masubuchi, Doug Perrine, Douglas D. Seifert, Mark Strickland and Peter Verhoog and also to the research team of Prof. Hans Fricke (Max Planck Institute of Ethology, Seewiesen, Germany) for some outstanding shots made from aboard the research submersible *Jago*. The abbreviation IV following some of the photographers' names in the captions below the pictures stands for Doug Perrine's renowned photo agency INNERSPACE VISIONS.

Thanks also to the authors of the various picture stories. They are uw-photographers, scientists or just people with some sort of special relationship towards cartilaginous fishes. Each of them is mentioned in the introductory paragraph of the story. It is these stories which make this guide so different from ordinary identification books by presenting first-hand information on elasmobranch behaviour or the people dealing with them.

I would also like to thank all those countless people, laymen and scientists alike, who have collected so much information and done research on cartilaginous fishes over the past decades and from whose books and papers most of the information presented in this book is taken. Their works are listed in the bibliography section at the end of this book.

Very special thanks go out to the elasmobranch specialist Dr. Peter Last for writing the foreword and adding some latest data as well as to the ray specialist Dr. Matthias Stehmann who was my first teacher in 'matters cartilaginous.' I am also grateful to Bernd Peyer for translating a few of the picture stories from the German original into English.

I want to dedicate this book to a few people who in one way or another made it possible for me to study my favourite vertebrates, the cartilaginous fishes: My parents Eleonora and Eberhard who always supported my weird interests, my grandma Theresia who taught me to respect nature and gave me first word of the megamouth shark shortly after it was discovered, and last but not least, 'my gal' Gerlinde who not only tolerates me.

Ralf M. Hennemann

FOREWORD

More than 1,000 sharks and rays are known to inhabit the rivers, coastal seas and open oceans of the world. They contain some of the largest and most feared, but poorly known, animals on the planet. Primitive members of this group extend back in time to the days of the dinosaurs. Since then, elasmobranchs have colonised all major aquatic habitats - from well inland in remote freshwater lakes and rivers to the depths of the oceans and abyssal floor - from warm tropical seas to the icy polar regions. However, despite their success in colonising aquatic environments, they remain particularly vulnerable to over-exploitation by humans.

Sharks are major predators of the sea but the long-held perception that all are dangerous man-eaters is grossly unjustified. Negative publicity created by this aura has partly led to their demise. Fishing, in particular the practise of shark finning, has seriously and wastefully depleted shark and ray populations in many parts of the world. Many species are targeted directly, while others are discarded dead in bycatch of numerous fisheries. So seriously damaging have these activities become, that some species have been almost eliminated in parts of their range.

Our perceptions of elasmobranchs needs to change. Guides such as this one are invaluable for teaching us about the diversity of the group and serve to rectify myths and misconceptions about their threat to humans - few species are genuinely dangerous. Most are a spectacle to behold in their natural environment. Divers, who have a unique opportunity to observe these fishes going about their daily lives, would agree that encounters with live sharks and rays is an unforgettable experience.

Underwater photography is challenging at the best of times but enters a new dimension when large, often fast moving animals such as sharks are involved. When taking portraits of large animals underwater, photographers need to be as close as possible to their subject. This poses obvious problems when attempting to capture images of hungry white sharks, basking sharks immersed in soupy plankton laden seas, or bull sharks in silty estuaries. Even dark, plain-coloured stingrays can appear featureless without thoughtfully planned imagery.

IKAN founder, Helmut Debelius, has delivered a selection of underwater guides which must surely rank amongst the best of their type ever produced. The high quality of the underwater images and their reproduction is beguiling to those with a knowledge of marine biota, and provides an unique insight for those less initiated. In this volume he teams with Ralf M. Hennemann to produce a worldwide treatment of these fishes. Ralf's text, which is well written and very informative, complements the amazing imagery - a rich diversity of ecological types, representing some of the sea's giants through to animals rarely seen, are beautifully depicted in their natural environment. Together they have produced a landmark reference that will be useful to a whole variety of interest groups including divers, fishers, scientists, as well as those who simply like to admire these magnificent animals.

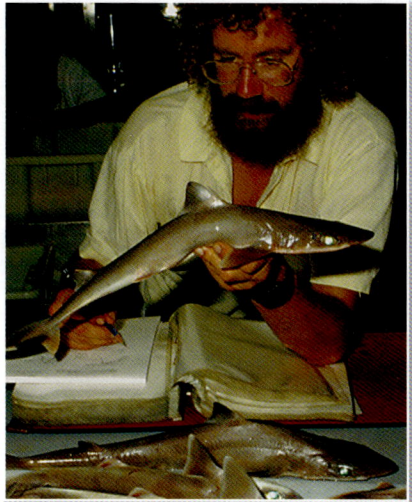

Dr. Peter R. Last
CSIRO Marine Laboratories
Hobart, Australia, February 2001

INTRODUCTION AND EXPLANATION OF TERMS

Of the approximately 1,000 known extant species of cartilaginous fishes (sharks, rays and chimaeras) actually none can be considered well-known. Additionally, many species are simply known to exist because one or a few have been collected by scientists who happened to be there when they were caught. Paradoxically, there seems to be little interest in investigating cartilaginous fishes while there is still so much to learn about them.

Besides lack of funding for such research, this paradox may also have ancient roots in people's minds. The first reaction of many people after mentioning sharks is one of horror, although they are standing on dry land, far away from any dark, deep body of water and have never seen a living shark except on TV or perhaps in an aquarium. You may ask psychologists about the 'Angst' phenomenon and anthropologists about our ancient relatives in the East African steppes during the dawn of mankind. They will tell you about hunters and the hunted, eating or being eaten and you may then be able to visualise the fear of fishermen being confronted with the maw of a large shark. But we have come a long way since and reality looks a bit different now.

Compared to other 'natural hazards' like lightning stroke or bee stings - you name it - sharks and rays, let alone chimaeras, never posed a serious threat to people. And vice versa, for most of the past few million years since the advent of man, they were not endangered by man at all. A few were probably caught and eaten, the skin of some ended up as sandpaper or fine leather goods, some made a career stuffed as curios in collections or a sailor's home and some were even worshipped as ancestral gods by the natives of small islands in the Pacific Ocean.

Today, the situation has dramatically changed: Not only since the beginning of the 'new millenium' are people on this planet counted by the billions, and all of

PETER KRAGH

them are hungry and want to eat. Techniques for catching highly mobile food in the sea have improved from stone age fishing spears to satellite-tracking of fish schools and GPS guidance for fleets of fishing vessels, each powered by thousands of kilowatts and literally scooping out the living resources of the oceans season after season.

We all have witnessed what has happened to the large whales: Their populations have been (and still are) decimated, probably beyond a reasonable minimum for the survival of the species. As mammals (and compared to cartilaginous fishes) they are our closer relatives and have at last found a lobby who wants to protect them. In the

View onto the deck of a fishing vessel at the Pacific coast of Costa Rica.

case of the archetypical monsters, however, no one really cares if sharks or rays are killed by targeted or accidental fishing by the hundreds of millions each year (the former often only for their fins which are sold to the profitable shark fin soup industry, the latter mainly in huge drift nets set for tuna and the like)! Considering their importance in the ecosystem, it is high time to stop this waste of life.

The only way to eliminate the fear of the unknown and thus the indifference towards the death of those perfectly adapted, age-old creatures is to educate people about them and their ways of life. A hard task, when keeping in mind just how little is known about that. Being a popular, albeit scientifically correct pictorial identification guide with lots of additional information, this book wants to make a small contribution towards a better understanding of one of the most interesting groups of marine animals which do deserve all our attention just as much as any whale or furry baby seal.

What are cartilaginous fishes?

The three groups of living cartilaginous fishes are the sharks, rays and chimaeras. Together they comprise what zoologists call the class of Chondrichthyes. The oldest known fossil members of this group are more than 400 million years old. Other than in the so-called bony fishes or teleosts (e.g. herring, tuna, salmon, eel) cartilaginous fishes have an internal skeleton made of cartilage, a flexible, light-weight material which may more or less be hardened by calcareous deposits but contains no true bone. A cartilaginous skeleton may help reduce the overall weight. Contrary to the typical bony fish, cartilaginous fishes lack a swimbladder and sink to the bottom if they do not swim. Other major differences between those two main groups of fishes are:

1) Cartilaginous fishes lack the bony scales of teleosts. Instead, their skin is covered with small dermal denticles which are indeed built like miniature teeth with root, crown and enamel.

2) Sharks and rays (collectively called elasmobranchs because of their arch-shaped gills) have - depending on species - five to seven pairs of gills and gill slits without a bony cover typical for teleosts; chimaeras, however, are an exception as their gills have a cover and only one pair of gill openings like bony fishes.

3) Mouth and nostrils of almost all cartilaginous fishes are located on the underside of the head (exceptions are the frilled shark, the megamouth shark and the manta ray with terminal mouths).

4) The males of cartilaginous fishes have a pair of so-called claspers which are located at the inner margin of the pelvic fins. One or (rarely) both of these intromittant organs are inserted into the genital opening (cloaca) of the female during copulation. This technique makes internal fertilisation of the generally few but large and yolk-rich eggs possible.

5) The jaw teeth of sharks and rays are embedded in the gums and continuously replaced (revolver dentition), but not attached to the jaws like those of bony fishes. Chimaeras' teeth are fused into continuously growing large, paired plates.

6) For osmosis control the blood of cartilaginous fishes contains urea, which breaks down into ammonia after the death of the animal. This results in an infernal smell of the decaying flesh and has prevented many people from eating these 'inedible' fishes (the trick is to prepare them fresh or deep-freeze them immediately after the catch).

The senses of cartilaginous fishes

All cartilaginous fishes are predators and many species also scavenge on dead animals. While most sharks and rays are opportunistic feeders on any kind of animal prey as long as it fits into the mouth and can be partitioned by the teeth, some species show preferences for certain types of food, at least during part of their life. Adult great white sharks, for instance, mostly feed on seals while young ones primarily prey on fish.

A few decades ago, sharks were still considered to be 'primitive' animals, unable to learn and instinctively reacting to external stimuli - just like robots controlled by a fixed computer program. Today we know that this is not true. Cartilaginous fishes have a comparatively large brain and an impressive set of sensory organs which are used to find a way through the oceans, locate prey and eventually join up with mating partners. Consequently, they also have a complex behaviour. Most of their sensory abilities - especially those of sharks - have been investigated only after the US Navy lost quite a number of servicemen during World War II because of shark attacks on the crews of sinking ships and ditched aircraft. Therefore, scientists tried desperately to find a reliable shark deterrent. Meanwhile a lot of knowledge about the behaviour of sharks and rays has been collected not only by scientists studying elasmobranchs in aquaria and the field, but also by myriads of divers and underwater photographers. Especially the latter category of elasmobranch enthusiasts contributed much on the topic of behaviour to this book in the numerous stories interspersed among the systematic pages depicting and describing the species.

The following list gives a short overview of the senses of cartilaginous fishes:

1) Smell (via the nostrils) plays an important role when detecting the scent of a distant food source, e.g. a wounded fish loosing blood.

2) Hearing is much better developed than once was thought and also is long-ranging as the sound travels almost four times faster under water than in air (reef sharks have been attracted by banging onto SCUBA divers' air tanks).

3) Vision is very acute in many sharks, some (e.g. lemon sharks) can activate an additional reflecting layer in their eyes to further enhance vision in the dark.

4) Touch is a close-range sense and becomes obvious when touching a whale shark or manta, they immediately react to a swimmer's hand. It also seems to play a role when checking a potential prey (many sharks nudge at their food before biting it).

5) Taste comes into play when prey items are taken into the mouth. If the results are not satisfactory the 'food' is spat out again (white sharks may 'test' a surfer or his board but almost never devour either. There is some evidence that they can taste certain fatty substances which are present in seals and whales. Humans are simply too skinny for them!

6) Besides these familiar senses cartilaginous fishes - like bony fishes - possess a lateral line system (a row of fine pores along each body side and branching on the head) which allows detection of vibrations over a distance of several metres, e.g. the aberrant movements of speared fish.

7) Via the so-called ampullae of Lorenzini (visible as large pores scattered over the head) they are also able to detect weak electric fields, for example those generated by muscular activity. Even if buried in sand, the heart and gills of a flounder are constantly active - an easy target for any bottom-living shark!

8) Additionally, cartilaginous fishes have a magnetic sense which may aid in spatial orientation and finding the right way during long migrations. But as in the case of bird migration research, the mysteries of this sense have not yet been lifted.

In summary, cartilaginous fishes are much better equipped sensorically than the average vertebrate, including humans. Their battery of senses is certainly one of the reasons for their successful survival through hundreds of millions of years.

Where do cartilaginous fishes live?

Today we know approximately 1,000 species of cartilaginous fishes compared to more than 20,000 fish species with a bony skeleton. While the latter are found in almost every body of water if it is not overly polluted, most cartilaginous fishes live in the saltwater of the oceans. Some species enter brackish water and rivers (e.g. bull shark, Ganges shark) and only a few spend their entire life in freshwater (freshwater stingrays, freshwater sawfish). The majority of species is found along the highly productive coasts and around islands in warm temperate and tropical seas. Most of the photographs

Silky shark surrounded by a school of juvenile jacks forming a natural 'bait ball', Cocos Island.

shown in this book come from habitats in these shallow waters. These habitats are familiar to most divers and include coral reefs, rocky bottoms, seagrass meadows, kelp beds, sand flats and the like, all of which are situated on the continental (or insular) shelf from the intertidal zone on the beach down to a depth of about 200 metres. Included are also some species that live in cold-water habitats less often frequented by divers, the most extreme example being the Greenland shark which was only recently photographed in its natural environment for the first time.

There are also many species - especially those of the dogfish shark family and a few exotic rays - which are true inhabitants of the deep sea and live on the continental slope from the rim of the shelf in 200 m down to a depth of about 2,000 m and deeper. This fish guide is unrivalled in presenting a great variety of excellenty photographed denizens of these deep waters. Some of the photographs have been made by skilful SCUBA divers using special breathing gas mixtures, others by remotely controlled underwater cameras on research submersibles. However, some of the species depicted in this book have never been seen other than after being caught by a research vessel's bottom or midwater trawl net. Hence, they are shown here in the lateral view (left or right side) or from the top (in the case of the rays) as is typical for scientific photographs of dead fishes that have been placed on a neutral background by digital photo processing.

How to use this book

This book is primarily a pictorial guide. When trying to identify a shark, ray or chimaera, the photographs - mostly from the natural habitat - are the main source of information. Photographer and photo location are given below each picture. The photos are supplemented by the accompanying text which gives further details on the depicted species. The latinised scientific name (genus and species) comes first and is followed by the popular name. Occasionally, a second (local or interesting) popular name is given. **Length** (L) gives the total length of the animal measured from the tip of the snout to the tip of the tail. **Width** (W) gives the width across the pectoral fins and body disc of those ray species in which this

9

measurement is more useful than total length because the long and thin tail may often be broken off. Max. stands for maximum length (or width) reliably recorded for the species. Additionally lengths at the onset of sexual maturity (or length ranges for mature specimens recorded) are given for either sex (if known). **Distribution** (Di) gives a short (and often generalised) list of geographical localities where to find the species. **Depth** (De) gives the known depth range of the species. **General** (G) gives additional information such as anatomical details for better recognising the species, varibility of coloration, habitat preferences, details on reproduction, prey preferences, behaviour including interactions with man, suitability for aquaria and the like.

The reader will find the species of the three major groups of cartilaginous fishes - sharks, rays and chimaeras - grouped into families according to generally accepted taxonomy. Popular and Latin family names are given in the coloured bar on top of each page. The same colour also appears on the contents pages at the beginninig of the book for easy reference. Families are grouped into orders which are appearing on the contents pages only. The orders reflect the relationships between the families of cartilaginous fishes. In order not to confuse the reader, most higher systematic categories as well as subfamilies and subspecies have deliberately been ignored. Classification and systematics of living organisms often depend on the view of the specialist(s) working on the group in question and should be of no importance to the user of this identification book.

When trying to identify a species, get familiar with the depicted ones first, compare their looks and the information on distribution and habitat with the specimen(s) observed by yourself and make your decision. Of course you will find not every shark or ray species at once and you might be lucky enough to see one not included in this guide. But at least you will come very close in most cases.

Special note

The scope of this book does not permit for further detailing the current endangerment of cartilaginous fishes by man other than mentioning it in the text at the beginning of the introduction, some of the picture stories and in selected species accounts where appropriate. It is the author's request to pass on that there are many enthusiastic people who work for the survival of sharks, rays and chimaeras as species on our planet. In order to learn more about these fantastic animals and hopefully also contribute to their protection from the greed of our own kind, some valuable sources of information are listed below. In our age of real-time telecommunication, the following homepages and website addresses of conservationist (and other) groups provide the most recent data on the status of elasmobranchs (cited from Elasmoskop 2(1), June 1997, and 3(1), December 1998, a publication of the Deutsche Elasmobranchier Gesellschaft e.V., Hamburg, Germany):

American Elasmobranch Society (AES): http://www.albion.edu/fac/biol/aes
Associaçao Portuguesa para o Estudo e Conservaçao de Elasmobranqueos (APECE): http://alfa.ist.utl.pt/~apece/htm-en/en-apece.htm
Deutsche Elasmobranchier Gesellschaft (DEG) e.V.: http://www.elasmo.de
European Elasmobranch Association (EEA): http://www.dholt.demon.co.uk/elasmo.htm
Shark Foundation: http://www.shark.ch (English); http://www.hai.ch (German)
Shark Info: http://www.shark-info.ch
Shark Trust: http://www.dialspace.dial.pipex.com/town/close/zg47
The Pelagic Shark Research Foundation: http://www.pelagic.org
Australian Shark Conservation Foundation (ASCF): e-mail address: 106447.2516@compuserve.com
Grupo Italiano Ricercatori sugli Squali (GRIS): e-mail address: wwfmiram@utsax3.univ.trieste.it

Obertshausen, January 2001   *Ralf M. Hennemann*

GLOSSARY

abdomen - belly

adult - sexually mature, able to reproduce

alar thorns - large dermal denticles on the outer pectoral fins (body disc) of most adult male skates

antitropical - distributed north and south of the tropics but not in the tropics

benthic - bottom-dwelling, living on or close to the substrate

cloaca - the common opening for gut, urinary and genital tracts of fishes

continental shelf - the rim of a continent that is covered by the sea, from the coast down to a depth of about 200 m where the seafloor starts to slope down to the deepsea plains in about 4,000 m depth

continental slope - the part of the seafloor that slopes down from the border of the continental shelf in a depth of about 200 m down to the deepsea plains in about 4,000 m depth

cosmopolitan - widely distributed in all oceans

dignathic - referring to both jaws

endemic - occurring in a restricted area of distribution

gestation period - time during which the embryos develop until birth

heterodont dentition - jaw teeth have different shapes

homodont dentition - jaw teeth have same shape

interdorsal ridge - a ridge on the skin between the first and second dorsal fin of certain sharks; its presence or absence is an external anatomical feature of requiem shark species

juvenile - young individual, not yet sexually mature

litter size - number of young per reproductive cycle, number usually increases with size of mother

mode of reproduction - the way in which young are supplied with nutrients and born

monognathic - referring to one jaw

nictitating membrane - a special 'lower eyelid' of certain sharks (e.g. catsharks, smoothhound, requiem and hammerhead sharks) that can be closed to protect the eye while feeding

ovary - the organ where the eggs ripen in adult females

oviduct - a tube-like organ through which eggs from the ovary are guided into the uterus

oviparous - 'egg-laying', a mode of reproduction in which the female deposits horny capsules on the seafloor each containing (usually) one fertilised egg; after several months - time depending on species - the fully developed young hatch to commence an independent life (e.g. catsharks, skates)

ovoviviparous - a mode of reproduction in which the fully developed young hatch already inside the mother's uterus from their thin egg capsule after having resorbed their yolk sac reserve; the young are born soon afterwards and are not supplied with additional nutrients from a placenta-like structure, hence this mode of reproduction is also called aplacental viviparous (e.g. dogfish sharks, whale shark, tiger shark, many rays)

pelagic - living in the open water as opposed to benthic

placental viviparous - a mode of reproduction in which the young are supplied with additional nutrients from a placenta-like structure inside their mother's uterus after having resorbed their yolk sac reserve; litter size is generally low but the fully developed young hatch at a relatively large size (e.g. hammerhead sharks, most requiem sharks)

plankton - 'the drifting', all of the (mostly minute) organisms which passively drift with the currents of the sea, e.g. unicellular algae, many diverse larvae of marine animals (including fishes), jellyfish, krill (euphausid shrimps)

planktonic - being part of the plankton

sympatrical - sharing the same range of distribution (or part thereof)

term size - size at birth

taxonomy - the science of correctly naming newly discovered species and placing them in the system of living organisms

term size - size at birth

BODY PARTS OF CARTILAGINOUS FISHES

1. TYPICAL (GENERALISED) SHARK

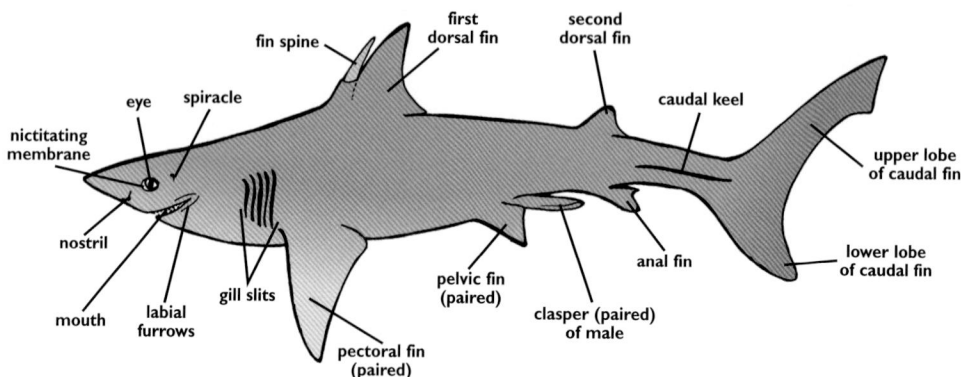

fin spine

first dorsal fin

second dorsal fin

caudal keel

eye

spiracle

nictitating membrane

upper lobe of caudal fin

nostril

lower lobe of caudal fin

mouth

labial furrows

gill slits

pelvic fin (paired)

anal fin

clasper (paired) of male

pectoral fin (paired)

2. TYPICAL (GENERALISED) RAY

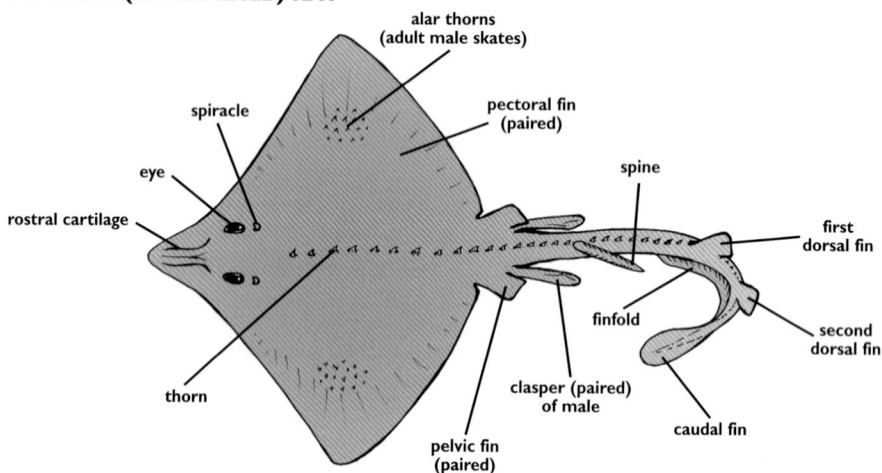

alar thorns (adult male skates)

spiracle

pectoral fin (paired)

eye

spine

rostral cartilage

first dorsal fin

finfold

second dorsal fin

thorn

clasper (paired) of male

caudal fin

pelvic fin (paired)

3. TYPICAL (GENERALISED) CHIMAERA

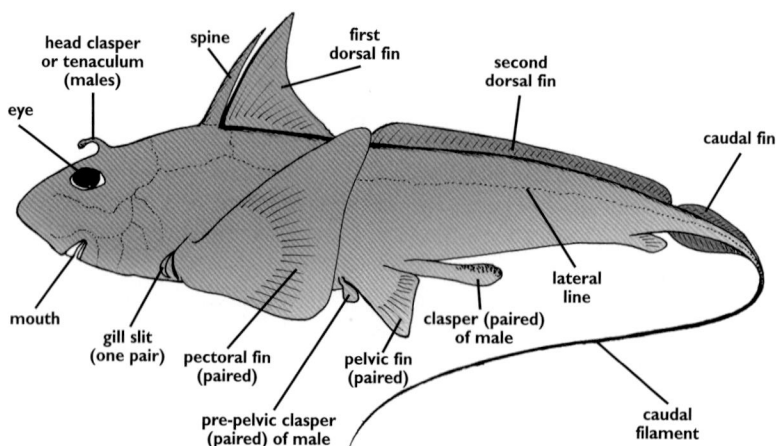

head clasper or tenaculum (males)

spine

first dorsal fin

second dorsal fin

eye

caudal fin

mouth

lateral line

gill slit (one pair)

pectoral fin (paired)

clasper (paired) of male

pelvic fin (paired)

pre-pelvic clasper (paired) of male

caudal filament

Hexanchus griseus   **British Columbia, Canada**

This family comprises five species, some of which may be observed inshore. The distinctive morphological characters of all family members include a single dorsal fin, six or seven pairs of large gill slits, broad-based, cockscomb- or sawblade-like teeth in the lower jaw and shorter, more hook-like teeth in the upper jaw. All sharks of this family are generally deepwater dwellers, except for the broadnose sevengill shark, which is most common in shallow, coastal habitats. In certain parts of its range, the bluntnose sixgill shark is regularly encountered by divers and is known to frequent shallow waters at night. All family members are ovoviviparous, the unborn young rely on a huge amount of yolk from an individual yolksac to reach term (parturition) size inside the mother's uterus. Large litter sizes of more than 100 young (in a large mother shark) let the hexanchids range among the most fecund of all sharks. Their large size and predilection for large prey makes them potentially dangerous to divers. The bluntnose sixgill shark is rarely housed in public aquaria, whereas the broadnose sevengill shark is frequently kept in temperate water aquaria and usually adapts readily to captivity. Both species are known to consume elasmobranchs (including other cow sharks) housed with them.

Saul Gonor

Hexanchus griseus
Bluntnose sixgill shark
L: At birth 65-70, max. 480, mature females 450-482 cm. Di: Wide ranging in all tropical and temperate seas. De: 25-1,875 m at least. G: Six gill slits. Snout broad, blunt. Coloration pale grey to black or brown above, lighter below. Eyes reflect light bright green. Most common in the abyssal depths, but observed regularly by divers in the summer at Hornby Island in the northern Strait of Georgia, west of Vancouver and at the mouth of the Alberni Inlet, Barkley Sound, on the west coast of Vancouver Island, in 25-70 m. With litters of 22-108 one of the most fecund sharks. Preys on a wide variety of vertebrates including hagfish, sharks (spiny dogfish, prickly sharks), rays, chimaeras (in the NW-Atlantic *Hydrolagus affinis* and *H. pallidus),* bony fishes (anglerfish, hake, marlin, gurnards, herring, ling), whale blubber and seals, but also on shrimp and crabs. Cannibalistic tendencies have been observed in captive specimens. Often with large copepod parasites on fins and body. Although not known to attack humans, its large size and varied diet make it a probable threat. Divers encountering this shark report some aggressive behaviour in situations where the shark has been provoked. Shows possible threat display of circular swimming with mouth agape.

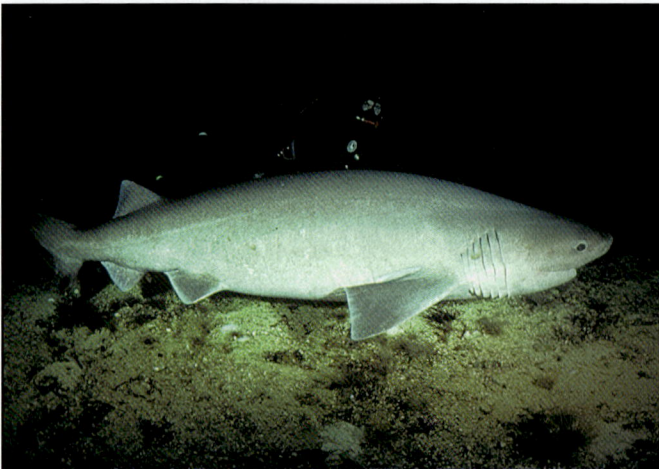

Neil McDaniel   both photos British Columbia, Canada

Notorhynchus cepedianus

Broadnose sevengill shark

L: At birth 50, max. 300 cm. Di: Most temperate seas. De: Intertidal to 50 m. G: Seven gills slits. Snout broad, blunt. Greyish brown with black spots above. Usually in shallow inshore waters over sand or near kelp. Litter size up to 82. Gestation period 12 months. Breeds in spring and summer in California. Nursery grounds in shallow bays (e.g. San Francisco Bay). Preys on octopuses, lampreys, sharks (conspecifics!) and rays, bony fishes (salmon, sturgeon, anchovies), gastropods, carrion and seals. Found near seal rookeries. Individuals have been observed hunting seals cooperatively. observed Females with mating scars, but also males, possibly a result of intrasexual aggression. Captive specimens continuously swim. Has bitten working divers in public aquaria and has behaved aggressively toward spearfishermen in the wild. Humans remains have been found in the stomach of one specimen.

Peter Verhoog   South Africa

Heptranchias perlo

Sharpnose sevengill shark

L: At birth 26, max. 137, mature females 98-93, mature males 85 cm. Di: In most tropical and temperate seas. Western Atlantic: North Carolina to Cuba, Brazil, Argentina. Eastern Atlantic: Mediterranean, Morocco to Angola. Indian Ocean: South Africa, S-Mozambique, Aldabra, SW-India. Western Pacific: S-Japan to China, Indonesia, Australia, New Zealand. Eastern Pacific: N-Chile. De: Intertidal to 1,000, usually 30-700 m. G: Seven gills slits. Snout narrow, pointed. Eyes large. Coloration greyish brown above, lighter below. No spots on body. Prominent black tips on dorsal fin and upper caudal fin lobe in juveniles, faded in adults. Deepwater species, strong swimmer. Active and aggressive when caught, readily bites, but harmless when not molested. Ovoviviparous. Litter size 9-20. Preys on bony fishes (hake) and squid. Bycatch of bottom trawl and longline fisheries, little utilised.

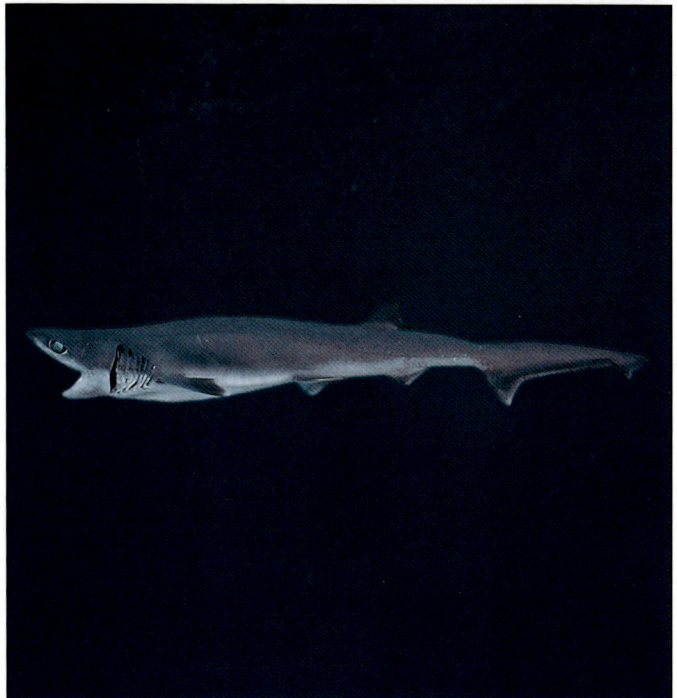

Pedro Nini Duarte / DOP   Azores, Atlantic

This family comprises a single genus and species. The frilled shark is a so-called 'living fossil' and one of the oldest shark types that have survived through the ages until today. It holds an isolated position within the system of elasmobranchs, it has no close, living relatives. On the other hand, its teeth closely resemble those of *Phoebodus,* one of the oldest sharks known. Specimens of this extinct genus were discovered in the Antarctic, where they thrived in Devonian times, about 380 million years ago! Also the teeth of *Thrinax baumgartneri* closely resemble those of the living frilled shark, but are of Middle Eocene age and thus 'only' about 35 million years old; they can be found in marine deposits of southern Germany and Austria.

Being a rare catch of the deep sea longline fishery, it is very hard to give reliable figures on the actual abundance of the peculiar frilled shark. Although described already in 1884, almost nothing is known about the biology and habits of this extraordinary species, mainly because its deep sea habitat is hardly accessible and the number of well documented and scientifically examined specimens is small. Its lifestyle is assumed to be mesopelagic and/or benthic. A (nocturnal?) encounter between frilled shark and diver thus would be a major sensation. The mode of reproduction is ovoviviparous, litter size about 4-6, size at birth about 60 cm. Because the eggs in the uteri of pregnant females are large and rich in yolk, the time for development of the young (gestation period) is assumed to last 1-2 years.

Chlamydoselachus anguineus   Frilled shark
L: At birth about 60, max. 200, mature females to 200, mature males to 150 cm. Di: Possibly worldwide in temperate seas. Eastern N-Atlantic: Northern Norway and Shetlands to Madeira, not known from the Mediterranean. Also South Africa, Japan, California, Australia. De: 120-1,100 m. G: An unmistakably long and slender, eel-like shark with almost terminal mouth, very wide gape and long, fringy appendages on the six pairs of gill slits, which make the entire gill region look like a frilled collar. The head is described as having a reptilian appearance (French popular name requin lézard = lizard shark). Coloration uniformly dark brown or grey, lighter below, with dark stripes on the flanks. The teeth are unique among living sharks; they have three main hook-like cusps (tricuspid) that serve in holding back slippery fish prey, look all alike (homodont dentition) and several successive rows are in function at a time. Compared to other shark dentitions, the tooth rows are widely spaced laterally.

all photos Rudie Kuiter   South Australia

Eight species of horn (or bullhead) sharks have been described and named. All have a pig-like snout, stout short spines in front of each of the two dorsal fins (spine never reaching tip of fin), large ridges over the eyes (the so-called supraorbital ridges) and a strongly heterodont dentition with pointed clutching teeth in the front and large molariform teeth in the back of the jaws. There is an ontogenetic change in tooth shape. In juveniles, the teeth are all of the clutching type, with 5-6 cusps in the case of the horn shark. But as the shark grows, the posterior teeth become more like molars, lose their cusps and are better for grinding up invertebrate exoskeletons, i.e. shells of molluscs and crustaceans. The large and muscular pectoral fins are used to 'walk' along the bottom. The spines in front of the fins are an effective device against predators, discouraging bony fishes and other sharks from eating smaller specimens (see family account on angel sharks). There is one report of a dead wobbegong found with a Port Jackson shark stuck in its jaws.

All family members are oviparous. The egg case is cone-shaped with an auger-like ridge spiralling around it. The cases of some species have tendrils at one end (the much flatter egg cases of catsharks usually have a tendril at two or all of their four corners). The horn shark egg case 'drills' by means of its auger-like flange into crevices where it stays locked motionlessly until it is eaten or the young shark hatches. Tendrils additionally serve in fastening the egg case to structures on the substrate.

Although generally harmless to divers, horn sharks have been known to bite if harassed. Most of the family members do well in large home and public aquaria. Several species mate in captivity.

Heterodontus francisci
California horn shark

L: At birth 15, max. 120 cm. Di: Eastern Pacific: Central California to Sea of Cortez, also possibly Ecuador and Peru. De: Intertidal to 50 m. Juveniles in shallow waters. G: Supraorbital crests moderately high. Spots on body small (less than one third length of eye), sometimes absent, no light bars between eyes.

Adults usually on rocky bottom or among kelp, juveniles on sand. Preys on sea urchins, worms, anemones, crabs and bony fishes. Utilise a small area during the day, often spending daylight hours at one spot, usually a cave or crevice. Adults move onto the reef at dusk and utilise a home range of about 1,000 square metres. Return to the same daytime resting sites at dawn. In deeper waters in winter. The spines in front of the dorsal fins are often worn in adults as a result of abrading them against rocks in caves. Usually solitary, sometimes in groups. Long lived (up to 25 years), mates in captivity.

Oviparous. Pairs of 10-cm-long egg cases are laid every 11-14 days, 1-2 weeks after copulation. After 7-9 months the eggs hatch. For copulation the male grasps the female's pectoral fin in his mouth and inserts one clasper into her cloaca for up to 40 minutes. Photo right shows two small males courting a large female.

Helmut Debelius

Mark Conlin   both photos California, Eastern Pacific

Helmut Debelius   Sea of Cortez, Mexico

Heterodontus mexicanus
Mexican horn shark
L: At birth 14, mature at above 55, max. 70 cm. Di: E-Pacific: Baja California to Peru, limits not well known. De: Intertidal to 50 m. Nursery area in Bahía Magdalena. G: Common in the Gulf of California.

Norbert Wu   Western Australia

Heterodontus zebra
Zebra horn shark
L: Max. 120, mature males 84 cm. Di: W-Pacific: Japan, Korea, China, Vietnam, Indonesia and Australia (recorded recently from the continental shelf of Western Australia in 150-200 m). De: Mostly shallower than 50 m, but also down to 200 m. G: Dark vertical bands on lighter background (zebra pattern), bands frequently extending onto fins. Supraorbital crests low, sloping away gradually behind the eyes. Dorsal fins very high, with rounded tips in juveniles, relatively lower in adults. Oviparous, but otherwise little is known of the biology and habits of this species. Probably preys on benthic invertebrates and small fishes.

Kazu Masubuchi   Izu Peninsula, Japan

Heterodontus japonicus
Japanese horn shark
L: At birth 18, max. 120 cm. Di: Japan to China and Taiwan. De: 6-37 m. G: Eye ridges low. Tan with darker bands (juv.) or saddles (ad.). Preys on crustaceans, molluscs, teleosts, sea urchins. Sometimes hydroids grow on teeth. Several females use same egg-laying site, up to 15 eggs in a single nest. Two eggs are laid at a time, 6-12 times during a season, hatch after 1 year. During courtship, male grasps one pectoral fin of female and wraps posterior part of body under her so he can insert a single clasper into her cloaca. Copulation lasts up to 15 minutes. Mates in aquaria.

Heterodontus galeatus
Crested horn shark
L: At birth 17-22, max. 150, mature females 70, mature males 60 cm. Di: Australia: Queensland, New South Wales. De: Intertidal to 93 m. G: Supraorbital crests very high. Brown with several dark bands; head band masks eyes. On reefs, among algae and seagrass. Shoves head between rocks in search of sea urchins, crustaceans, molluscs and small fishes. Jaw teeth often stained purple from eating sea urchins. Spiral-flanged egg cases 11 cm, tendrils up to 2 m long, deposited all year round, incubated 8 months. Does well in captivity.

David Fleetham / IV   New South Wales, Australia

Heterodontus portusjacksoni   Port Jackson shark
L: At birth 24, max. 170 cm. Di: Queensland to Western Australia. De: Intertidal to 172 m. G: Harness pattern distinct. Female lays 10-16 egg cases (15 cm) in 1-30 m in winter, moves to deeper water in spring. Mating as in *H. japonicus*. Juveniles in bays and estuaries (nurseries) for several years, then segregate by sex and move offshore. Makes long migrations (850 km) along coast. Groups rest in caves. Preys at night on sea urchins, sea stars, molluscs, crustaceans, benthic bony fishes and own egg cases. Uncovers buried prey by rapidly pumping water and sand through the gill slits. Reproduces in aquaria.

Helmut Debelius   Western Australia

Heterodontus quoyi
Galapagos horn shark
L: At birth 17, max. 59, mature males 48 cm. Di: E-Pacific: Coasts and offshore islands of Peru and Galapagos Islands. De: Intertidal to 50 m. G: Supraorbital crests low. Greyish brown with large black spots. Common, but little known. Preys on crabs. Below: Typical horn shark egg with hatchling emerging.

David Fleetham / IV   Galapagos, Eastern Pacific

19

Ed Robinson   Hawaii, Central Pacific

Mark Strickland   Andaman Sea

Helmut Debelius   Sea of Cortez, Mexico

Rhincodon typus
Whale shark

L: Up to 12 m; there are questionable, unconfirmed reports of sightings of specimens up to 18 m in length. Di: Circumtropical. De: 1 - 130 m. G: The largest living fish species is easily recognised by its immense size and a two-tone pattern of light spots and lines on a dark back. Unlike that of most other sharks, its enormous mouth is terminal in position and can be opened wide to filter large amounts of water for small fish, squid, crustaceans, and other plankton organisms (bottom).

This shark species is ovoviviparous and not oviparous as has been thought previously for a long time. In 1995 an 8-m-female was harpooned off Taiwan. The pregnant shark contained about 300 embryos (the largest litter size of all sharks!) in different stages of development. Some had already hatched from their egg shells: They were about 70 cm long, probably the size at birth.

Only very little is known about the habits of the whale shark. Subadults live in small groups, but these are seen only very rarely. Adults are usually solitary, but no details are known about the way these giants find each other, or how and where they mate.

Whale sharks are very popular with divers and in spite of their random occurrence encounters are guaranteed off the Seychelles, Thailand (Andaman Sea), Christmas Island, and tropical Western Australia (Ningaloo Reef) at certain times of the year.

The whale shark is often accompanied by other fishes, the largest is the cobia *Rachycentron canadum* (centre and facing page).

See also the following pages and WHALE SHARK HUNTERS on pp.42-45.

Rhincodon typus   **Andaman Sea**

LARGEST SMILE
ON THE PLANET

On a wall of the National Aquarium in Baltimore a quote is carved in stone that lyrically sums up the romantic fancies and imaginative yearnings of those who have an interest in the sea. But when the poet/naturalist Loren Eisely wrote, "if there is magic on this planet, it is contained in water," he was describing the interactions in a freshwater puddle that he experienced from a wholly terrestrial point of view. If Eisely had had the opportunity to enter the Indian Ocean off the coast of Western Australia during the months of March, April and May of each year, he would have realised the fullness of his observation and the magnitude of its understatement. Douglas David Seifert was there...

The elements that combine to produce this magical alchemy are a phase of the moon in the Australian autumnal season; the reproductive cycle of a colonial organism; the occurrence and short life of a counter-current; an aggregation of predators and predated that forms a tangible strata of the food chain pyramid; and the arrival, in large numbers, of the world's largest fish.

Regular occurrences of natural phenomena, even when we don't entirely understand them, can serve to put obscure places into the forefront of our consciousness. The annual migration of polar bears to Churchill, Canada; of grey whales to Bahía Magdalena, Baja California; and of the monarch butterfly to the Sierra Madre Mountain Range of Mexico are the most well-known examples. Joining this catalog of locations is the small country town of Exmouth, located on the Northwest Cape of Western Australia. It is a remote place, situated along a virtually uninhabited desert coastline, some 1,200 kilometres north of Perth, the most isolated metropolis on the planet. Exmouth's attraction is that it is the closest settlement to Ningaloo Reef, and from March through the end of May, it is, without even a close second, the most reliable spot on the planet to encounter and observe the largest fish of all the oceans of the world.

If you're a scuba diver of more than novice experience or an armchair naturalist with an inquiring mind, you may know from periodicals or books or documentaries that there are some animals living in the sea which possess such grandeur - such mere physical presence - that a firsthand encounter with them is an experience unparalleled underwater. Among the big animals, whales and dolphins have captured the public imagination by their apparent similarity to humans and all the qualities associated with the better part of human nature: Intelligence, playfulness, tolerance toward strangers, non-aggression, family-based social structure, and an expressive language. Sharks also appeal to the public imagination, but for the opposite reasons: A tacit nod to humanity's least desirable attributes. Also, on occasion they have been known to eat the odd human, a direct path to our self-interest.

But the largest shark - which has been accurately measured at 12.18 metres in length and has been less accurately reported at eighteen metres - is completely harmless to human beings. It is the whale shark (Rhincodon typus), known by a common name based entirely upon its size. But - other than a largely planktonic diet shared with baleen whales - there all similarity to a whale ends. It is a water-breathing, cold-blooded, cartilaginous fish, certainly the largest cold-blooded animal in the world. The denticled skin of an adult whale shark may attain a thickness of 10 cm, limiting its possible predators to great white

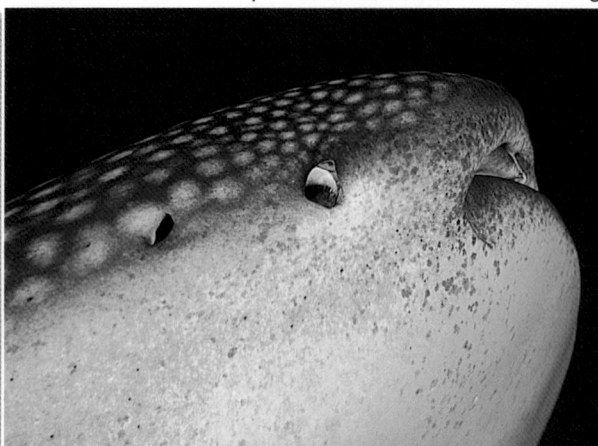

MARK STRICKLAND

sharks, tiger sharks, and orcas, and though none have been observed attacking a whale shark, some individuals show scars and bite marks from some apex predator. Humans are by far the whale shark's biggest enemy: Mortality is caused by shipping collisions and by the harpoon fisheries of India and Asian countries like the Philippines.

At first glance, the whale shark exhibits the classic, easily recognisable shark profile: Spindle-shaped body, long pectoral fins, tall dorsal fin, and great sweeping tail. But the whale shark departs from the standard shark form with its broad, blunt head; three prominent ridges running transversely along its body; and - most distinctly - an improbable

The whale shark's unique eye rolls inward and is covered by a lid for protection.

combination of coloration and markings that seem designed by an unabashedly eccentric abstract painter. The shark's upper surface has a base coloration ranging from gunmetal grey to black to dusky blue, mocha brown, or bronze. It is cross-hatched with thin, chalk-striped horizontal and vertical lines further broken up by a bold pattern of yellowish or creamy white spots, up to 5 cm in diameter, that can only be described as polka-dots. This coloration and marking pattern alone identify the whale shark at a glance, even more readily than its potential size, to such a degree that even a juvenile less than one metre long cannot possibly be confused with any other shark - or any other animal, for that matter. Underneath, its belly and lower surface coloration is a snowy white.

Scientists call the contrasting coloration of a dark upper surface and a light lower surface countershading, a trait indispensable for stealth and protection. Marine animals are believed to be unable to see in true colour, only in the contrasts between black and white. Viewed from below, the white of the whale

Observed head-on underwater, a feeding whale shark's gaping maw appears to be the inspiration for the Biblical story of Jonah. The mouth shape of the whale shark modifies itself, as the jaws widen and distend outward, to form an enormous funnel - sizable enough to hold a beer barrel or two.

shark's underbelly blends with the pale, reflective coloration of the surface; viewed from the surface looking downward, the dark base pigmentation blends with the blue of deep water, and the seemingly random swirl and spot patterns mimic the patterns of coral growth, jellyfish swarms, fish schools, or a planktonic haze. In front of the pectoral fins and gill slits, at the border where the polka-dotted, dark upper surface joins the snowy underbelly, are two round recessed orifices on either side of the head. The larger of the orifices houses an unblinking eye the size of a golf ball. The eye shows no expression other than the wide-eyed stare of a large black pupil that swivels to follow any activity going on around it. If an object approaches the eye too directly, the eyeball swivels backward into the cavity, and a tough flap of skin slides forward to cover the opening to the vulnerable sensory organ. Once the shark perceives that the threat has passed, the skin flap retracts and the eye returns to its wide-open stare. Many other large shark species possess a similar shielding eyelid (called a nictitating membrane) which slides forward to protect the eye, but the whale shark's sight protection adaptation is unique among all species and is as yet unclassified.

A slightly smaller opening spaced back from the eye and ahead of the gill slits marks the site of the spiracle chamber. A spiracle functions as a vestigial first gill slit, allowing water to be taken in and oxygen extracted, whereupon the oxygenated blood is pumped directly to the eye and the brain via a separate blood vessel. More commonly associated with sedentary, demersal sharks, sawfish, guitarfish, and stingrays, the spiracle in this non-bottom-dwelling shark was one of the criteria that led taxonomists to make the whale shark the sole representative of the family Rhincodontidae in the Orectolobiformes order - an order comprised of nurse, zebra, and carpet sharks. In addition to spiracles, whale sharks and other orectolobiforms also have five pairs of gill slits, two dorsal fins, a single anal fin, and a mouth terminating in front of the eyes.

In the order of Orectolobiformes, reproduction takes one of two forms. Nurse sharks reproduce ovoviviparously, which means the embryo forms within an egg retained in the mother's womb. The yolk supplying the embryo with nutrients is not connected to the mother after the egg case is formed. When the fetus reaches term and has exhausted the yolk sac, the juvenile breaks free of the egg case and is delivered out of the womb via the cloaca. The zebra shark reproduces oviparously, by laying an egg case containing fetus and yolk sac, and development occurs wholly outside the mother's body. Until a very recent

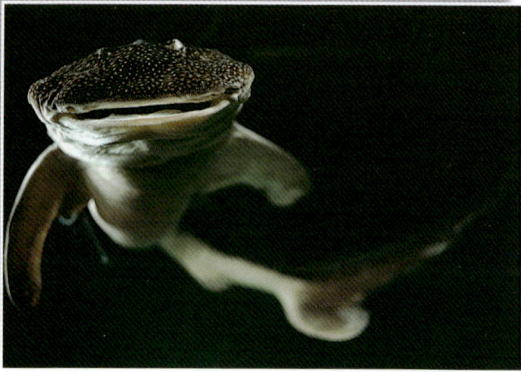
This 45-cm-long embryonic whale shark already resembles the adult.

NORBERT WU

discovery in 1995 off Taiwan that a whale shark does indeed give birth to live young - up to an astonishing 300 young from two uteri - there had been considerable debate over whether the whale shark was ovoviviparous or oviparous. This controversy stemmed from the discovery of a single, near-term whale shark egg case trawled up from 80 metres of water in the Gulf of Mexico in 1953. The egg case was the size of a football, but uncharacteristically thin-walled and without attachment tendrils (essential in attaching an egg case to the bottom) as on other oviparous shark egg cases. Since no other egg cases had been discovered, it was generally accepted that this particular egg case was a miscarried or aborted fetus.

A further connection to other close shark family body types is found in the whale shark's mouth. The mouth extends across the width of the flat, blunt snout and terminates well before the eyes. But a mouth running the width of a whale shark's head can reach tremendous proportions: The 12.18-metre specimen's mouth measured 1.36 metres across, in the slack repose of death; alive and in the process of feeding, it may have extended two or three times that width. Like bottom-dwelling sharks, the whale shark also has twin nasal slits and sensory barbels - a trait it shares with the nurse shark - in front of its blunt snout on the upper surface above the mouth. In sharks, the nasal openings do not lead to the mouth nor do they play any part in respiration. Their purpose is to augment the shark's sense of smell, enabling it to detect prey without relying on sight. The majority of other shark species have a ventrally located mouth and an elongated snout, which provides a greater surface area for the receptors of the ampullae of Lorenzini - the sensory organs which detect weak electric fields and provide sharks with sensory information when their vision is hampered. Placed as they are, far forward and extending over the mouth, the whale shark barbels augment the ampullae of Lorenzini and signal the location of prey, guiding the whale shark's mouth toward its reward.

When the whale shark swims passively, in a non-feeding mode, its jaws hang slack, with the mouth 15 to 20 cm open, allowing water to pass through to the gills for respiration. From a head-on perspective, the oval shape of an approaching whale shark looks like a very large "Have a Nice Day" button from the 1970s; their faces appear to have, without a doubt, the largest smile on the planet. Yet, as a whale shark feeds, the smile disappears entirely. This behaviour is classified as passive filter-feeding, and the whale shark shares it with two other large sharks: The temperate-water-dwelling basking shark (Cetorhinus maximus) and the recently discovered, deep-water-dwelling megamouth shark (Megachasma pelagios). In passive filter feeding, a shark's forward motion works in conjunction with the exaggerated shape of the open mouth to create a continuous intake of large volumes of water. This flowing channel of water serves two functions: Nutrition and respiration. Tiny fish, crustaceans, invertebrates, and planktonic organisms are suspended in the water, and the water itself contains dissolved oxygen. The flow of water passes through to the cartilaginous gill rakers, situated at the rear of the mouth, which act as a fine sieve to trap and filter out food items before water is passed on over the gills - where respiration occurs - and out the gill slits. When the gill rakers have trapped enough food, the whale shark closes its mouth and swallows in a discernible gulping movement; then the mouth opens wide again and the process continues.

When larger prey such as anchovies or krill are in dense concentrations at the surface, the whale shark - unlike the basking and megamouth sharks - may also feed via a process known as active suction filter-feeding. In this behaviour, it isn't necessary for the whale shark to propel itself forward to take in its prey. Encountering an abundance of food, whale sharks may either hang vertically in the water, swing their heads from side to side, or make a rushing lunge at the aggregations of prey, all the while 'vacuuming' in large quantities of water and food, gulping and swallowing with abandon, creating a froth of trapped air bubbles that exit their gilt slits as they take advantage of the amassed protein. This combination of lunging and vacuuming is thought to allow the whale shark to capture the more agile fish and krill that might easily elude the slow-moving, trawl-like funnel of passive filter feeding.

Although the whale shark's mouth is wide and capacious, it shows no evidence of the terrible teeth of other sharks. In fact, the whale shark has 300 rows of minute teeth in each jaw, each tooth less than three millimetres long and rasp-shaped. These teeth are the reason for its original generic classification, Rhiniodon. Rhiniodon means rasp tooth; the name came originally from Dr. Andrew Smith of South Africa in 1828. The current term, Rhincodon, is the unfortunate result of a long-ago printer's error - a 'c' typeset instead of an 'i'. Some scientists believe the whale shark's teeth to be vestigial. since their function has not been observed, but it seems probable that they are used in processing larger food items, such as the small albacore, tuna, and mackerel that are occasionally captured but are too large to pass easily through the whale shark's narrow gullet. Whale sharks have also been observed to evert their stomachs when the contents are not fully digestible or have clogged the gullet. In gastric eversion, the shark vomits the stomach out from its mouth, essentially turning it inside out like a sock; once the passage has been cleared, the shark swallows the stomach again and continues about its business. Though both passive filter feeding and active suction filter feeding have been documented and were previously considered characteristic behaviour, recent studies now indicate that whale sharks feed primarily at night and at depth, because it is under the cover of darkness that the deep scattering layer of planktonic and nektonic prey move up the water column in the densest concentrations. For most of the year, at least during the day, the amount of food taken in during subsurface cruising is equivalent to snacking, while the main meal comes after dark in deeper water.

While whale sharks are widely distributed in a band around the equator ranging from thirty degrees north to thirty-five degrees south latitudes and are present in all the oceans of the world except the Southern and Arctic Oceans and the Mediterranean Sea, they are infrequently encountered. So infrequently that the preeminent whale shark scholar, the late Dr. Fay Wolfson, never had an opportunity to dive with one, and Dr. Eugenie Clark dived for over twenty years before she had her first encounter. Most reported sightings came from fishermen or divers who encountered them sporadically - usually solitary individuals. They have been seen around the Caribbean, the Gulf of Mexico, the Eastern Pacific Islands of the Northern Galapagos, Malpelo, Cocos, the Revilla Gigedo Islands, the Sea of Cortez, Hawaii, Papua New Guinea, Indonesia, the Andaman Sea and the Maldives, the Red Sea, the Seychelles, and Natal Province, South Africa. As scientists cannot easily study what they cannot find, scientific research in the past focused on accounts of brief chance encounters and the autopsies of dead specimens that were either entangled in fishing nets or the victims of shipping collisions. For scientists and keen scuba divers, the possibility of encountering a whale shark in the wild came along once in the proverbial blue moon.

While the largest fish on earth is present in almost all the oceans of the world, encountering a whale shark is a rare event.

Although science gives little credence to cliches, blue moons do exist - even if they're nothing but an optical illusion. More importantly every month has a full moon, and recently, under just such a full moon, it appeared that something extraordinary was happening in die Indian Ocean. Reports of whale shark sightings in the waters off the Northwest Cape in Western Australia began to increase in regularity in the early 1980s as the numbers of fishermen working the area - both commercial prawn trawlers and sport-fishing boats - multiplied both in time at sea and frequency of excursions. The Northwest Cape is ideal for fishing activity: Not long enough to be classified a peninsula, the land juts out from the Western Australian coastline to form a fat finger pointing into the Indian Ocean on its western face and forming a sheltered, mangrove-lined backwater known as Exmouth Gulf to the east. The waters of Exmouth Gulf are bountiful, supporting a large population of western king prawns and their attendant fleet and industry. The Northwest Cape is fronted by a fringing coral formation called Ningaloo Reef and it is unique. It is not only the only sizable coral reef, some 260 kilometres in length, occurring on the west coast of any continent, it is also the longest fringing reef in the world. The reef runs parallel to the coastline, at some points as close as 30 metres offshore to as far out as 5 kilometres, and a shallow lagoon lies between the breakers of the reef front and the sandy shore. As intriguing as the frequent whale shark sightings off Ningaloo

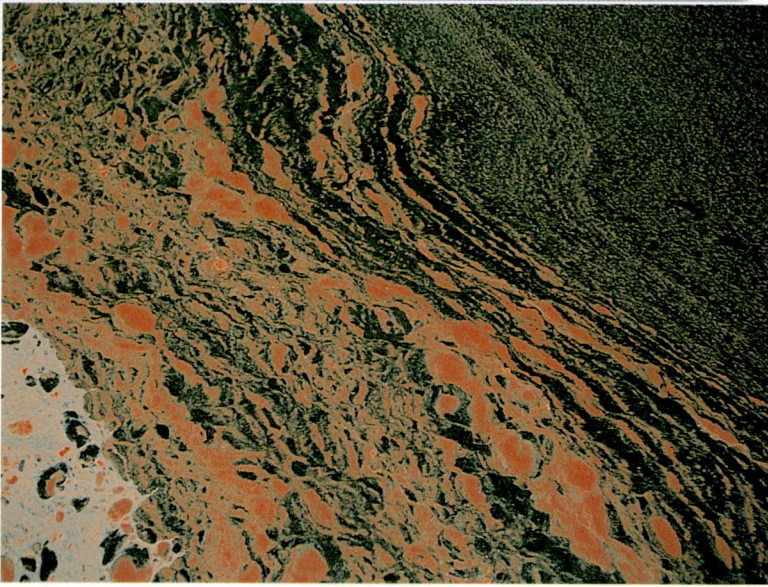

Red coral spawn slick covers the water surface at Ningaloo Reef late in March each year.

GEOFF TAYLOR

Reef were the accounts relayed by local fishermen of a puzzling, pink-coloured slick which would suddenly materialise upon the surface of the ocean in the latter part of March each year. The descriptions of the slicks attracted the attention of coral reef biologists.

Prior to 1982, no one knew that the various hard coral species might have a common biological clock. Yet, during a series of nocturnal observations on the Great Barrier Reef in November of that year, scientists discovered that for more than 130 species of scleractinian (stony) corals - all types disposed to releasing eggs and/or sperm into the water for external fertilisation - mass coral spawnings are an annual, predictable event in the life of the reef. The trigger for the mass coral spawn is unrelated to water temperature; the determining factors are the phase of the moon in the month of November in conjunction with the phase of the tide on a particular night. The corals spawn under cover of darkness during the third quarter of the moon on neap or ebb tides, though how the polyps know that the conditions are at their optimum is still a mystery. The polyps begin preparation months in advance: Each polyp transforms from asexual to hermaphroditic, with the creation and maturation of sex organs. The eggs are formed first, followed by the testes, and both are gathered in

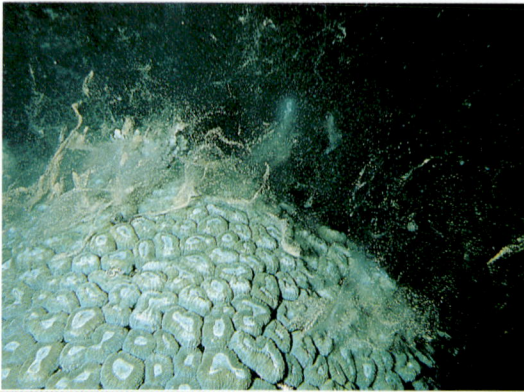

HELMUT DEBELIUS

a pink-coloured package and stored in readiness beneath the mouth of the polyp. When time and tide signal that the moment has arrived, the polyps release their egg and sperm packages. The packages are buoyant and float slowly upward from their sessile parents, creating a scene that looks like a vigorously shaken-up souvenir snowstorm diorama, albeit in pink, as millions of packages are cast to their fates. Upon reaching the surface, the packages break open, releasing the eggs to drift and the sperm to stream out

Left: Spawn and sperm rise from a *Favia* stony coral.
Below: Tropical krill (*Pseudeuphausia latifrons*).

and try to collide with an eligible egg. The surface of the water becomes stained with an oily pink slick, signalling on a grand scale the microscopic spectacle occurring below.

Although in a different season and a different ocean, the slicks seen off Ningaloo Reef seemed virtually identical to those in the Great Barrier Reef, and investigation in 1984 confirmed the fact. Australian scientists examined this phenomenon and discovered that at Ningaloo Reef up to 102 species of scleractinian corals do indeed spawn simultaneously three to four hours after dark seven to nine days after the full moon during the

GEOFF TAYLOR

The formation of 'baitballs' - made up of small pelagic fishes - by several whale sharks working in collaboration is an outstanding phenomenon in the world of sharks and reminds one of the bubble screen hunting technique of humpback whales.

neap ebb tides in the month of March. Subsequent research also discovered that every third year, there is a second coral spawn following the same pattern after the full moon in the month of April. Not only were the scleractinian corals reproducing, but so too were polychaete worms, sea stars, molluscs, sea urchins, and other invertebrates. This supersaturation of planktonic and nektonic larvae in a concentrated area served to attract a higher level of predators - krill and a variety of species of jellyfish from the deep waters of the continental shelf and large schools of small fish, such as anchovies and juvenile mackerel, from the open ocean. They, in turn, served to attract the whale sharks, manta and mobula rays, and other predators, such as the resident bronze whalers, silvertips, blacktips, and tiger sharks, as well as tuna and a variety of cetaceans, including even Bryde's whales. In the months of March through May, off Ningaloo Reef, the Indian Ocean dinner bell is ringing in abundance.

As lucky observers soon found out, not only do whale sharks passively filter-feed on the superabundance of coral and invertebrate larvae, but they have a special taste, previously unsuspected, for larger prey. Scientists now believe that while the whale sharks do consume vast clouds of plankton, it is the concentrations of larger prey feeding upon the plankton that serve to draw them in such numbers. Whale sharks become frenzied upon encountering krill (Pseudeuphausia latifrons), small, shrimp-like crustaceans, about 1 cm long. Their temperate-water cousins are the principal diet of baleen whales, and they are most responsible for the large size that filter-feeding megafauna attain. When they find clouds of krill, whale sharks engage in lunge feeding, and the surface of the water is churned with each rushing gulp. One of the reasons for the frenzying, which is consistent with other shark species' behaviour, is competition. When one whale shark detects an abundance of food, other sharks are soon on their way as well; whatever sensory detectors bring a whale shark to its prey seem to send out signals that draw others nearby. And with an estimated 200 to 400 visiting whale sharks in the area during the spawning season, it isn't uncommon to see several feeding together whenever large enough concentrations of krill are found.

Among the most spectacular phenomena observed off Ningaloo Reef is the formation of 'baitballs' (also called 'fishballs' or 'fishboils'). Smaller pelagic fish species - such as anchovies, sardines, and mackerel - travel together in large schools made up of thousands of individuals as they move inshore from the open ocean and migrate south along the coast. These small fish are drawn by the planktonic larvae and coral spawn, which they feed upon avidly. Their excitement as they feed translates into pulsed vibrations, sending out a signal to larger predators, such as tuna, trevally, and bronze whalers and silvertips. When schooling baitfish are detected, lone sharks appear, initially making wide passes at the school, until the sharks' numbers increase to a cooperative group. The amassed sharks then make rushing charges at the schooling fish. Schooling behaviour among fish is a defensive 'confusion-by-sheer-numbers' adaptation; with so many similar fish, a predator's attack system is overwhelmed, and it has difficulty focusing its attention on a specific individual. Yet as the pack of sharks continues to press, the school is consolidated, its ranks concentrated as spacing between individuals is decreased until the school becomes a tightly packed

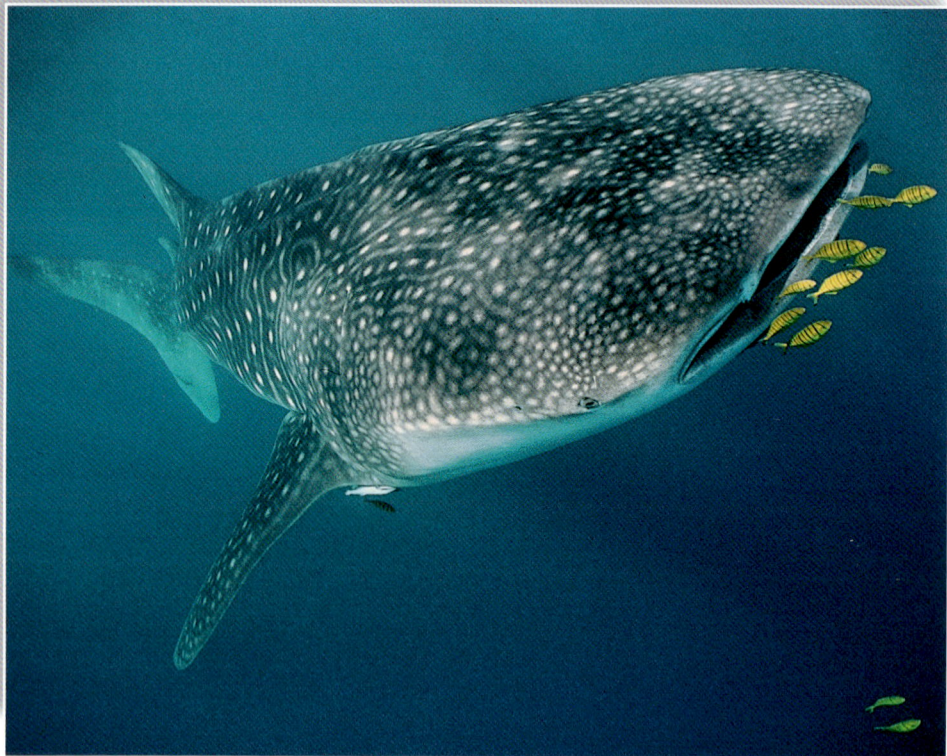

A whale shark in the company of a group of pilot jacks, only one of the fish species frequently seen near these gentle giants.

ball of fish. The sharks work in collaboration, patrolling the perimeter of the ball and rushing it intermittently to keep the school closely contained. Individual sharks alternate attacks by charging through the baitballs, snapping at the dense quantity of prey without seeking individual targets.

Besides taking advantage of the abundance of food, whale sharks may also engage in mating while in the area. Although the majority of whale sharks observed at the beginning of the season have been immature males, there is growing evidence that mature males and females appear toward the end of the season. And while no one thus far has observed whale sharks mating - or at least there isn't any photographic evidence - observers have seen claspers (the dual male sex organs) raw and scarred, presumably from recent use in the area. There have also been reports of whale sharks trailing newborns - like a gaggle of baby ducks - seen from the air and by boats. Only time and continued research will reveal the answers to these exciting possibilities.

Above left: The paired claspers (intromittant organs) of a male whale shark. Above right: The slit-shaped cloaca of the female.

Stegostoma fasciatum
Zebra shark

L: At birth 25, max. 350 cm. Di: Red Sea to South Africa and Indonesia, Australia, Samoa and north to Japan. De: Intertidal to 65 m. G: The family zebra sharks contains one monotypic genus (meaning it has only one species). The zebra shark inhabits reefs in tropical to warm temperate regions. Unmistakable shark with a tail fin almost as long as its body, ridges on its flanks and back and a characteristic colour pattern. Specimens longer than about 90 cm have dark spots on yellow to cream background; smaller inidividuals have yellow stripes and spots on a dark brown background (name). At least one albino specimen has been recorded. The long, flexible body and tail enable this species to enter reef crevices and caves to capture concealed prey (gastropods, clams, crabs, shrimps, bony fishes).

The zebra shark is a sluggish bottom dweller that is mainly active at night when it hunts over the reef and adjacent sandy areas. By day it rather rests on the bottom, often 'standing' on its pectoral fins with mouth open and facing towards the current, a pose which obviously facilitates breathing (small photo below). Usually seen solitary, rarely in aggregations, in lagoons, reef channels and reef faces. Often in the company of discfishes and cleaner wrasses. Reproduction oviparous, the purple to brown egg cases are about 20 cm long, lack tendrils, but are adhesive. Up to four eggs are laid at a time. Juveniles are only rarely seen, they probably inhabit depths greater than 50 m. Courtship has been observed in the wild (photo at right showing male on top of female while clutching at her right pectoral fin).

A hardy species, does well in public aquaria. Inoffensive, there are no reports of attacks on humans. At Phuket Island, Thailand, they are commonly hand fed and some individuals allow divers to handle them and rub their bellies, sometimes for extended periods of time. Also commonly encountered near Lady Elliot Island, Great Barrier Reef.

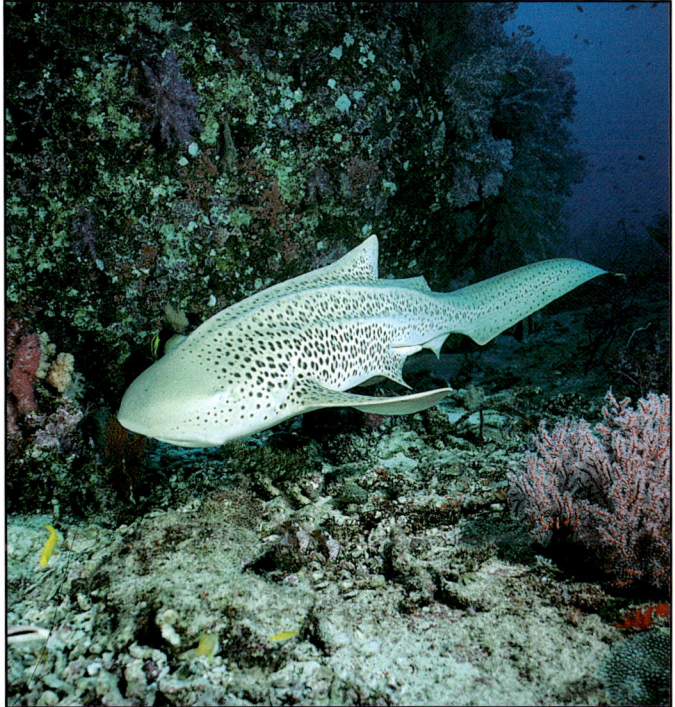

Helmut Debelius   **Maldives, Indian Ocean**

Peter Verhoog   **Egypt, Red Sea**

This family of small sharks comprises seven species placed in two genera, one of which (*Parascyllium* with four species) is restricted (endemic) to the more temperate waters of southern and western Australia. All family members have long, slender bodies with a relatively narrow head and an anal fin well in front of the lower tail lobe and partially in front of the end of the second dorsal fin. The colour patterns are important for identifying the different species. However, little is known about the biology of these crepuscular elasmobranchs. Some of them are known to be oviparous and deposit tendril-bearing egg cases. As far as is known, all family members prey on small bottom-dwelling invertebrates such as worms and crustaceans but occasionally also small bony fishes are taken. The narrow heads and slender bodies of collared carpet sharks enable them to hunt and capture prey that hides in cracks and crevices in the varied reef environment. They are nocturnal hunters and themselves hide in caves and among large kelp during the day, thus going unnoticed by most divers. These sharks are harmless to divers, do well in aquaria and readily breed in captivity.

Rudie Kuiter   Tasmania, Australia

Parascyllium ferrugineum
Rusty carpet shark

Length: At birth 15, max. 80 cm.
Distribution: Australia: Victoria and Tasmania.
Depth: 5-55 m.
General: Coloration brownish grey with a few dark saddles, a dark collar around neck and dark spots on fins and body. Has fewer spots than the otherwise similarly coloured endemic Tasmanian carpet shark *P. multimaculatum*. Found in shallow water around southern Tasmania, in deeper water off Victoria, among kelp or sea grass where hunting benthic crustaceans at night. Hides under ledges and in caves during the day.

Rudie Kuiter   Victoria, Australia

Parascyllium collare
Collared carpet shark
L: 86 cm. Di: Australia: Queensl. to Victoria. De: 20-160 m. G: Living demersal on the shelf, oviparous.
Below: **Necklace carpet shark,** *P. variolatum*, length at birth 16, max. 91 cm. Australia: NSW to Victoria and Tasmania; common around King Island. Intertidal to 165 m. Nocturnal among kelp and sea grass.

This family comprises two genera (*Chiloscyllium,* the bamboo sharks, and *Hemiscyllium,* the epaulette sharks) with at present about 15 valid species, at least 8 of which occur in reef habitats. They have small mouths placed in front of the eyes, no dermal flaps on head and a low rounded anal fin that looks different from the caudal fin. Their muscular paired fins enable them to 'walk' along the ground, their slender bodies allow them to slip between coral branches and into narrow reef cracks. Some of the species are oviparous. Juveniles are rarely observed in the wild. They apparently live deep in coral crevices and dense staghorn corals where they hide from predators, including other sharks. The colour patterns of most family members change during growth. Juvenile bamboo sharks usually have distinct bands that fade or disappear with age, juvenile epaulette sharks have bands that dissolve into spots during their growth. The colour patterns of epaulette sharks provide an effective camouflage while the eye-spots may serve to intimidate enemies by mimicking the head-on perspective of a piscivorous fish. Since many predators would approach an epaulette shark resting on the ground from above they would first see both ocelli when looking down! These sharks are usually harmless, but nevertheless may bite if harassed. They are very well adapted to live in confined places; as a consequence they do well in public and large home aquaria. At least three family members are known to regularly reproduce in captivity.

Hemiscyllium freycineti
Freycinet's epaulette shark

L: Max. 72, adult 61 cm. Di: Irian Jaya, Papua New Guinea. De: Shallow waters. G: Juveniles with alternating dark brown and tan bands that later break up into spots. Body of specimens from 30 cm in length with a honeycomb of rusty-brown spots on a cream background and an inconspicuous ocellus over each pectoral fin. Spots of about eye size in front of eyes. Inhabits shallow coral reefs. Hides in crevices during the day, sometimes with only the tail sticking out. Little is known about its biology. Forages at night on the reef and adjacent rubble, sea grass and sand areas. Prey/reproduction probably similar to that of its relatives.

Hemiscyllium hallstromi
Papuan epaulette shark

Length: Max. at least 75 cm. Distribution: Known only from Irian Jaya and1 Papua New Guinea in the western South Pacific. Depth: Shallow inshore waters. General: A bottom-dwelling shark that is found in the shallow waters of coral reefs. Colour pattern of scattered small and large dark spots. A conspicuous large black, white-ringed ocellus over each pectoral fin. Each ocellus surrounded by additional smaller black spots. Nothing is known about the biology of this small shark. Prey and reproduction probably similar to that of its relatives.

both photos Bob Halstead   Papua New Guinea

Bob Halstead   Great Barrier Reef, Australia

Hemiscyllium ocellatum
Ocellated epaulette shark

L: At birth 15, max. 107 cm. Di: Australia (Queensl. to West. Australia), New Guinea. De: 1-10 m. G: Large black, white-ringed ocellus behind each pectoral and numerous smaller spots. Two colour morphs: golden with more, darker, smaller spots and distinct ocelli; tan with fewer, larger spots, ocelli less distinct. In coral reefs, tidepools. Eats worms, shrimp, crabs (dug out or sucked from crevices). Solitary, nocturnal. Hides with head stuck under a ledge. Mating lasts 2 minutes, male grasps one pectoral fin of female with mouth and inserts one clasper. At night, two tendril-less eggs are deposited at a time.

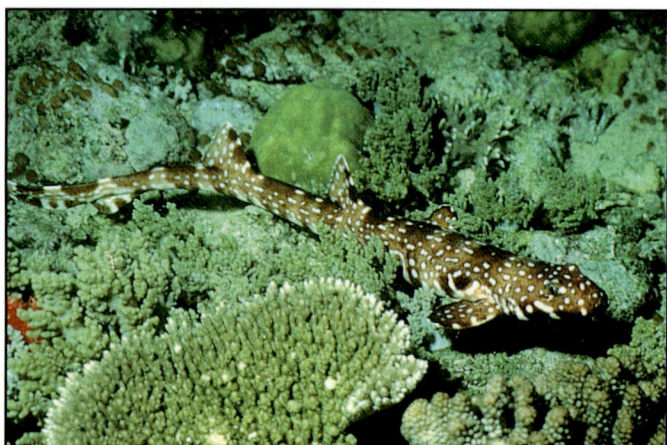

Scott Michael   Papua New Guinea

Hemiscyllium strahani
Hooded epaulette shark

L: Max. known 80 cm, one 75-cm-male still immature. Di: Indonesia, Papua New Guinea. De: 3-18 m. G: Ventral surface of head with dark markings, giving a 'hooded' appearance to the entire head of some individuals. Ocelli present, but not very conspicuous. Nocturnally active coral reef inhabitant. Reportedly fairly common along the northeastern coast of Papua New Guinea where encountered only at night. Found on reef flats and reef faces in areas with abundant stony corals. Hides during the day in crevices and under table corals. Does well in captivity, one specimen survived for over seven years. Harmless.

Rudie Kuiter   Great Barrier Reef, Australia

Hemiscyllium trispeculare
Speckled epaulette shark

L: Max. 65, mature males 53-64 cm. Di: Australia (Western Australia to Queensland). De: 1-20 m. G: Common. With a large black ocellus (bordered in white) above each pectoral fin and two or three larger spots behind each ocellus. Background coloration yellowish, upper surfaces with numerous very small brown spots. Brownish bands on body and tail. Underside pale. In the shallow water of coral reefs, including tide pools. Specimens from Australia's Northern Territory have much smaller and denser spots than those from Western Australia. Biology hardly known; oviparous, probably preys on benthic invertebrates.

Chiloscyllium punctatum
Brownbanded bamboo shark
Length: At birth 16, max. 104 cm.
Distribution: India to Indonesia,
Australia (Queensland around
north coast to Western Australia)
to Japan. Depth: Intertidal to
60 m. General: Posterior edge of
pectoral fins concave, not rounded. Colour pattern of juveniles
dark bands on light background,
(top) the bands being faint or
absent in adults (bottom). Inhabits
coral reefs, tidepools, tidal flats
and reef faces. A solitary, nocturnally active shark that is very
secretive, especially when young,
hiding deep in crevices at the base
of coral heads during the day. The
eggs have no tendrils. Reproduces
in captivity. The male bites one
pectoral fin of the female before
and during copulation, and may
also bite gill area during courtship.
The invertebrate prey is 'dug up'
with the snout from coral rubble
and sand. May bite if provoked.
Other family members:
Grey bamboo shark, *C. griseum,*
max. 74 cm, Arabian Gulf to
Papua New Guinea and north to
southern Japan. Posterior edge of
pectoral fins rounded, not concave. Base of first dorsal longer
than that of second dorsal.
Colour pattern of juveniles prominent transverse dark bands on
light background, adults uniformly
light brown. A common inshore
benthic shark. Little is known
about its habits and biology.
Oviparous, small oval egg cases
are deposited on the ground. Prey
probably consists of invertebrates.
Whitespotted bamboo shark,
C. plagiosum, at birth 15, max.
95 cm, India to Thailand, Indonesia, Philippines, Taiwan to Japan.
Posterior edge of pectoral fins
rounded, not concave. Colour
pattern of grey to dark brown
background with lighter bands and
white spots. Found in tropical reef
areas. Hides in crevices during the
day, comes out at night to hunt
and feed on fishes and crustaceans. A hardy species in captivity,
reported to survive as long as 25
years in public aquaria.
Arabian bamboo shark, *C. ara-
bicum,* at birth 12, max. 78 cm.
Arabian Gulf to India. Posterior
edge of pectoral fins rounded, not
concave. Base of second dorsal
longer than that of first dorsal.
Coloration of juveniles and adults
uniformly brownish. In caves and
crevices of coral reefs.

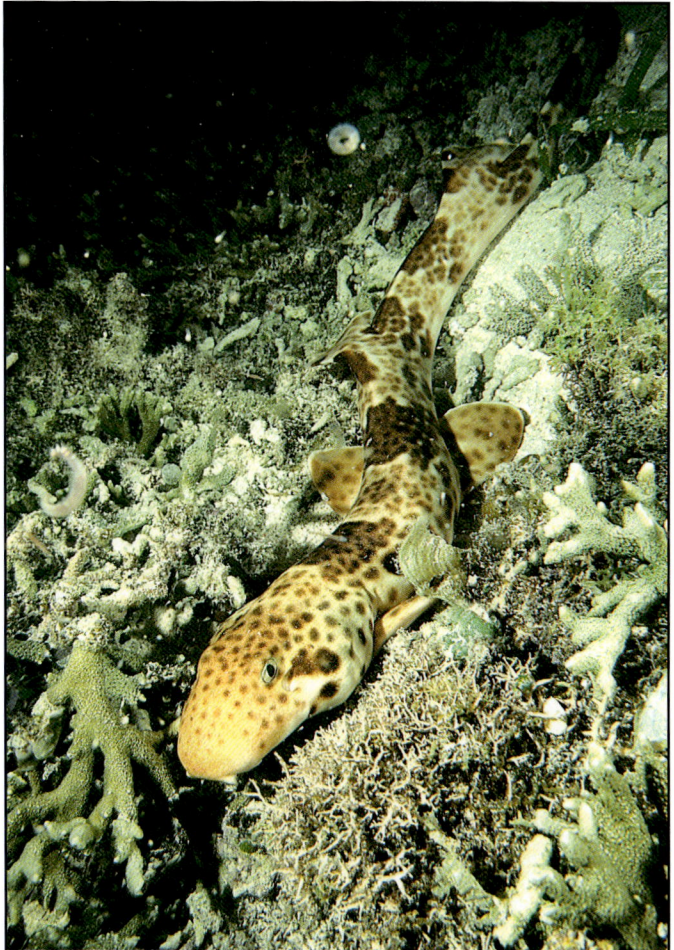

Peter Verhoog   Irian Jaya, Indonesia

Rudie Kuiter   Great Barrier Reef, Australia

This family comprises three genera with six described and several undescribed species. All of these unusual sharks inhabit coral and/or rocky reefs. Wobbegongs are characterised by a flattened body, a very broad head with a huge terminal mouth, long barbels and skin flaps around the mouth. Coloration, presence or absence of tubercles on the back and number of skin flaps in the upper lip area are features serving in distinguishing species. Their ornate colour patterns help to camouflage them among coral, rocks and algae, where they lie in repose. From these daytime resting sites, they ambush prey that comes into striking range. They hold large prey items until they are subdued, then manipulate the catch so it can be swalloed head first. At night they actively stalk octopuses, squid, crabs, sharks (including other wobbegongs), rays and bony reef fishes. Their flattened front part enables wobbegongs to shove their heads into crevices and extract hiding prey. All species are ovoviviparous (yolk sac viviparity) as far as is known. Juveniles are cryptic and live in deep crevices, under rocks and in tide pools. They may be reclusive to avoid larger congeners. One species gives birth in shallow areas away from the paths frequented by adults. Unlike some of their close relatives, wobbegongs do not crawl on their pectoral fins, instead they swim in a sinuous fashion. Individuals have been observed moving from one tide pool to another in water so shallow their backs were exposed.

Wobbegongs have a dubious reputation. Most divers consider them harmless unless provoked, but there are reports of wobbegongs biting divers for no apparent reason. These sharks obviously have poor visual acuity and will strike at any movement near their head, therefore it is not recommended to place a hand or foot near the mouth of any (even a smaller) wobbegong. Wobbegongs readily adapt to a life in captivity and two species are known to have mated in public aquaria.

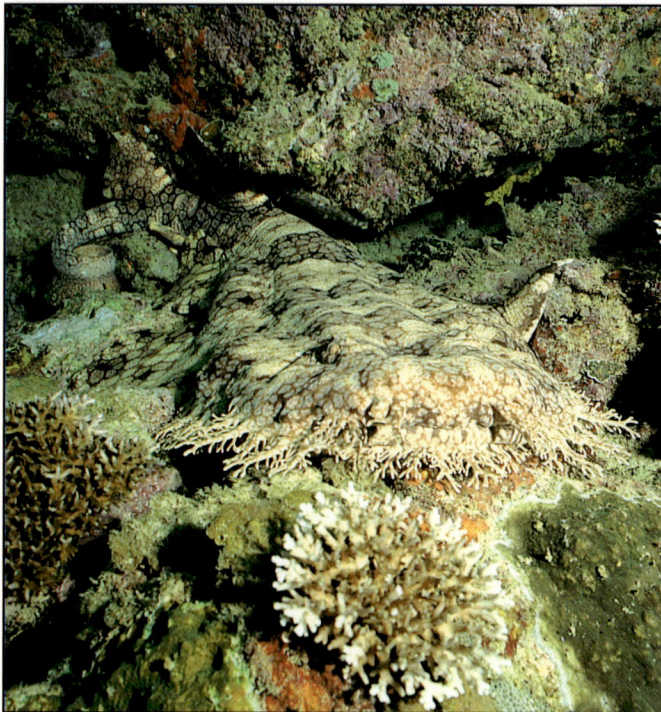

Jim Black   Papua New Guinea

Eucrossorhinus dasypogon
Tasselled wobbegong
L: At birth 22, max. 300+ cm. Di: Tropical Indo-West Pacific: Indonesia, Papua New Guinea, Australia (Queensland along northern coast to Western Australia). De: 1-40+ m. G: Head very broad, entire margin covered with dermal flaps resembling a beard. Unlike in other family members, dermal lobes also present on lower jaw. Coloration of camouflaging reticulated pattern of dark dots and lines on a lighter background. There is one report of a pair mating in a cave at night. Obligatory coral reef dweller. Found solitarily in reef channels and on reef faces. Usually seen resting in caves and under ledges during the day, leaves these sites to hunt bony fishes (squirrelfish, soldierfish, sweepers) in the reef at night. Also preys during the day on these nocturnally active teleosts in the caves sheltering both predator and prey. Uses several daytime resting sites within a small home range. Rests on bottom with tail curled up. Cleaner shrimp live on and pick at its skin. Aggressive, has attacked divers without being provoked. See also facing page.

Eucrossorhinus dasypogon   **Irian Jaya, Indonesia**

Rudie Kuiter   New South Wales, Australia

Orectolobus maculatus
Spotted wobbegong
L: At birth 20, max. 320 cm. Di: Australia: NSW to W. Australia. Possibly also Japan and South China Sea, but may be a different species. De: 1-100 m. G: Each side of head with 8-10 skin flaps. Light ocellus-like markings on yellowish-brown with darker bands. In temperate waters on rocky reefs and in coastal bays, juveniles in estuaries or sea grass beds. By day often resting in caves, sometimes in aggregations of a dozen or more individuals. Litter size up to 37. Prey includes crabs, lobsters, octopuses, bony fishes, other sharks (incl. conspecifics) and rays. Aggressive, attacks on divers reported (see also family account).

Hiro Tatsuuma / IV   Shizuoka, Japan

Orectolobus japonicus
Japanese wobbegong
L: At birth 20, max. 100+, mature males 103 cm. Di: Japan to Philippines. De: Inshore waters. G: Nocturnal. Litter size up to 27. Preys on benthic fishes. Mating observed.

Rudie Kuiter   New South Wales, Australia

Orectolobus ornatus
Ornate wobbegong
L: At birth 20, max. 290 cm. Di: Australia, PNG, Indonesia and possibly Japan. De: Intertidal to 30 m. G: 5-6 dermal lobes on each side of head, dark and light areas between saddles on back that have spots with light centres. In reefs of temperate and tropical waters. Nocturnal. Also in aggregations, sometimes piled on top of one another, resting in caves during the day. Ovoviviparous, litter size at least 12. Preys on bony fishes, sharks, rays, cephalopods and crustaceans. Ambush feeder. Aggressive when provoked, large males are more aggressive during the breeding season. Said to attack waders and fishermen in tidepools.

Sutorectus tentaculatus
Cobbler wobbegong

Length: At birth about 20, max. recorded 92, possibly to 200 or 300 cm.
Distribution: Australia: Western and south Australia.
Depth: Inshore waters.
General: Differs from other family members by tubercles (large dermal denticles) on head and body which are present in juveniles and adults, long unbranched nasal barbels and a colour pattern of dark saddles with jagged edges, interspaced with light areas with irregular dark spots. There are two rows of large fang-like teeth in the upper jaw, but three such rows in the lower jaw (see photo centre right).

Little is known about the habits and biology of this shark which is common off southern Australia. The cobbler wobbegong is found living benthically in rocky and coral reefs of temperate continental waters. Reproduction is presumably ovoviviparous like in other wobbegongs. A litter of near term young was reported to have still sizeable yolk sacs; these young were 18 cm long.

Prey items have not yet been reported but most probably include benthic invertebrates and bony fishes. As other wobbegongs eat elasmobranchs including their own kin, this species too may prey on other sharks as well as rays.

Brachaelurus waddi
Blind shark

L: At birth 15-18, max. 122, mature at 62-66 cm. Di: Australia: Northern Territory to NSW. De: 0-150 m. G: Long nasal barbels. Juveniles greyish with black bands, adults brown with inconspicuous bands, often with white flecks. Juveniles in crevices in shallow, surgy waters. Adults by day in caves, prey at night on anemones, cuttlefish, crabs, shrimp. Named for its habit of its closing eyelids when removed from the water where it can survive for up to 18 hours; can survive in tide pool until next high tide. Does well in capitivity. Ovoviviparous. Litter size up to 8.

all photos Rudie Kuiter   New South Wales, Australia

This family comprises three species, all of which may be encountered by divers or snorkelers in reef habitats. All family members share the following characters: Nasal barbels, nasoral grooves, small mouths positioned in front of the eyes, fourth and fifth gill slits very close together, second dorsal fin originates well in front of anal fin origin, long tail. The reproductive regime of nurse sharks is yolk sac viviparity, the reproductive behaviour of one species is well-known. By day nurse sharks rest on the ground, during the night they hunt bony fishes and invertebrates in the reef. In places where only restricted swimming movements are possible they can use their pectoral fins to 'walk' forward or backward. After locating prey hiding in rubble or a crevice, they place their small mouth near it and litterally inhale it by rapidly expanding the pharynx. This technique has been termed suction feeding. The paired barbles just in front of the mouth are covered with taste receptors and help locating prey. Fish prey is held with many multicuspid teeth, hard shelled invertebrates (molluscs, crustaceans) are crushed between the powerful jaws. Generally these large animals are considered harmless and even serve as riding platform for the youngsters of certain Pacific islands. But reports of several unprovoked attacks on divers by nurse sharks also exist. Often they are reluctant to release a victim from the powerful grip of their jaws after an attack. As they spent most of their time resting on the ground, especially when they do not have to hunt for their food, nurse sharks adapt well to captivity in large public aquaria; some specimens have survived for more than 25 years.

The photo sequence on these two pages shows a pair of the Atlantic nurse shark *Ginglymostoma cirratum* engaged in courtship and mating. Bottom: To overwhelm his partner, the male has to get a firm grip of one of the female's pectoral fins in order to control her movements. Facing page, top: After accomplishing this task, the male flips the female over and aligns his body parallel to her's - a tricky maneuver; bottom: Only now can he insert one of his two penis-like organs (claspers) into her cloaca. The copulation itself can last for up to two minutes.

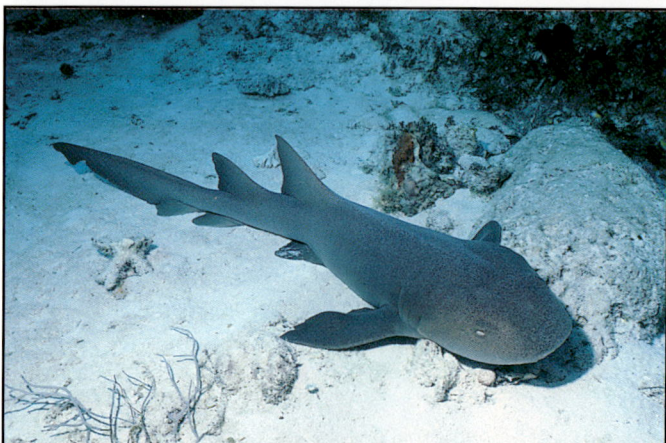

Paul Humann   Caribbean

Ginglymostoma cirratum
Atlantic nurse shark

L: At birth 27, max. 430 cm. Di: W-Atlantic: Rhode Island to Brazil. E-Atlantic: Senegal to Gabon. E-Pacific: Baja California to Peru. De: 1-50 m. G: Dorsals rounded, barbels long. Brown, juveniles with scattered dark spots. On reefs, in mangroves. By day, often resting in small groups on bottom. At night, hunts benthic invertebrates (spiny lobsters, crabs, octopuses, sea urchins), but also stingrays and bony fishes. Digs under coral debris for concealed prey. Territorial, tagged individuals recaptured in same area after 4 years. Ovoviviparous, litter size up to 28. The West African nurse shark *G. brevicaudatum* is the second genus member.

Mark Strickland   Myanmar

Nebrius ferrugineus
Tawny nurse shark

L: At birth 40, max. 320 cm. Di: Red Sea and South Africa to Australia and Society Islands. De: Intertidal to 70 m. G: Dorsal and pectoral fins sharply pointed, pectoral fins sickle-shaped. Uniformly tan, juveniles (bottom) with spots. Sluggish inhabitant of coral reefs. Juveniles are frequently seen in shallow lagoons, where they hide in reef crevices. Adults are found in a variety of habitats including lagoons, reef flats, channels, and reef faces. Larger individuals are often seen moving over the reef or resting in caves during the day. Sometimes they aggregate on the ground, with individuals in a pile, resting on top of or in close contact next to each other. This shark is territorial and has a delimited home range, individuals usually return to the same area every day. At night, it leaves its daytime resting place to hunt in the reef. Prey includes invertebrates like octopus, squid, crabs, lobsters, sea urchins, but also bony fishes and occasionally sea snakes. Generally considered harmless, but several unprovoked attacks have been reported and bites with remarkable tenacity. Ovoviviparous, litter size up to eight. *Nebrius concolor* is a junior synonym.

Helmut Debelius   Mauritius, Indian Ocean

Nurse sharks are among the favourites of divers as they are benign and seem to enjoy the contact with people.

MARK STRICKLAND

Generally considered sluggish animals, even nurse sharks get into a 'feeding frenzy' when food is made available.

WHALE SHARK HUNTERS

Pando Operio and his crew have been bobbing up and down for three days now on their boat *Ocean Wolf* in the Bohol Sea in the Philippines. They are in search of the largest fish in the oceans, the whale shark. For many hours, they ride against the *habagat,* the south-westerly wind. The boat rocks dangerously under the midday heat and only the wind and sprays of sea water brings cooling relief to the crew. At the small town of Jagna, they turn the boat and sail back, always along the coast, always watching out for a moving shadow in the water. Pablo, the spotter, sits sleepily on the short mast of the boat. Suddenly he jumps up, points at a shadow in the water and shouts *"Balilan! Balilan!"* Photographer Jürgen Freund spent two weeks with the fishermen of the Bohol Sea.

One of the whale shark hunters jumps into the water with a large steel hook in his hand.

The coral and plankton bloom take place in the tropical waters of the Philippines at full and new moon every year between January and May. During this time, whale sharks come up to the water surface to feed on plankton. This is the hunting season for the fishermen of Pamilacan Island! For more than a hundred years, the residents of this small island located 20 kilometres south of Bohol are famous as hunters of the ocean. The word *pamilac* in their local dialect means to hunt, and *Pamilacan* means the place where they hunt. In the olden times, they caught gigantic manta rays, which they call *sangga,* with self-made steel hooks. When the number of rays dwindled, they switched to whales and whale sharks. After the Philippines joined the worldwide ban on the killing of whales and dolphins, the fishermen started to concentrate on the largest of all sharks. Since the islanders found it too complicated and expensive to make harpoons, they invented simpler steel hooks and became masters in this craft. Only the most courageous and skillful fishermen can perform the job of the 'hooker' or 'human harpoon'. The hook hero these days is Miguel Valeroso. Aged 28, he has many scars which tell of his adventures. To date, his injuries include a broken jaw, rib and bone fractures and a big scar on his arm where a hook went in while he was fighting with a Bryde's whale.

However, there are times when no whale sharks can be hunted. Then the islanders depend on tuna. Sometimes, big foreign fish trawlers empty the sea around Pamilacan Island, making food scarce for the islanders. When this happens, the people of Pamilacan have only rice, sea urchins and sea weed for food.

The Pamilacan population of whale sharks have been wiped out and the hunters now have to travel relatively far to Jagna in Southern Bohol where there are more sightings of their preferred 'prey'. The nights will be spent in Jagna to spare the long return to Pamilacan. The hunting is planned to last three days. Just before the catching trip begins at sunrise, the six hunters light candles in a small chapel nearby. All are very religious and ask their saint, *Santo Niño,* for a good hunt and a safe return.

Usually it takes longer than one hour to drag the hooked whale shark to the boat.

Only after fighting for hours, the huge catch of the fishermen can be tied to the boat with ropes and towed home.

With only one canister of freshwater, some cooked rice and some canned sardines for food, the hunters sail to Jagna. It takes them 8 hours to get there. The first day passes without any sightings. The monotonous ride is only interrupted by a short lunch break. The second day also passes by uneventfully. Luck comes to them on the third day.

Pablo spots a shadow far away in the water. He screams *"Balilan! Balilan!"* and jumps down from the bamboo mast of the boat. Tonio, the hook man, runs to the bow while his brother Pando accelerates the engine. Suddenly, everything happens very fast. With a huge steel hook in his hand, Tonio jumps into the water head first and hooks the approaching whale shark on its back. He rams the hook into ten-centimetre-thick skin. Simultaneously, Pando turns off the engine. The sudden silence is interrupted by the loud flaps of the shark's enormous tail. The water seems to boil. Tonio disappears underwater. All of a sudden, he appears again, grinning: "We are lucky today. The shark is 14 metres long."

Tonio quickly climbs into the boat through the outrigger. There is no time to waste. The injured whale shark immediately heads downward. The animal's descent is only stopped by the end of the 150-metre-long and 2.5-cm-thick nylon rope tied to the hook. The end of the rope is strapped firmly to the boat's mast. While Tonio climbs aboard, the rest of the crew tries to stop the fast reeling of the rope. A rough tug-of-war starts between the giant and the six fishermen. Slowly, the boat closes in on the shark metre by metre. With every wave, the men pull in the rope. But the fight does not seem to be lost for the shark yet. Quite often, the fish manages to rip the rope out of the crew's hands. But finally, after two despairing hours, the ten-ton-shark is exhausted and hangs

Holes have been cut into the living animal in order to secure it to the boat.

motionless underneath the boat, held only by 12 arms. Now, the hunters prepare to weaken the animal further. Tonio jumps back into the water, this time with a *bolo,* a very long knife. It is his task now, to cut the spinal chord of the whale shark.

The biggest problem is to transport the 14-metre-whale shark back home to Pamilacan Island. The shark hangs head down but is still alive. Equipped with knives, the entire crew dive to the shark's head to cut holes as large as dinner plates into the jaws. The holes are made to tie a rope to the shark on which the catch will be dragged back home. Despite it's cut spinal cord, the shark is still able to fight. Again and again the animal rears and tries to shake off its torturers. The situation is extremely dangerous for all men involved. Sometimes, the struggling shark is hidden from view by thick clouds of plankton only to become visible again just before it can hit one of the hunters. Finally, after two hours, the shark is ready for transportation.

The boat drags the bulky catch at a very slow speed. Only now, after four hours of hard work, the men get a break. They stop at a little coastal village to refuel and to buy *Tanduay* rum to celebrate. Not all hunters are as successful as the crew the *Ocean Wolf.* Just a few days ago, the captain of the smaller *Sta. Cruz* misjudged the size of the whale shark his men had hooked. The whale shark was able to drag the boat below the surface of the sea. Even the crew's attempt to cut the rope failed because their knives broke. Luckily, after floating in the sea for over an hour, all men - the youngest merely being 15 years old - were picked up by another boat.

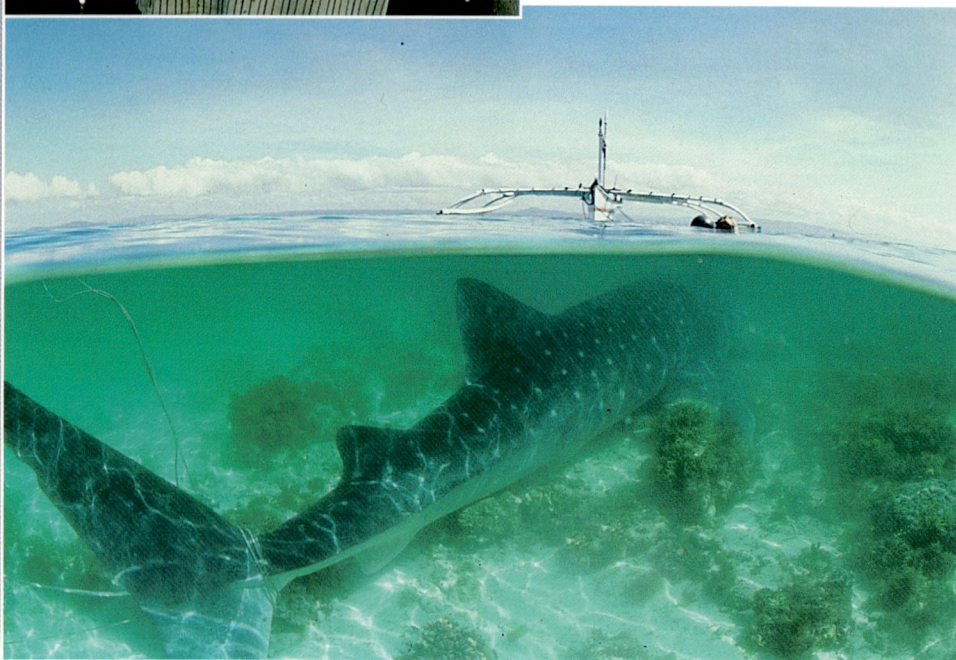

Top: Since 60 years now, the men of one Pamilacan family forge the massive steel hooks. Above: The catch has been brought home!

In the shallow water at the beach of Tondo, the whale shark is butchered and partitioned by the fishermen and their helpers.

Late in the evening, the *Ocean Wolf* returns to Pamilacan with its precious catch. The animal is still alive and pulled into shallow water. Early in the morning, the hunters and their helpers come out to pile up large Styrofoam boxes along the shore which are filled with ice. After one hour of slaughtering, only the head and parts of the cartilaginous spinal column of the 14-metre-shark are left. The shark's white meat is cut into rectangular pieces and placed inside the ice boxes. Skin and fins are hung on bamboo racks for drying. The head, the cartilaginous skeleton and the gigantic liver are dumped into the sea. A few hours later, the beach looks as if nothing has happened. The only sign of the slaughter is the animal's red blood dispersed in the shallow water and on the beach. On this very same day, a shipload of fresh whale shark meat will be sent to Taiwan.

In the past, whale shark meat was cut into strips and dried in the sun. These strips were sold on the local markets in Bohol. Several families were engaged in the butchering and marketing of shark meat. But the recent demand of the Taiwanese for the fresh meat wrought havoc on the social structure of the islanders. The price of whale shark meat suddenly went up and the hunting effort was multiplied. In recent years, the number of whale sharks drastically went down. In 1996, a total of about 100 whale sharks were caught in the region and slaughtered for the Taiwanese market. In 1997/98, the catch was only about 50 whale sharks. Under the pressure of a Philippine daily newspaper and the Philippine section of the WWF (World Wildlife Fund), former President Fidel Ramos enforced a ban on the killing of whale sharks and manta rays in March 25, 1998. The following figures show that occasional whale shark slaughter still continues until today. December 1998: 812 kilograms of whale shark meat confiscated at the Ninoy International Airport (NAIA); May 1999: Many kilograms confiscated in the port of Cebu; January 2000: 2,000 kilograms confiscated in NAIA; January 2000: 829 kilograms confiscated in Puerto Princesa Airport. What will be in the future for the largest fish in the oceans, the majestic, yet always benign whale shark?

A few years ago, each villager got his share of the catch; today, almost all of it is sold to other countries. The photo shows the transsected thick white skin and the pink flesh.

This family comprises 5 species, i.e. great white shark, longfin and shortfin mako, mackerel and salmon shark. All have conical snouts, one or two keels laterally on the caudal peduncle, a lunate (moon crest-shaped) caudal fin, five pairs of long gill slits and tiny second dorsal and anal fins. All family members maintain a body temperature several degrees above that of the surrounding seawater by conserving metabolic heat with the help of special networks of blood vessels (rete mirabile, Latin for wonder net) set deep in their musculature. Additional 'wonder nets' are associated with diverse organs (brain, eye retina, gut) which in that way receive extra metabolic heat and can 'work' faster and more efficient than in other sharks. The white shark's muscle is 3-5°C, that of the shortfin mako 7-10°C warmer than the ambient water. The prey of these superpredators includes bony fishes, other elasmobranchs and - in the case of the great white shark - marine mammals. Mako sharks occasionally hunt billfish, but these extremely fast swimmers are not without defence: One shortfin mako had the fractured bill of a small sailfish embedded in its eye, a longfin mako (Isurus paucus) was found with a swordfish bill broken off in its belly! All family members are aplacental ovoviviparous and oophagous like their cousins, the thresher and sandtiger sharks. White shark and shortfin mako are known to attack people in the sea. Unfortunately, even large public aquaria are not able to maintain these magnificent, ever-swimming and active sharks for sustained periods of time. Singular trials with young white sharks in San Francisco ('Sandy') and Japan resulted in the release of the animals which were not at all doing well in captivity.

Carcharodon carcharias
Great white shark

Length: At birth 110-130, max. at least 650, mature females 400-650, mature males 240-550 cm.
Distribution: Worldwide in most temperate and some tropical regions. Coastal and mostly amphitemperate. E-Atlantic: Mediterranean and Madeira to South Africa. W-Atlantic: Newfoundland to Cuba and northern Gulf of Mexico, Brazil to Argentina. Indian Ocean: South Africa, Seychelles. W-Pacific: Siberia to Philippines, Australia to New Zealand. Central Pacific: Marshall and Hawaiian Islands. E-Pacific: Gulf of Alaska to Gulf of California, Panama to Chile.
Depth: 0-1,280 m.
General: A large, unmistakable shark characterised by the following features: Caudal peduncle with a pronounced keel on either side. Teeth of adult specimens very large (up to 8 cm high), triangular, with coarsely serrated edges, but without secondary cusplets. Gill slits long. Coloration black, grey or white above, usually becoming lighter with age, white below; the border between the dark dorsal and the light ventral colour is sharply marked. Often with a distinctive black spot at the axil of the pectoral fin, undersides of pectoral fins with black tips.
    Great white sharks are found in coastal areas, around rocky reefs and in kelp fields, only very rarely near coral reefs. An old record from the warm, coral-rich Red Sea is most probably erroneous. Adults are more common in warmer waters than juveniles. The white shark is usually encountered singly, sometimes also in pairs. The observation of feeding aggregations of up to about twenty individuals at major food sources like dead whales is a rare event.
    The white shark's mode of reproduction is aplacental ovoviviparity, litter size is 7-9. Correct data on pregnant females have been reported only a few years ago. The rarity of pregnant females may be explained by spatial separation from other white sharks during pregnancy, their sheer size which precludes capture by most fishing gear, and by a possibly low fecundity with relatively few adult females in the stock. Like observed in other family members, the embryos of this species gorge themselves with eggs released from the ovaries into the maternal uteri. Near term, the embryos' stomachs are grotesquely swollen with yolk from these eggs. However, no signs for intrauterine cannibalism (or adelphophagy, as seen in mako, salmon and mackerel sharks) have been found so far. This practice usually allows only a few (the strongest) young to survive in each uterus.
    Prey species preference changes with age, young individuals (<2 m) mainly eat fish, while larger white sharks (>3 m) have shifted toward fat-rich marine mammals, i.e. seals and sea lions as main dish. Vertebrate prey includes a wide range of bony fishes (sturgeon, menhaden, pilchards, salmon, hake, halibut, rockfish, mackerels and tunas among others), other sharks (Galeorhinus, Mustelus, Carcharhinus, Rhizoprionodon, Sphyrna, Squalus spp.), stingrays, eagle rays, chimaeras (ratfishes), sea turtles, sea birds and marine mammals (harbour porpoise, dolphins, harbour seal, northern elephant seal, Steller's sea lion, California sea lion, South African fur seal among others, and probably sea otter). Carrion from near-shore slaughter-houses, dead baleen whales and other large cetaceans, fish caught on lines and the basking shark Cetorhinus maximus are also eaten. Pinnipeds may be an important prey where they occur together with this shark species, but in tropically warm areas without these mammals the white shark can subsist on other sharks, teleosts, turtles and whales.
    Following is an excellent account b Chris Fallows from the University of Cape Town on white shark-pinniped interaction at Seal Island, South Africa. He was the first to come up with the idea of using seal decoys to elicit the breaching behaviour in white sharks (see also FLYING WHITES on pp.295-300).

Facing page: A perfect example for 'predatory breaching'. This white shark just has grabbed a Cape fur seal in mid air.

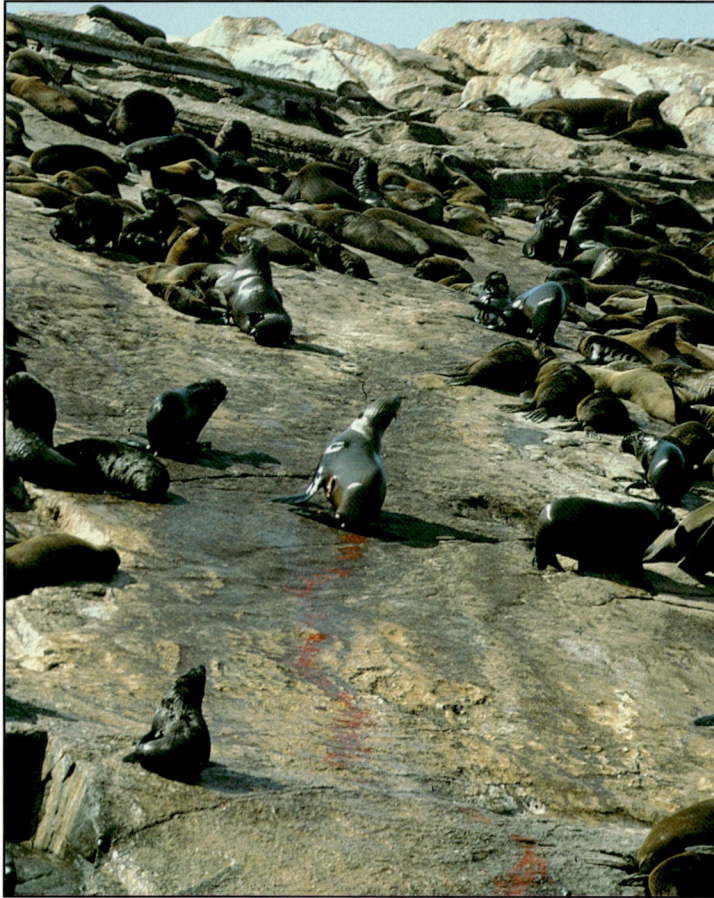

Cape fur seal bleeding from predatory bite of great white shark on Seal Island, South Africa.

C. & M. FALLOWS / IV

Seal Island lies on 34°08'18" S and 18°35'00" E, which is only 13 km off Cape Town, South Africa. The island itself is around 450 m long, give or take a few metres depending on the tide. The water around the island on the western seaboard is deeper than on the eastern side that is characterised by a gentle sloping shoal. The depth of the island on the South and West quickly drops to 25 m and it is this depth that holds the key for the surprise attack strategy that the white sharks use. This terrain makes it the ideal hunting area for the great white that uses the deeper water as a means to vertically attack the seals which live there, thereby achieving maximum camouflage and speed. This attack strategy leads to the unique and spectacular behaviour of Seal Island known as 'predatory breaching'. This spectacular form of predatory adaptation is what makes Seal Island unique in terms of great white shark behaviour world wide. Nowhere else in the world does the great white show such regular intensity in pursuit of it's prey. Although at the Farallon Islands off San Francisco the great whites do breach as well as at Mossel Bay and occasionally at Dyer Island in South Africa, the sheer regularity of this breaching and hunting at Seal Island is not seen anywhere else in the world.

Seal Island has a long history of man-made destruction and was a seal-clubbing and guano-collecting site up until the early 1970s. The area is home to 64,000 Cape fur seals (Arctocephalus pusillus). They are everywhere and seem to pride themselves on being able to climb rounded boulders that many humans would struggle to mount. They comically cavort up and down the island playing, fighting and mostly sleeping. Anyone who has ever spent time amongst seals will tell you that their yawning, sighing, barking and coughing leave you with no doubt that this animal is highly social, intelligent and just generally great to be around.

Of the many interesting behaviour patterns the great white displays around the island none are more visually spectacular than predatory breaches. These bursts of flight have been witnessed on several hundred occasions and every one is unique. We have been involved with various research projects over the years. Shape discrimination has been the most exciting. Often these huge sharks launch themselves skyward merely 25 m away on shapes that resemble there natural food source, whilst in the background the seals watch thankful that it is not one of their own kind being catapulted into space. One of the research projects we undertook in 1999 (based on our results from 1996) was using a cut-out in the form of a square and another one in the form of a seal. When the targets where stationary there was unquestionably a clear preference for the seal, however, when motion was applied to both little discrimination occurred provided the shapes fell within a certain size range.

Other research projects have included colour-coded tagging where we use various colours on our tags to identify the various individual animals. In this way we have discovered that all the great white sharks have unique and clearly defined personalities. We have seen certain sharks one year after the next displaying the same traits that add to their individual appeal. We have one shark known as 'Black-White-Black' that has 'visited' us 24 times. This shark is very confident and is a highly successful hunter, having been seen to make three kills in a week. Another shark known as 'Rasta' is a miracle of success in the shark world and goes against the adage that only the fittest survive. She is so slow and laid back around the boat that we all wonder how she ever catches a seal. We have seen her on a dozen occasions and every time she appears to be hardly moving, but rather just drifting wherever the currents takes her. Besides conducting the above projects we collect all climatic and environmental data on a daily basis.

This behaviour is undoubtedly what makes Seal Island the world's premier natural feeding ground for the great white.

A great white shark launches a 'flying' attack on a Cape fur seal in False Bay, South Africa.

It has been estimated that at least about every second attack by experienced white sharks on seals is successful.

The success rate is phenomenal. The rate of actual successful kills being in the region of 45%, which amongst large predators is very high. Certain individuals are, as mentioned, far more successful and determined than others and the success is probably closer to 80%. The seals themselves are highly resilient to these attacks and have their own defences such as sharp teeth and claws with which they inflict often-serious cuts to the sharks' snouts and eyes. Often when in mid-breach firmly clamped between the jaws of the shark, these never-say-die seals are still trying to bite their powerful adversary that relentlessly hunts them. The seals that do sometimes escape the vice-like grip of the jaws often are left with the most horrific bites and it astounds us time and again to see these seals actually recover from such bites. It is always the seals that are the first indicator of the abundance of sharks around the island. The relaxed state or lack thereof is always an indicator of just what is happening in the world under the waves. Often when the great white sharks are absent the seals cavort with gay abandon forming large rafts many metres offshore but when the sharks are present the seals hug the island ready to climb out of the water at a mere glance of the great fish. The seals do have a respite, however, as like at most other great white shark sites around the world the presence of the shark is seasonal. The sharks are most prevalent at Seal Island during the colder months when the warm water has moved out of False Bay and so too have the migratory fish that the great white shows preference for. It is only logical from a predator's point of view to prefer to hunt a food source that does not bite back or potentially claw your eyes thereby possibly incurring a life-threatening injury. This is why we believe the great white prefers to hunt the larger fish species when they are available. It is also during the colder months that the seals that are still young of the year must go out and hunt for the first time. At this stage they are small and inexperienced often porpoising slowly in high risk areas making them easy prey for the ever alert sharks.

The great white shark has for centuries captured the interest of all those fascinated by the ocean and even those who aren't. Having being privileged to have worked with this magnificent predator for many years we can only but try to convey the necessity to protect it from the ignorance that is still prevalent today. The animal is not a scourge but rather a great asset to any country fortunate enough to have it occurring along their shores. In South Africa and around the world we must do everything possible to conserve the population we have and provide a safe refuge for this lonely lord of the seas.

M. & M. KAZMERS / IV

After succesfully hunting seals, this large white shark languidly floats at the surface near Dyer Island, S. Africa.

OUTSIDE THE CAGE

Almost anything you could think of, pertaining to wildlife, producer Don Meier had made into an episode of the TV series *Wild Kingdom*. In the mid-1980's, he was willing to entertain even the most outrageous ideas for underwater shows with one qualification: The show had to have sharks. So Howard Hall did shows about sharks: Blue sharks, mako sharks, hammerhead sharks, angel sharks, tiger sharks, lemon sharks - you name it. But he had never done a show about the great white. Don Meier considered the expedition to the great whites off South Australia simply too expensive. But how did the great white appear in *Wild Kingdom* then?

A 4.2-metre great white shark at Guadalupe Island. This was the very first great white any of us had seen underwater!

The film proposal would go something like this: I would call Don and ask, "How about a film on Guadalupe Island?" "What's it got?" Don would ask. "Great white sharks". "Have you seen 'em?" "Nope." "Have you ever been there?" "Nope." "Do you think you can find 'em?" "I dunno." "Well, what do they do?" Don would ask finally. "They eat people now and again," I'd reply confidently. That was about all it took. I'd sit around the house dreaming up trips like that and then book up a whole year of 'work' with a few ten-minute phone calls. They actually paid me to do this stuff!

Then, early in September of 1985, we loaded up the 17-metre motor vessel *Mirage* and headed south from San Diego, California nearly three hundred miles to Guadalupe Island which lies isolated one hundred and eighty miles off the coast of Baja California. Our crew consisted of two cameramen: Marty Snyderman and myself, and two assistants: Tom Allen and Jeremiah Sullivan. Upon arriving at the island, we decided to spend four days scouting locations and filming sequences we could use to make a show in the likely event that the sharks failed to show up. The water was clear, cold, and the island dropped off quickly into deep water. We filmed sea lions and Guadalupe fur seals and at the end of each dive we would hang on the surface looking down into the bottomless, cobalt-blue water and wonder if we were being considered for a meal. In preceding years two divers had been attacked by great whites in the exact spot where we were swimming. One died. I thought about that a bit as we drifted on the surface, like six drunken seals, for nearly an hour one day while anxiously waiting for the boat to pick us up.

Following double page: We also realised that this was the first time anyone was outside a cage with the great white shark.

The underwater visibility at Guadalupe Island is often excellent.

The next day we put bait in the water at the very same spot where we had drifted for an hour the previous afternoon. An hour and forty-five minutes later we attracted a monster. To say that it was a huge great white still seems like understatement, even now, more than a decade later. In the years that followed, I made five lengthy expeditions to South Australia to film great whites, some of these sharks were certainly 4.8 metres long. And still I have yet to see a shark that was anything like the size of the Guadalupe monster. I won't guess at how long it was since my best guess would seem an exaggeration. But I will say that instead of dashing to the crane to lower our shark cage, our entire crew stood on the upper deck of the *Mirage* and watched, slack-jawed, as this thing circled the boat. It circled twice then disappeared and was gone. Later that day and all of the next we took shifts standing in the shark cage as bits of tuna flesh and coagulated blood drifted through our hair, and waited for the shark to come back. It didn't return. By the afternoon of the third day, we were getting bored.

Tom Allen and I were in the shark cage twiddling our thumbs when a small mako shark showed up. We decided that footage of any kind of shark was better than no sharks. After all, Don would probably call the film 'The Sharks of Guadalupe Island' whether we had sharks in the film or not. I swam out of the cage and up to the surface. "We've got a mako. Tom and I are going to film it," I yelled to Marty. Marty waved and began suiting up with Jeremiah. Tom left the cage and we drifted downcurrent with the mako. Tom loaded his shark tagging spear and I hoped to get a shot of him placing a tag in the mako's dorsal fin. We were about fifteen metres from the cage when Marty jumped in with Jeremiah. Jeremiah swam to the cage to dispense some extra bait while Marty joined Tom and me as we drifted away with the mako.

After ten minutes or so, the mako suddenly left. We had drifted beyond sight of the cage but I could still see bits of fish scraps drifting downstream in the chum line which helped define our course back to the cage. We were just about to head back when we heard a series of tremendous bangs. Someone was pounding on the shark cage. Tom, Marty, and I swam back toward the cage against the current, each of us occasionally looking back over our shoulder to see if the mako was following us up the chum line; the mako or something worse. When we got close enough to the shark cage to see what was making the loud banging noise, we all stopped in shock. The same two words passed through each of our minds: "Oh, shit!"

Jeremiah was making the noise in a desperate attempt to get our attention. The instrument he employed for this purpose was a twenty-five pound, partially frozen albacore tuna. Jeremiah was pounding on the bars of the shark cage with a frozen fish. Of course, a side effect of pounding on the cage with a partially frozen tuna was to create a great cloud of chum in the vicinity of the cage. And circling the cage in frustration was a 4.2-metre great white shark. It was the first great white any of us had seen underwater. Marty, Tom, and I each also realised that this may well be the first time anyone has swam outside the cage with the great white, certainly this far outside the cage. However, at the moment none of us felt our chests swell with pride. Instead, each of us felt like a horse's ass. Marty later confessed to issuing a silent prayer for salvation. "Dear God, if you just let me get out of this mess alive I promise I will never do anything as dumb as this again as long as I live." It was a lie Marty rather routinely told his maker.

But we were all experienced shark divers and our hesitation lasted only the briefest of moments. We all glanced at each other and in that moment, as our eyes made contact and without the benefit of oral communication, we agreed on a strategy for survival. It was a moment between men; men who routinely dive with sharks and who, in the complete absence of any form of threat, are entirely fearless. It was not necessary to discuss our strategy for survival in that fleeting moment when our eyes made contact. The strategy was almost shouted to one another telepathically. "Every man for himself!"

Marty made a mad dash for the swim step of the boat, no doubt thinking that once on board he could better fulfil his promise to God which he had supplemented with an additional promise to quit diving forever. The swim step didn't look good to me since there was a bait basket hanging beside the ladder and I instinctively knew that, even if I made the swim step in one piece, there would be several unacceptably long moments of uncertainty as I climbed the ladder leaving my legs dangling below. Marty almost certainly expected Tom and me to also go for the swim step and, having a head start, was thinking that he didn't have to worry about his dangling legs since Tom and I would be behind him and in a position to satiate the shark's appetite should it attack.

But for Tom and me the cage looked closer. We hesitated a moment as the shark circled toward the back of the cage then, with the cage between the shark and ourselves, rushed to the cage door and dived in on top of Jeremiah. It was only a two person cage, but under the circumstances, I doubt ten divers would have found it uncomfortably cramped. Things mellowed considerably after that. The shark continued to circle the cage rather lethargically in the manner typical of great whites. Tom and I, having survived our first moments outside the cage, soon decided that the experience hadn't been so bad after all. In fact, the shark never showed much interest in any of us. So as the afternoon passed, Tom and I left the cage several more times as Tom tried and finally succeeded in placing a tag on the shark's dorsal fin while I succeeded in capturing the process on film.

Back at home a week or so later, I received the inevitable phone call from Don Meier. Don, having reviewed the footage, was prepared to render his critical evaluation.

"Well, I looked at the footage," Don began.

"Whaddaya think?" I asked.

"Well, I think you got a show," he said. That was the whole ball of wax for me. Our underwater crew was still batting 1,000. After sixteen shows, we were still the only *Wild Kingdom* crew which had never failed to bring back a show.

"What did ya think of the shark?" I asked.

"Well, they don't do much, do they?" Don replied.

Ah, praise indeed!

Even the most experienced underwater cameramen had to learn that the great white does not behave according to its image.

Lamna ditropis
Salmon shark

Length: At birth about 70, max. 360, mature females 210-360, mature males 180-240 cm. Average size range 200-250 cm.

Distribution: North Pacific: Japan, Korea, Sea of Okhotsk to Bering Sea, southward to southern California and possibly Baja California, Mexico.

Depth: 1-150 m.

General: A heavy, spindle-shaped shark with a short conical pointed snout, medium-sized blade-like teeth with lateral cusplets and a secondary keel on the lower lobe of the caudal fin below the main lateral keel on each side of the caudal peduncle.

Coloration uniformly dark grey above, underside white with dusky blotches. Free rear tip of first dorsal fin of same colour as rest of fin (abruptly white in the related mackerel shark L. nasus). A coastal and pelagic shark with a pronounced preference for cold to cool temperate waters, it is common in continental offshore waters but also ranges close inshore. When following their salmon prey, these sharks are also abundant in the open ocean at times. The predominant large predatory fish in the North Pacific hunts singly, in small groups or large schools. Like its lamnid relatives, the streamlined salmon shark maintains a body temperature well above that of the surrounding sea water, enabling it to be a high-speed swimmer and active hunter.

Reproduction is ovoviviparous, litter size up to 5, young are uterine cannibals. The gestation period is not known, but may be around 9 months, as mating occurs in late summer and parturition in spring. Males mature at an age of about 5 years and 180 cm TL, females at 8-10 years of age and approximately 210 cm TL.

Prey includes primarily Pacific salmon (Oncorhynchus spp.), but also diverse other cold water bony fish species. Salmon sharks are direct competitors for commercially important species like salmon and herring in the Northwestern Pacific. During the 1990s their number in Alaska waters rose dramatically and triggered a number of catching, big game angling, filming and scientific activities.

Salmon sharks are well-known to commercial fishermen in Alaska who do not particularly like these fish because they damage gill nets and 'steal' fish and gear. Such encounters are usually costly to the fishermen and deadly for the sharks. Salmon sharks are most common there during their seasonal surface migrations into southern Alaskan waters from July to September. They associate with the roe herring fishery in spring from April to May and the fall herring bait fishery from September to October. In the Prince William Sound area they are most abundant during July and August which corresponds with the return of adult salmon from the open sea.

Because of its size and powerful appearance this shark has been regarded as potentially dangerous, but has been photographed (filmed) without any notion of attack in Alaska waters already in the 1980s (1990s).

Photo below from Prince William Sound, Alaska, North Pacific.

MARK STRICKLAND

Lamna nasus
Mackerel shark, porbeagle

Length: At birth 60-75, max. possibly 370, mature females 152-219 (possibly 370), mature males 219-262 cm. Distribution: Amphitemperate. Northern hemisphere: North Atlantic: South Carolina to Iceland, western Barents Sea, Morocco and Mediterranean. Southern hemisphere: Chile, Argentina, South Africa, southern Australia, Tasmania, New Zealand. Depth: 1-366 m. General: A heavy, spindle-shaped shark with a short conical pointed snout, medium-sized blade-like teeth with lateral cusplets and a secondary keel on the lower lobe of the caudal fin below the main lateral keel on each side of the caudal peduncle. Coloration dark grey above, underside white, without dusky blotches, free rear tip of first dorsal fin white (other than in the related salmon shark *L. ditropis*).

This shark is most common on the continental shelves but also found far offshore. It prefers water temperatures below 18°C and is fished commercially by longlining. The flesh is sold as shark steak in Europe. Also caught by anglers, especially at the British Isles, but described as sluggish when compared to the shortfin mako, does not leap clear of the water. Catch data of European fisheries indicate segregation by sex and size.

Reproduction is ovoviviparous: Only 1-3 large embryos develop in each of the mother's uteri where they feed on fertilised eggs (oophagy) and smaller siblings (adelphophagy, uterine cannibalism), thus developing grotesquely swollen bellies and enlarged branchial regions. In European waters they are born after an estimated gestation period of about 8 months. On the other side of the Atlantic, off the U.S. coast from Massachusetts to Maine, there also are breeding populations with pregnant females present throughout the year. Sexual maturity is reached after approximately 5 years, maximum age for this species probably exceeds 30 years.

The mackerel shark feeds on a wide range of pelagic and demersal schooling bony fishes such as mackerels, herring and cod, but also on tope *(Galeorhinus galeus)*, various dogfish sharks and squid. Because of its size and powerful appearance this shark has been regarded as potentially dangerous, however, there are no confirmed reports on attacks by this particular species. Its cold water habitat makes meetings with divers unlikely.

During the early 19th century, the porbeagle was in great demand for its liver oil which was used in the leather tanning business. During the 1960s, it was fished in Canadian waters by Norwegian fishermen; the annual catch was in the range of 4,000 tons, all destined for the food markets in Europe. Due to overfishing, this fishery declined. The porbeagle is usually caught on hook and line put out for cod. Baits should be deeply set and the wire leader on the line should be long as this shark has the habit of rolling up on the wire which often causes the line to abrade and break.

Photo below from Guernsey, an island in the British Channel, Northeastern Atlantic.

Isurus oxyrinchus
Shortfin mako

Length: At birth 60-70, max. 400, mature females 280-400, mature males 195-284 cm. Distribution: Temperate and tropical coastal and oceanic waters worldwide (photo below from California; bottom photo from Costa Calida, Spain, Mediterranean). Depth: 1-152 m. General: This magnificently streamlined shark is identified by a single keel on each side of the caudal peduncle, large black eyes, a sharply pointed long conical snout, very characteristic large narrow curved teeth without lateral cusplets, a moon crescent-shaped caudal fin and pectoral fins shorter than the head (longer than head in the otherwise similar longfin mako *I. paucus*). Coloration brilliant blue above, underside abruptly white. One of the fastest and most widespread sharks, has been measured to reach a speed of 70 km/h. It is found mainly in waters with temperatures above 16°C. Like in all family members, the body temperature of makos is higher than that of the surrounding seawater, permitting a higher level of activity. The pelagic species is only occasionally encountered in coastal waters, but then sometimes close inshore, just off the surf zone. The author examined a subadult male specimen caught in only three metres of water, just off the beach near Elat, northern Red Sea. Makos readily jump out of the water to get rid of crustacean skin parasites (specialised copepods) that can be seen trailing from the rear margins of fins and gill slits in many specimens. They are also well-known for their power, endurance and fighting capabilities including jumps clear of the surface after being hooked by anglers. Literature Nobel prize winner Ernest Hemingway held the all tackle world record for shortfin mako in the 1930s and wrote about this extraordinary shark in several of his novels, e.g. 'The Old Man and The Sea'. The mako's mode of reproduction is ovoviviparous, litter size 4-16, young are uterine cannibals. Prey includes primarily bony and cartilaginous fishes (mackerel, tuna, bonito, anchovy, herring, grunt, cod, ling, whiting, horse mackerel, sea bass, swordfish, blue sharks, requiem shark, hammerhead shark), but also turtles, dolphins, squid and rarely marine mammals (the latter are found solely in large makos which have more flattened, blade-like upper teeth than smaller individuals). The shortfin mako is known to have attacked people and boats, especially after being hooked. Unfortunately, there also are a few reliable reports on unprovoked attacks on swimmers, at least one of which resulted in the death of the female victim after she suffered from the loss of a limb and post-traumatic depression. The flesh of the mako is excellent eating, with a natural lemon aroma in cases, but only seldom utilised throughout the wide range of this impressive species.

DAVID HALL

JUAN C. CALVIN

Mitsukurina owstoni
Goblin shark

Length: Max. 335, mature males 264-322 cm. Distribution: Western Atlantic: French Guiana. Eastern Atlantic: France, Madeira, Portugal, South Africa. Western Indian Ocean: South Africa. Western Pacific: Japan, Australia. Depth: Shallow waters (only occasionally), down to at least 550 m.

General: A bizarre, unmistakable deep sea shark with a very much elongated, flat, blade-like rostrum (snout) and long, protrusible jaws which bear very slender teeth with long cusps. Body soft and flabby, eyes tiny, caudal fin very long and straight, without lower lobe. Coloration pinkish-white in life.

The goblin shark is an uncommon benthic shark whose habits and biology are only very poorly known. It lives on the outer continental shelf and upper slope region, only rarely occurring in shallow inshore waters. The long caudal fin, the soft consistence of the body and the small paired and unpaired fins suggest that this shark is a slow swimmer with a density close to that of seawater. The blade-like snout superficially resembles those of the teleost (not related!) paddlefishes of North American and Chinese freshwater streams. The paddlefishes of the family Polyodontidae are related to sturgeons and use their elongated snouts to detect prey items. The goblin shark may well use its snout in the same manner. Its pick-like anterior and lateral teeth are very long and have slender cusps, the posterior teeth are short and stubby. It is suggested that this shark picks up relatively small, soft-bodied prey such as fishes, shrimp and squid with its slender front teeth and subsequently crushes such prey between its posteriors, much like the sandtiger shark does. The goblin shark's jaws are highly specialised and can rapidly be projected from the head. Such an elaborate jaw anatomy suggests that the jaws are used in snapping up small, but fast-moving prey items, much in the same way certain mesopelagic teleosts catch their food.

There is a lot left to be learned about this spooky-looking denizen of the deep which is closely related to the Upper Cretaceous fossil shark genus *Scapanorhynchus*. Of the latter, complete fossils are known which have a very similar rostrum, but differ in some details and hence are placed in a separate family, Scapanorhynchidae, by some scientists.

Because this species is not common and usually found in deep waters, the interest to fisheries is minimal. The goblin shark is taken as bycatch in the deepwater trawl fishery and is only occasionally caught with deepwater longlines, gillnets or purse seines. Its flesh is utilised dried and salted.

DAVID SHEN / IV

Megachasma pelagios
Megamouth shark

Length: At birth: Unknown, max. known 515, mature females unknown, mature males 400 cm.
Distribution: Widely distributed in tropical to temperate waters of all oceans. The locations of the 14 specimens found up to date include Japan, Philippines, Indonesia, Southwest Australia, Hawaii, California, South Brazil and Senegal.
Depth: 150-1,000 m.
General: The first specimen of this extraordinary shark has been discovered in 1976; since then only 14 specimens have been found. Being so extraordinary, the megamouth shark was placed in its own family (Megachasmidae) of the order Lamniformes. It is easily recognised by its large size and its large rounded head with a short snout and a large terminal mouth which exceeds 1 m in width in adults. The jaws are strongly protrusible and lined with numerous small hook-like teeth which are set in over 100 rows in each jaw. All fins are small with the exception of the pectoral fins and the caudal fin which has a strongly elongated upper lobe. The caudal peduncle lacks keels. Coloration is dark grey to bluish black on the back, fading to a lighter colour on the sides; the underside is pale grey. Some specimens show a faint pattern of spots on the lower jaw. The posterior margins of most fins and the tips of the pectoral fins and the pelvic fins are white. The inner lining of the mouth is silvery in life and becomes black after preservation.

This oceanic shark lives pelagically in deep water where it feeds on planktonic organisms such as euphausid shrimps, copepods and comb jellies. It is assumed that the silvery interior of its mouth cavity is luminous and may serve to attract crustacean prey. The megamouth shark itself is possibly preyed upon by other large sharks; it is also one of the victims of the cookiecutter shark as characteristic wounds have been found on its skin. Its soft flabby body and relatively small gills suggest that this shark is much less active than the other two large planktivorous sharks, the (related) basking shark and the orectolobid whale shark. Its large liver is rich in low density squalene (a fatty compound) and - together with the poorly calcified cartilaginous skeleton and the watery flesh - probably helps to achieve neutral buoyancy in the water. The mode of reproduction is yet unknown because the largest female found was still immature.

The few known specimens of the megamouth shark were retrieved from a Navy vessel's parachute-like drift anchor, from set nets, a longline and after stranding on beaches. Because of its preference for planktonic food, chances are low that this shark is caught on hook and line; therefore all finds have been more or less haphazardly and further specimens are badly wanted by scientists in order to reveal more about the biology and habits of this most unusual creature. Some specimens have been conserved and are on display in Japan, California and Australia. A few specimens were found alive and returned to the ocean, one of them carrying two ultrasonic transmitters (one with depth sensor). The shark was tracked for 50.5 hours, during which it made distinct vertical moves at each dawn and dusk transition, staying shallow at night (depth 11-22 m) and deeper during the days (120-166 m, well above the 700-850 m bottom). These moves appeared to be triggered by light changes. Horizontally, the shark moved slowly (0.95 km/h), on a relatively straight path (Nelson et al., 1991). The photo below shows the first specimen found alive, off California. See also following pages.

TOM HAIGHT

MEGAMOUTH

What lives in the sea, is about five metres long, weighs about a tonne and was totally unknown to science until 1976? You have guessed right, it is the giant megamouth shark. Its relatively late discovery was a sensation for experts and interested lay people alike. So far only thirteen specimens of this extraordinary and apparently rather rare shark species have been seen, including several live ones. Helmut Debelius reports:

To this day this was a unique encounter between a diver and a live megamouth, which was released with a transmitter attached.

It happened on 15 November 1976. Only about 25 miles offshore from Oahu (Hawaiian Islands), on a bright and sunny day, a United States Navy ship rolls in the ocean swell. It is under orders to search for and recover lost practice ('dummy') torpedoes. However, as the ship is about to change position, a dead marine animal drags down its parachute-like drift anchor. A giant fish – of a type never seen before by man – is pulled from the sea. It is five metres long, of brownish colour and has a conspicuously large mouth, the jaws bearing thousands of small teeth.

Soon it became obvious that a new shark species had been discovered. Yet, this single specimen was so unusual that it took another seven years for it to be described scientifically as *Megachasma pelagios*, the only species in its own family Megachasmidae (Order Lamniformes). At that time, shark expert Dr. John McCosker considered the possibility of finding another specimen of this species as being extremely unlikely. However, eight years later, and only a short time after the publication of the original species description, another (dead) specimen was brought up in a fishing net off the coast of California. Four more years down the track the same events occurred again in the vicinity of the West Australian city of Perth. Finally, in October 1990 the sensational event happened: A fisherman, working off Los Angeles, caught a live megamouth shark in his net. He tied the shark to his boat and telephoned scientists. Dr. Robert Lavenberg, who coordinated the subsequent investigations, referred to it as a living fossil, which had undergone hardly any changes during the last ten million years. He telephoned around the world in order to find a suitably large aquarium for this rare find, but his efforts were in vain. There was simply no aquarium large enough to hold this megamouth shark. Consequently, tissue samples were taken at great haste from the defenceless animal; it was measured, photographed and filmed. The general interest in this exceptional shark was at least as large as its giant mouth. The underwater photographer Tom Haight exclaimed enthusiastically: 'While I was swimming around the shark it turned its head and its eyes were following my every movement'.

So far, scientific investigations have revealed the following picture: These sharks have about 236 rows of needlesharp, small teeth in their giant mouth. They live at depths from 200 to 1,000 metres, and their eyes are large and of a brown colour. The cartilaginous skeleton is soft and pliable, and the rubbery skin encloses loosely a weak musculature; in essence then, this is not a shark that swims at great speed through the ocean and tears its prey apart with overpowering strength. Instead, giant megamouth sharks push themselves along slowly and leisurely and at great depth. They feed by filling their large mouth with water, straining it through their teeth and then swallow whatever is being retained: Small sea jellies, juvenile fishes and various crustaceans. In addition, these filter feeders have a strip of tissue along their lips that glows in the dark and is to attract prey organisms. The understandable curiosity among shark experts about sex-specific differences and clues to the reproductive behaviour of megamouth sharks was finally satisfied in 1994, when a dead megamouth female was washed ashore on the coast of Japan. The carcass (471 cm, 790 kg) was in good condition and so it was preserved frozen for subsequent scientific studies.

Based on the locations where megamouth sharks had been found up to 1995, it had been assumed that the distribution of these sharks was confined to temperate waters of the Indo-Pacific region. But the puzzle was still not complete yet.

This dream of every shark enthusiast only came true in 1990.

In May 1995, off Dakar (coast of Senegal), the captain of the French tuna catching vessel *Le Bougainville* was surprised when he pulled his net out of the water and sorted through the catch to find a strange, 180 cm long fish among all the skipjack tuna *(Katsuwonus pelamis)*, filling his net to the brim. He was familiar with the manta rays as well as three other sharks in the catch, but then he suddenly becomes suspicious and quickly makes the right decisions: He freezes the smallest of all megamouth shark specimens ever found. Simultaneously, he also records relevant data, such as the unusually high ocean temperature of 22.5° Celsius.

The fact that the third-largest plankton-feeding shark is not quite as rare in the Atlantic Ocean as had been previously assumed, is supported by the capture of a megamouth shark off the coast of Brazil in September 1995. This then confirms definitively that the new star of cryptozoology does occur in all oceans.

Dr. Barry Hutchins, ichthyologist and Curator of Fishes at the *Western Australian Museum* (Perth) was an eyewitness to one such find and reports first-hand: On that particular morning I received a telephone call. A large, very unusual animal was reported to lie on the beach, something nobody had ever seen before. It appeared to be a cross between a shark and a whale. I was supposed to come out there and take a look at this extraordinary creature. Just what I needed, I thought...merely because of a rotting whale! But the caller was adamant and somehow the unusual details made me curious. And so I raced the 50 km to Mandurah, south of Perth (Western Australia). When I got there a large crowd was already waiting for me. Helicopters from various media organisations had landed on the beach and there were reporters everywhere. When I finally got to the animal, it was (with 5.2 m) longer than I had anticipated. It had fins and gills, just like those of sharks. But the shape of the head was totally new to me; it almost resembled a baby whale's head ! With that then there was no doubt; I was indeed looking at a megamouth shark.

Questioning the finder of this rare sea creature revealed, that surfers a day earlier had sighted it at the surface near the beach. At that time it was assumed to be a whale. Since whales are sometimes known to kill themselves by swimming onto a beach, and to prevent this from happening, the animal was towed back out to the open sea, but without success. Next morning, the animal was found dead on the beach. In the meantime teachers and their students started to arrive, to get a first-hand look at the animal. On that day I counted about 4,000 people, gathering around the megamouth shark. Therefore, it was quite difficult to transport the 700 kg giant to our museum, where it was preserved in formaldehyde and where it is now on public display.

All of us have a dream to do something nobody else has ever done before. Tom Haight, a SCUBA diver and underwater photographer got his chance on 22 October 1990: An unusually warm current had attracted many unexpected visitors that summer to the coast of California. Never before had I seen so many snappers, trigger fishes or dorados in the home waters of our State Fish, the Garibaldi, off Los Angeles, and I was taking photographs of these fishes with great enthusiasm. Just as I was mounting all this slide material, I received a telephone call from a fisherman friend. He recounted to me with enormous excitement that he may have caught a specimen of the rare megamouth shark, and when recognising this fact he had towed the live fish into the Port of Los Angeles with his boat. Right now it was swimming below the boat. Quickly, I packed my NIKONOS with a 15 mm lens and then raced down to the harbour. Indeed, it was a megamouth shark, as confirmed by a biologist who had arrived in the meantime. Hurriedly an operational plan was devised on how to acquire optimum scientific data from the healthy-appearing shark. A transmitter was to be attached painlessly to its dorsal fin, before it was to be released the next day. Filming the animal was to document the swimming movements of this deepsea denizen. The next day I got into the water with great anticipation and enthusiasm in order to photograph the five-metre-long shark. The animal was calm so that we could record all details as required, while it was still tied up. Biologists took the measurements, and after the transmitter was attached, the rope was untied. Slowly the large shark descended into the blue depth below. To see this extraordinary animal and to be part of all these activities was probably the most exhilarating event in my 28 years as a diver! To this day it was a genuinely unique encounter between a diver and a live megamouth shark, which was allowed to return to freedom with a transmitter attached. This is a dream every shark enthusiast has and it only came true in 1990.

This family contains one monotypic genus (meaning it has only one species). The basking shark is a schooling, plankton feeding species, and the second largest shark (after the whale shark) and fish in the world. It has a conical snout (often grotesquely pointed and deformed in young individuals), extremely long gill slits, densely packed stacks of gill rakers (modified dermal denticles, that consist of the same hard materials as the regular teeth), pronounced keels on the caudal peduncle and a lunate (half moon-shaped, more or less symmetrical) tail. Each jaw is lined with several rows of minute teeth, the entire dentition resembling bands of tiny hooks. The teeth may hold back small prey items, but the bulk of the daily food ration is gathered in a unique way: Basking sharks swim through the water with their large mouths wide open in order to filter as much water as possible over their erect gill rakers for plankton. The latter adheres to the mucus covered rakers and gill arches and is squeezed into the esophagus when the gill rakers are collapsed while the mouth is closed. The gill rakers are shed and replaced approximately every 4-5 months. Whether the sharks stop feeding and retire to deeper waters during this period is not clear. Individuals without rakers have been caught near the surface, allegedly feeding. Attempts to track the migration paths of individual basking sharks with satellite transmitters attached to them has not yet revealed the entire story of "Where do they go when they do not feed?" The liver is huge and makes this shark almost neutrally buoyant when it cruises oceanic or inshore habitats. It is highly migratory and seasonally abundant in certain areas. For example, it is most common off the coast of Scotland in spring and summer; off California it appears during fall and winter. Little is known of the reproduction, although the species since a long time has been (and still is) the target of commercial fisheries for oil and meat, e.g. around the Hebrides, Scotland, off the coast of Norway and in Japan. As a result of overfishing during the last decades a dramatic drop in population numbers occurred. The species is presumably ovoviviparous and oophagous, fecundity is low; all that makes the stocks vulnerable to overfishing. Females have only one functional ovary which may contain up to 6 million tiny eggs (5 mm). Often crustacean parasites (copepods), sometimes also lampreys, adhere to the skin of basking sharks.

Cetorhinus maximus
Basking shark

Length: At birth 170 or less, max. at least 980 (doubtfully reported to 1,220 and 1,520), mature females 810-980, mature males 400-500 cm. Distribution: Circumtemperate. Depth: Surface to possibly 200 m and deeper.

General: The characteristically long gill slits almost meet on the ventral surface of the head. Coloration dorsally brown to slate grey or black, often mottled with lighter areas. Snout of young individuals longer and more pointed. Occurs in bays and fjords, near offshore islands and in the open ocean. Observed singly, in small groups or large aggregations of over 100 individuals. They often swim, and feed, with their first dorsal and upper caudal fins projecting from the water's surface. Two, three or more individuals may swim in tandem and have repeatedly been mistaken for a single huge 'sea serpent' in the past. Dead basking sharks are sometimes stranded on a beach, in a state of decay and partly dismembered from rolling in the surf; more than once such carcasses have been misidentified as 'sea monsters', particularly so-called 'sea horses', which is due to the mane-like appearance of the exposed decaying gill filaments of the shark.

Mating has not been observed, but courtship may involve males following females. After mating, both sexes show abrasions near the vent region caused by their rough skin rubbing against each other. From scar patterns in the vagina it has been deduced that females mate more than once in a breeding season. In parts of the range (Scotland) mating occurs in inshore waters in the early summer. After mating, females move offshore when they give birth to their young. Reproduction seems to include a gestation period of more than a year and ovoviviparous development of 2 - 6 young, but evidence is scarce since females caught during the summer fishing season are not pregnant.

Prey mainly consists of minute planktonic crustaceans, pteropods and fish eggs. Divers may swim unharmed among a school of filtering basking sharks that often are encountered immediately at the surface in a basking fashion (hence the name). See also following pages. **Howard Hall**

California, Eastern Pacific

MOUTHS WIDE OPEN

The port of Peel is located on the West Coast of the Isle of Man, a small island - 50 km long and 20 km wide - geographically centred in the middle of the Irish Sea. Peel lies at the base of a vast rolling hill and in the shadow of the ramparts of a thousand-year-old castle. High, rocky cliffs rise up on either side and the jade green waters of the Irish Sea lay before a long, tall cement breakwater that shelters the entrance to Peel harbour. On a clear day, the headlands of Scotland and Ireland can be seen looming upon the horizon. What an impression for journalist Douglas David Seifert, who came straight from the South Pacific to write a report on a giant shark.

The basking shark, the second largest shark (and fish), is a gentle giant that filter-feeds on plankton with its mouth wide open.

The richness of the Irish Sea is made abundantly clear as all manner of fishing boats unload their catches at quayside: Boxes upon boxes of crabs, lobsters, scallops, and langoustines are hoisted from their decks and stacked onto waiting trucks. Cranes strain to lift overstuffed nets full of small silver fish from the vessels' holds; the fish flow out like a quicksilver river as the catch is poured into vast containers. A profusion of sea birds squawk and compete for fish scraps. Cars come and go amongst wandering crowds out for a morning stroll. Amateur anglers cast their lines for plaice and mackerel. Jellyfish can be seen bobbing and blobbing just beneath the water's surface. The scene is quite lively.

Amidst the activities of this working pier, a dozen school children are making their way down a cement stairway and boarding an old, well-used but quite seaworthy, wooden fishing boat. They are on a school field trip to explore a unique phenomenon that occurs in their own backyard. A tall, lean man with a moustache and salt and pepper hair helps the children on board with a smile and welcomes them. He is Ken Watterson and the pupils, including his daughter, are his guests, part of his commitment to education. The lines are cast off and the vessel, *Jasmine*, captained by the ruggedly able Brian Maddrell, slowly motors out to sea.

In other parts of the world, it is increasingly more common to find whale-watching excursions. On the Isle of Man, people depart for shark watching. Basking shark watching. It is completely unique to this tiny island and to this small town and it is wholly the creation of Ken Watterson. This commercial enterprise serves multiple purposes: To raise awareness and educate people about basking sharks and an attempt to provide some kind of funding and an operational platform to study the basking shark. Ken Watterson founded 'The Basking Shark Society' as a registered charitable trust in the early 1980's. It remains self-funded, even under-funded, without any support from the local Manx government or from the United Kingdom. The monies raised by the shark watching trips are an attempt to offset the costs of research and the expense of a boat, a captain, and sea time. Watterson left a successful career in banking to study the basking shark and receives no salary. The entire operation runs on a shoestring budget with long days and little sleep for Watterson and Maddrell during the five month basking shark season. The only support they receive is a generous donation of time and assistance by befriended volunteers on their days off from work.

The basking shark that fascinates Watterson so and tempts the day-trippers' imaginations is the second largest fish in the sea, which can reach a length of nine metres and weigh in excess of 3,500 kilograms. In the

Ken Watterson, founder of the 'The Basking Shark Society', with students aboard the *Jasmine*.

shark family, the basking shark *(Cetorhinus maximus)* is most closely related to the great white shark, but fortunately, like its larger, distant cousin the whale shark and decidedly unlike the great white shark, it is a harmless filter-feeder who eats only plankton. Both the basking shark and the whale shark are found in all oceans, but the whale shark is restricted to tropical waters and the basking shark is found mostly in temperate waters. And, as far as anyone knows, nowhere is the basking shark found more regularly and in greater abundance than in the British Isles during the summer months.

From May to September - for reasons unknown but the subject of considerable speculation - basking sharks are particularly plentiful in the waters within five kilometres of the west coast of the Isle of Man. The sharks may be in other places as well, but one of the difficulties in studying sharks - as opposed to say, marine mammals, which have a marvellous trait of having to breathe air (a clear indicator of whether they are around or not) - is that unless sharks come to the surface, they remain invisible and unknowable. They could be stacked as thick as cordwood but if they are more than a few metres beneath the surface, there is nothing to denote their presence. Fortunately, basking sharks need to eat an enormous quantity of plankton and will follow that food source wherever it occurs.

Plankton is the basic building block from which the inverted pyramid of the food chain rises. Phytoplankton are drifting microscopic plants, and as plants, require sunlight to nourish themselves and to reproduce, hence they are often found near the surface. They are, in turn, fed upon by zooplankton, tiny invertebrate animals. Many species of zooplankton are more capable of independent movement than their drifting, suspended prey and are known to make vertical migrations in the water column as they feed upon the phytoplankton.

During the summer, a combination of water temperature and currents, possibly connected to the ten-metre tidal changes in the relatively shallow Irish

Whenever Ken has the opportunity he snorkels with the basking sharks.

Sea, seems to provide favourable conditions for a population explosion of phyto- and zooplankton. And with this population explosion comes a massing of plankton predators, from herring and mackerel, to jellyfish and basking sharks. When this explosion occurs at the surface, it often brings the basking sharks within range of human observation. And since whatever phenomenon in nature signals a basking shark that food is in a certain localised area, it may also signal a number of basking sharks that food is available. Basking sharks may occur en masse at such times, with gatherings of eight individuals to 200 sharks not uncommon.

As *Jasmine* motors along, Watterson gives a talk to the school children about the sea life around the Isle of Man and shares his knowledge of basking sharks. Watterson is soft-spoken and earnest, exuding a warmth and dedication to which the children can relate. He explains that basking sharks were once hunted for their livers, which accounts for one-third of their weight and contains oil high in concentrations of squalene, used in the lubrication of engines, for cosmetics and pharmaceutical products. Synthetics have replaced sharks as a source of squalene, but they are still hunted, mostly by the Norwegians, for their fins, which are sold to the Asian markets. The fishing industry in Scotland was so effective in hunting basking sharks that the population abruptly crashed and has not recovered in the decades since the industry closed.

The answer that humans eat sharks provokes the question of what and how much do the sharks eat. Watterson explains that like in many of the baleen whale species, the basking shark's preferred food is copepods, tiny planktonic crustaceans. Pity the poor copepod: It is estimated that in order for a ten-metre basking shark to meet its energy requirements, it needs to consume more than 500 litres of zooplankton per day. It achieves this by swimming at a speed of 2 - 3 knots and filtering around 1,800 cubic metres of water through its mouth per hour, which is roughly the equivalent of the volume of an Olympic-sized swimming pool, per hour! The shark swims with its mouth open wide - the opening larger than a metre across - and the water flows over stacks of flattened, rod-like strainers called gill rakers, before continuing over the gills and out the gill slits. The shark swims and filters until the gill rakers have collected a satisfactory amount of copepods, then the shark closes its mouth, forces the remaining water out its gills and swallows the copepods. Then it begins again until the supply of plankton is exhausted.

Captain Maddrell has the binoculars to his eyes and scans the horizon. The sea is flat, with little wind to stir the waters, excellent conditions for finding basking sharks. He is looking for anything out of the ordinary breaking the featureless plain of the sea's surface. Something catches his eye. It is a fishing buoy. But the weather is variable - extremely variable - during the summer, from Wrath of God storms to the appearance of a millpond. It is the afternoons that usually bring the calmest weather.

Watterson concludes his talk and excuses himself to climb to the crow's nest platform high upon the forward mast. As he climbs the tall ladder, he calls out: "Fin!" All eyes are directed to a black dot on the horizon. The boat heads towards the spot and as the distance closes, the black spot is seen to be a dorsal fin. It is a big fin, projecting almost a metre in height above the water's surface. It is grey, with cream coloured speckles and it moves slowly, almost randomly across the water's surface. Unlike the movie shark's fin, this fin isn't straight and rigid. It seems to flop, from on side to another, as it moves.

Above: Schools of up to 200 basking sharks were common in earlier days. Today, seeing more than a few is a rare exception.
Facing page: The black openings of the gill arches give the inside of the basking shark's white mouth a ribbed appearance.

The big first dorsal fin of the basking shark projects almost a metre in height above the water's surface, moving slowly across it.

As the boat is manœuvered closer still, the pupils can see the entirety of the basking shark swimming just below the water's surface. The shark's body is a wrinkled, brownish-grey and its mouth is open at full gape, distended, the throat flared wide from the shark's body. The shark's nose seems sharply pointed and its eye is seen above the arch of its open jaw. A tiny sliver of pink gill peeks out along each of the five gill slits on either side of the shark's throat. In contrast with the cloudy green water, the inside of the mouth is white, so white that it glares. The black openings of the gill arches give the mouth a ribbed appearance. Jellyfish dot the waters all around the shark, but it seems to avoid ingesting them with a delicacy and precision that can only be by the shark's choice. It passes close to the bow, seemingly non-plussed by the boat's presence amidst its feeding area, essentially where it is performing the aquatic equivalent of grazing.

The shark's lack of concern about a twenty-metre vessel immediately next to it makes it easy to understand how fishermen could approach basking sharks to harpoon them. Whenever Watterson has the opportunity he snorkels with the basking sharks. The waters of the Irish Sea are warm for the British Isles, thanks to the influence of the Gulf Stream. Although he used to wear a wetsuit, Watterson now uses a drysuit when he is in with the sharks. Most of the scientific data that is known about basking sharks comes from dead specimens. Watterson is trying to change that. He does as much in water observation as he can, attaching identification tags to the sharks, noting any unique physical characteristics, determining their sexes, photographing and videotaping them and observing behavioural patterns. So far, he has successfully attached one satellite tag which should help unlock the mystery of basking shark migration.

There are two schools of thought in basking shark migration. One belief is that the sharks make a long distance migration marked by their appearance off Spain in the spring and continuing along the coast of Brittany and into British Waters, ending ultimately with the disappearance of the sharks off the coast of Norway in the fall. It is assumed the sharks make their migration back to the south in deep waters out of the sight of man during the winter. The other theory of shark migration holds that the sharks are localised in each of the areas they are seen: Spain, Brittany, the Irish Sea, and Norway and they do not travel such long distances. This theory contends the sharks move offshore into deeper waters in the winter and are only seen in the summer when they have moved inshore. The inshore/offshore theory seems more plausible than the long distance migration theory because of the scarcity of basking sharks off the coast of Scotland where they were heavily fished. It seems logical that if there was a long distance migration, those numbers would remain unaffected, but their scarcity indicates an eradication of a localised population. Until Watterson can satellite tag more basking sharks and follow their movements with high technology, this mystery will remain.

Jasmine turns back towards the coast and returns to Peel. The school children have had a terrific day, and are all abuzz about seeing the world's second largest fish and right in their own backyard. For Watterson and Maddrell, it is back to port for a quick turnaround and another evening trip out to see the basking sharks. No time off today. The staff from a local bank has chartered the boat for the evening cruise and there is hope that some of the funding 'The Basking Shark Society' so desperately needs might be sponsored by the bank. Until then, Watterson will continue to work sixteen-hour-days, under-funded, and having to make compromises that set back the cause of their research in order to pay the bills, unable to spend much of their waking hours with their families during the long summer of basking sharks.

This family of unusual sharks comprises three species all of which are shown here. All are characterised by a long upper caudal fin lobe that is nearly as long as the body. This exaggerated tail is used to herd and stun schooling fishes when these sharks hunt in small packs. An old report even states that a common thresher slapped a loon that was sitting on the water's surface. All family members are known to be ovoviviparous and oophagous like their lamnid cousins.

Alopias superciliosus
Bigeye thresher shark
Length: At birth 64-106, max. 460, mature females 355-430+, mature males 270-400 cm. Distribution: Oceanic and coastal, circumtropical. Depth: 0-150 m. General: This thresher shark is easily distinguished from its congeners by two deep grooves on top and sides of the head and its comparatively very large eyes situated high on the head. The grey coloration of the back gradually fades to white on the belly; there is no white on the bases of the pectoral and pelvic fins as in the common thresher shark *A. vulpinus* (below). Mainly found in the open ocean, although it is also caught nearshore, e.g. in the Azores (photo). Prey includes small pelagic fish and squid which are grasped with about 30 rows of small unicuspid teeth in each jaw.

RICARDO SANTOS

Alopias vulpinus
Common thresher shark
L: At birth 114-150, usually to 400, max. possibly to 610, mature females 376-550, mature males 319-420 cm. Di: Oceanic and coastal, circumglobal in warm seas. Atlantic area records are from the North Sea and Mediterranean south to Gulf of Guinea, and from South Africa. De: 1-366 m. G: Distinguished from the pelagic thresher (following page) by overall coloration (white of belly extending up and over the pectoral fin bases into the grey of the back) and shape of pectoral fin tips (rounded in pelagic, pointed in common thresher). A fast swimming shark with a body temperature several degrees higher than that of the surrounding seawater, a physiological adaptation also found in great white, mako, salmon and mackerel sharks. Shy, difficult to approach underwater. The photo was taken in 20 m depth in the Visayan Sea, Philippines, where several of these sharks gathered at a cleaning station on a seamount.

SCOTT TUASON

Alopias pelagicus
Pelagic thresher shark

L: At birth 96, max. 330, mature females 264-330, mature males 276 cm. Di: Wide ranging in the Indo-Pacific: Sea of Cortez and Galapagos Islands to Red Sea and South Africa. De: 0-150 m. G: Pectoral fins straight with rounded tips. Flanks dark, white of belly not extending up and over the pectoral fin bases. Eyes clearly smaller than in *A. superciliosus*, compare photos. Oceanic, epipelagic, occasionally observed by divers around seamounts. Prey includes bony fishes and squid. Often leaps out of the water, up to five leaps in sequence have been observed. A shy species, difficult to approach underwater. Ovoviviparous and oophagous, litter size 2.

The two Brother Islands (in Arabic: El Akhawein) are located about 65 miles SE of the Red Sea port of Safaga. Weather conditions have to be good in order to make the 8-hour-trip from the mainland to the remote islands. Big Brother is famous for its shipwrecks, but Little Brother is preferred by diver Thomas Reich who was lucky to find an extraordinary sight there: "On one of the dives I met a female thresher shark of about 3 metres total length at a depth of 30 metres where it was swimming at an angle in a head-up stance in the downward current near the boundary layer. It kept its almost stationary position by slowly moving its long caudal fin. When I approached with my camera, it left its position, circled me once and disappeared into the blue water."

THOMAS REICH

70

This family comprises four species. All have long, dagger-like teeth that protrude from the jaws. Anal and dorsal fins are large, the pectoral fins are small. The eyes are quite small and lack a nictitating membrane. Family members are considered sluggish, but capable of quick lunges at prey. Sandtiger sharks have a peculiar mode of reproduction: They are ovoviviparous, but only one embryo in each uterus can grow to term as the unborn shark eats its siblings (embryophagy, adelphophagy) and fertilised or unfertilised eggs (ovophagy) inside the uterus (intrauterine cannibalism). As a consequence the term size, at which the young hatch at the end of their gestation period, is relatively large, but litter size and fecundity are low when compared to that of other sharks. Mode of reproduction of other family members unknown, but probably similar to that of the common sandtiger.

The bigeye sandtiger shark *Odontaspis noronhai* has been described in 1955 from Madeira, Portugal, where the holotype is displayed in the museum at Funchal. The holotype from the Atlantic is the only specimen known, but a set of jaws from the Seychelles in the Indian Ocean confirms the validity of this species.

The teeth of extinct sandtiger species are certainly one of the most numerous macro-fossils and can be found in many locations all over the world where Tertiary marine sediments are exposed. The extinct species differ in tooth morphology and hence are placed in different genera like *Synodontaspis* and *Striatolamia*.

Odontaspis ferox
Smalltooth sandtiger shark

Length: At birth about 100, max. 360 (females), males mature at 275 cm. Distribution: Eastern Atlantic: Gulf of Gasconey into Mediterranean. Western Indian Ocean: South Africa, Madagascar. Pacific: Japan, Australia, New Zealand, Kermadec Islands (where observed infrequently by divers), Hawaii, southern California, Sea of Cortez, Isla Malpelo (Colombia). Depth: 13-420 m.
General: First dorsal fin larger than second dorsal. Coloration grey, belly often with blotches, some individuals have small reddish spots scattered over entire body. In the deeper layers of temperate and tropical waters. Has been observed by divers on coral reefs near drop-offs. Like the sandtiger shark has been reported to aggregate and swim amid large schools of baitfish. Prey includes bony fishes, squid and shrimp. Not aggressive towards divers, but encounters with this species are very rare. See also the following pages.
The photos shown here are the first from the natural habitat published in an identification book.

Clay Bryce   **Malpelo Island, Eastern Pacific**

MALPELO SAND TIGERS - PUSSYCATS OR SHEEP?

Many people, even if they are not really keen on sharks, know the awesome-looking 'all-teeth' portraits of the common sand tiger shark. But even for many elasmobranch enthusiasts the small-toothed sand tiger shark is still just one more name on the list of sharks they would like to see. Here they are! Clay Bryce risked a cold to photograph a rare member of the sand tiger family at Isla Malpelo, Colombia, in the central eastern Pacific.

PHOTOS: CLAY BRYCE

By all rights you should be able to see them as you descend into the depths surrounding Isla Malpelo – but you don't. The surface layer of water is warm and rich in plankton, giving it a soupy quality. At a depth of around twenty metres everything changes; the brain-numbing cold hits you like a slushy snowball and the water clarity expands your visual horizons to what seems like limitless vistas. As the bottom comes into view, so far down still, small shapes can be seen languidly moving about the flat rubble bottom. Nitrogen narcosis kicks in, you

Above: Seeing small-toothed sandtiger sharks - and that as far south as Isla Malpelo, Colombia - is not business as usual, even for shark specialists.
Right: Head-on portrait of a Malpelo small-toothed sandtiger. Its large, dark eyes and placid habits may remind one of more familiar, woolly creatures roaming the countryside back home.

If one gets close enough, *O. ferox* can be distinguished from the more familiar *C. taurus* by counting its more numerous upper jaw teeth.

hope your camera isn't flooding– better check!, doubly hoping the new flash units aren't quietly filling with water as well and then you arrive - bottom at 60 metres!

Odontaspis ferox or the small-toothed sand tiger shark seems to blend well with the dark waters as it quietly cruises the bottom. The clarity plays tricks with the eyes and the shark doesn't seem so big. As it approaches those famous 'small' teeth (I don't know which idiot named it small-toothed!) hang out of its mouth as it appears to grin. Its approach is without menace as it does a half turn to face you head-on. A stout body trails behind a seemingly small head with a comically pointed snout, strong pectoral fins angle downward and with that toothy grin your point of view suddenly changes as it swims straight at you.

It didn't seem so close through the camera lens. *O. ferox* swam right over the top of me, as if I didn't exist. It made several passes and when it went to swim away, a fellow diver herded it back. An amazing sight to see a diver swim beside one of these 4.5-metre-long toothy pussycats, gently herding it this way and that – or are they more like very large sheep?

O. ferox was previously only known to occur as far south as southern California. The Isla Malpelo record, first reported by Captain Heinz Buchbinder, significantly increases its known range. This species is similar to the common sand tiger shark *Carcharias taurus* with long sharp teeth used to impale prey. *O. ferox* also seems to congregate, we sighted up to five individuals on one dive, and probably has communal habits. Their slow sluggish movements are also similar to the common sand tiger and divers should be wary as their habits could easily give way to complacency. The small-toothed sand tiger shark is big and bulky (lots of muscle) and it does have those 'small' teeth (lots of them) for a reason!

The snout of the small-toothed sand tiger is conical, that of the common sand tiger is flattened.

LEVITATING SHARKS

What do you do with a shark that doesn't move? After all, sharks are supposed to "endlessly cruise the world's oceans, their dorsal fins ominously slicing the water." That may all be well and good for blue sharks, but in North Carolina, they have another kind of shark that prefers to levitate a foot off the bottom, as motionless and implacable as an Indian fakir. Known as 'raggies,' or ragged-tooth sharks, in South Africa, and as 'grey nurse' sharks in Australia, they are popularly referred to as 'sand tigers' in the U.S.A. Underwater photographer Doug Perrine expectantly went into their 'arena' to catch a few 'tigers' on film.

Sand tigers at the wreck of the *Papoose,* an oil tanker which was torpedoed off North Carolina by a German U-boat in 1942.

In scientific parlance, common sand tiger sharks are *Carcharias taurus, Eugomphodus taurus* or *Odontaspis taurus,* depending on scientists' latest view of sand tiger nomenclature. By any name, these are distinctive sharks. They have a beautiful golden sheen to their skin, which is often spotted, and a high arched back sloping down to a narrow snout. Protruding from the snout are the nastiest looking recurved teeth you have ever seen. It is these hook-like exposed teeth that really endear these animals to photographers. They are perfectly adapted for seizing and holding fish, and are quite different from the triangular serrated cutting blades sported by species such as great whites, which specialise in dismembering large food items like mammals. Sand tigers feed primarily by night, and are fairly inactive by day. From a photographer's point of view, it's a perfect shark - one that looks like a man-eater in the photos, but is actually quite unaggressive and nearly harmless.

Lest you be fooled by their docile daytime demeanour, keep in mind that each sand tiger you see is the survivor of pre-natal sibling rivalry in its most vicious form. The largest embryo in each of the two uteri (wombs) of a female sand tiger consumes all the other embryos in that uterus, as well as unfertilised eggs which the mother produces to nourish it. With this steady supply of high-energy food, and no siblings to share it, the embryo grows rapidly to a large size. After 9-10 months gestation, two pups are born at a size of about one metre - among the largest of any sharks at birth. In addition to the advantage of size, the pups are born with fully developed teeth, and experience in pre-

Above: If you can sit still in the stiff current near the *Papoose,* the 'tigers' will swim right up in your face. Below: By night, sand tigers feed ferociously on fish; by day, however, they don't show much interest in the schools of fish swirling around them.

dation. By night, sand tigers feed ferociously, primarily on fish such as drums and croakers, which live on open sandy bottoms. They can detect prey in the dark using their advanced sensory systems, which register even minute electrical fields, including those produced by living organisms. Introducing artificial stimuli into their environment can confuse these systems, as film maker John McKenney found out when a sand tiger that he was filming at night with the aid of bright movie lights suddenly turned and snapped off the metal arm supporting the lights.

By day, however, these sharks don't seem to show much interest in either the schools of fish swirling around them, or in divers that enter their domain. The way in which they hover in place is uncanny and a little unnerving. Don't they know that sharks don't have swim bladders? The lack of this little gas-filled organ, which keeps bony fish afloat, means that all sharks and rays have to either swim constantly, or rest on the bottom. All, that is, except sand tigers. These sharks, and no others, have learned the neat little trick of swimming to the surface and gulping air. When hooked by fishermen they sometimes expel the air loudly, giving them one of their other common names: 'Belching sharks.' A big bubble of air in the belly acts like a sort of internal buoyancy control, and gives these night shift workers the chance to hang around and doze during the day. Which brings us back to our original question. What do you do with a shark that doesn't move?

If you want to get a picture of a shark, you usually either try to place yourself in its swimming path, or put out some bait, and wait for the shark to swim to you. If it doesn't swim, there's not much to do except try to sneak up close enough for a picture without disturbing it. For this purpose, a rebreather, which doesn't release bubbles, is an obvious tool. Rebreathers have been used to great effect to film hammerhead sharks, which are extremely shy and bolt at the sound of a diver's exhalation. I knew that Howard Hall and Bob Cranston, who pioneered the use of rebreathers with hammerheads, had also used them to film the sand tiger sharks which aggregate on shipwrecks off the coast of North Carolina. So when a friend invited me to join him on a charter to photograph these sharks, I packed up my MK 15.5 rebreather and shipped it ahead to Morehead City.

The coast of North Carolina is often referred to as the 'Graveyard of the Atlantic,' due to the over 2,000 ships that have sunk there since the 16th century. At Cape Hatteras, the warm Gulf Stream collides with the cold Labrador Current, producing unpredictable currents and frequent heavy fog. Hurricanes and lesser storms, pirates, and war have created additional hazards to shipping. During WW II, this coast became a 'shooting gallery' for German U-boats, which sank 5 ships in one night in 1942. Some of the U-boats themselves were sunk as U.S. forces responded. Divers visit over 120 of the wrecks, and more are discovered each year. Only a few of the historically most significant, e.g. the Civil War *Monitor* and Blackbeard's pirate ship, *Queen Ann's Revenge*, are off limits to divers.

For years divers have been flocking to this coast to collect artifacts, take photographs, and spear the large fish which take shelter in and around the wrecks. It was not until 1980 that diving with sharks became a big draw. In the early 1980s, a new wreck was discovered, the U.S.S. *Tarpon*, a World War II submarine with a resident population of sand tigers. Male sand tigers migrate along the east coast from Florida to Cape Cod, stopping to mate in spring to early summer, but females are resident for most of the year. Just as word was getting out, the price of shark fins went sky high, and longlining boats moved in. By 1991, when federal management of the shark fishery began, the sharks on the *Tarpon* had been completely wiped out, as had most of the sharks in the area. The sharks never moved back onto the *Tarpon,* but as shark fishing quotas were reduced, they began to be seen on other wrecks, particularly the *Papoose*.

The *Papoose* is a 124-metre-long oil tanker which was torpedoed by a U-boat in 1942, and sank upside down in water 40 metres deep. Because of the sharks, and other marine life which it attracts, and because the water is generally very clear, the *Papoose* has become the most-dived shipwreck off the North Carolina coast. In 1998, complete protection was granted (in federal waters) for sand tigers, guaranteeing that divers will continue to be able to see sharks on the *Papoose,* and other wrecks in the vicinity.

The other two most-dived wrecks are the *Schurz* and the *U-352*. The U.S.S. *Schurz* is a 90-metre sail/steam cruiser, built in 1894. It collided with a tanker in a heavy fog in 1918 and sank. Today it consists mostly of scattered debris in 33 metres of water. The site offers generally clear water, the chance of recovering artifacts, and an abundance of marine life. The *U-352* is a 65-metre German sub which made the mistake of attacking the U.S. Coast Guard cutter *Icarus*. The first torpedoes fired missed and exploded on the bottom. The U-boat tried vainly to hide in the cloud of sediment, but the *Icarus* steamed right over it and dropped depth charges. The sub now rests on its side in 35 metres of water.

We were here for sharks, not history, so we made only a token dive on the *U-352* and concentrated our efforts on - you guessed it - the *Caribsea*. O.K., you didn't guess it. Sharks are free and don't have to stay where the brochure says they'll be. The reports we got on arrival were that there were more sharks hanging around the *Caribsea* than the *Papoose,* and that the visibility was unusually good for this site, which typically is not as clear as on the *Papoose*.

The *Caribsea* is a 75-metre-long steamer, built in 1919. It was carrying cargo from Cuba to Virginia in March of 1942, when it was struck by two torpedoes fired by *U-158*. It sank within minutes, and landed upright on the sand at a depth of 25-27 metres. The shallower depth offers significant advantages in bottom time over the *Papoose,* especially with nitrox. A closed-circuit rebreather mixes its own nitrox, so is able to optimise the mixture for any depth. More important was the stealth advantage I was expecting with my rebreather, which loops the exhaust gases back to a scrubber canister. The exhaled carbondioxide is removed in the scrubber, and the gas goes back to the breathing bag. No bubbles, no troubles! I was ready to scoot up to a dozing shark.

A closed circuit rebreather is nearly silent. It is far from invisible, though. Sharks have good vision, and can also sense water movement through their lateral line organ. They must have been amused by my attempts to crawl up close enough for a picture, which inevitably resulted in the shark gliding slowly away just before I got within camera range. I found, however, that a number of the sharks were cruising around slowly, and if I just sat still in the sand, eventually some would pass right by me. After three dives with the rebreather, mechanical problems forced me to go back to open circuit. With noisy bubbles forming a visual marker over my head, there was a change in the behaviour of the sharks. Instead of merely cruising by, as if I

The big, nasty-looking teeth made for impaling small fish and seen here in the mouth of a large female, are the reason why so many sand tigers are seen on magazine covers.

wasn't there, some of them would actually change course slightly to come investigate me.

A change in the weather forced another revision of our plan. The visibility dropped dramatically at the *Caribsea,* so we switched to the *Papoose.* There were fewer sharks on the *Papoose,* and our bottom time would be cut in half. The visibility was dropping, but was still better than what we had left. When we reached the bottom and located the sharks, we were confronted with another difference in their behaviour. These sharks were bigger, were swimming in tight patterns, and were nearly fearless. They had obviously become habituated to the large number of divers which visit the *Papoose.* All we had to do was pick a spot in their swimming pattern and sit motionless, and they would come right up to us.

The next day the wind shifted, and made it impossible to reach the *Papoose.* The following day, the wind had moderated a bit, and we were able to dive the *Papoose,* but there was a stiff current and surge on the bottom, and visibility had dropped to 7 metres. As we left Morehead City the next day, the weather was gorgeous, but a hurricane was headed up the coast. This is the dilemma of diving North Carolina. One day conditions can be spectacular, and the next they can be miserable. The weather odds are best during the summer, but it's still a matter of luck.

At least one variable has been removed from the equation. Unless the government re-opens fishing for sand tigers, the sharks will be there. They will have big nasty looking teeth hanging out of their mouths. And if you sit still, they will swim right up in your face. Are you feeling lucky today?

both photos Herwarth Voigtmann   **Aliwal Shoals, South Africa**

Christoph Gerigk   **Cape Verde Islands, Eastern Atlantic**

Carcharias taurus
Common sandtiger shark
Length: At birth 100, max. 330, mature females 220-330, mature males 220-260 cm. Distribution: Both sides of the Atlantic, western Indian Ocean, western Pacific. Depth: 1-200 m. General: Both dorsal fins set back on body, both of about the same size. Anal fin large, its size similar to that of dorsal fins. Coloration brown, often with spots. A wide ranging, locally common coastal species, occurring solitarily or in aggregations in tropical to temperate waters. In the Mediterranean found only in deeper waters out of reach of divers. In some areas follows certain fish species such as mullet and baitfish along their migration paths. It is a voracious feeder on a wide variety of bony fishes (including herrings, croakers, bluefish, eels, remoras, wrasses, mullets, sea robins, groupers, jacks, flatfishes) as well as small sharks (sharpnose and dusky sharks), rays (spotted eagle ray), squid, crabs and lobsters. This shark is denser than water, but it uses its stomach as a buoyancy compensator, swallows air at the water's surface and retains it in its stomach to maintain neutral buoyancy. Often floating motionlessly in the water column. Able to produce sounds by burping out air.

Aggregates in specific areas for mating, where the sharks loose quite a few teeth during courtship and mating bites. These teeth fall to the sea floor, accumulate there and indicate recent mating activities in the area. Sandtiger sharks are commonly encountered in reef channels off New South Wales, Australia, near rocky reefs off South Africa and around ship wrecks off North Carolina. Courtship and copulation have been observed in captive specimens. The male bites and holds the female's gill region, both maintain a parallel orientation during copulation. The females do not feed during their pregnancy, hydroid polyps grow on their teeth while they do not use them! Due to intrauterine embryophagy litter size is only 2; only one embryo per uterus survives because it thrives on its siblings and unfertilised eggs after its relatively small yolk supply is exhausted.

One of the most fearsome looking sharks due to the jagged appearance of the teeth in both its jaws, but not regarded as aggressive towards divers. Adult males are reported to be more aggressive during the mating season. See also pages 74-77.

This is one of the largest shark families (at present including 17 genera with about 80 species), but most species are deepwater or oceanic forms that do not frequent shallow waters. Family members are very diverse in shape and size, but most have a distinctive defensive spine at the front margin of each of their two dorsal fins; this feature is otherwise only present in members of the horn shark family Heterodontidae. An anal fin is lacking, the first dorsal fin originates in front of the pelvic fins. Some species (e.g. the lantern sharks of the genus *Etmopterus* and the pygmy sharks of the genus *Squaliolus*) have light-producing organs known as photophores, that may be used to attract prey or to communicate with conspecifics, especially in species which are known to live in huge (often unisexual) schools. Sizes range from the diminutive lanternsharks, some of which reach maturity at less than 20 cm, to the titanic sleeper sharks that attain lengths of 6 m or more. They are ovoviviparous (yolksac viviparity), with the spiny dogfish holding the record for the longest gestation period of any known vertebrate animal, i.e. 24 months. Some species in this family are known to hunt in large packs and several species are despised by commercial fishermen due to the destruction they cause to catch and equipment. They feed on a wide variety of bony fishes and invertebrates. A few species (cookiecutter sharks, probably also gulper sharks) even parasitise marine mammals and oceanic fishes by biting out small plugs of flesh, although they mainly feed on 'regular' prey. Dogfishes themselves often fall prey to larger shark species, like the bignose shark or the white shark (the stomach of one white shark contained eight dogfishes). In captivity, they often ram into the sides of the aquarium and cause sever, often lethal, injuries. However, a few dogfish species (those frequenting the upper reaches of the vertical distribution of the family) have been maintained in large public aquaria with some success.

Some taxonomists presently split the family Squalidae into several families each including related forms. For the sake of clarity only one family is recognised here.

The photo below shows the largest squaloid shark (and one of the largest sharks in general), the Greenland or sleeper shark *Somniosus microcephalus;* the depicted individual was captured through sea ice at Baffin Island, Canada, Western Atlantic.

GEORGE W. BENZ / IV

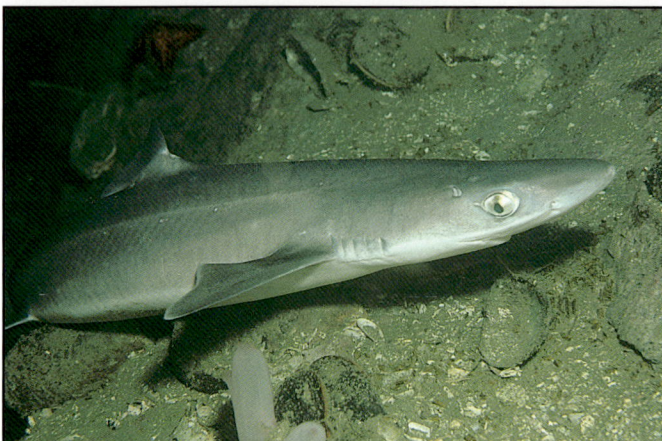

Florian Graner   **Norway, Eastern Atlantic**

Ingo Vollmer   **Canada, Western Atlantic**

Squalus acanthias
Common spiny dogfish
L: At birth 22-33, max. (exceptionally) 160, mature females 70-124, mature males 83-100 cm. Di: Worldwide, antitropical. De: 1-900 m. G: Dorsal fins with spines. Brown to grey above, usually with several scattered white spots, lighter below. Common in a variety of habitats, including rocky bottoms. Litter size 1-20, gestation period 18-24 months. Preys on bony fishes (herring, hake, cod, flatfishes), fish eggs, squid, octopuses, crabs, krill, worms, gastropods, salps and jellyfish. Adults feed more on fish than juveniles. Estimated annual food consumption rate for juveniles is 5 times the body weight, for adults 2.5 times. Often in schools which are segregated by sex and size. Forms mixed schools with leopard sharks and smoothhounds off California. Migrates in response to warming temperatures, some move into cooler, deeper waters in spring. One tagged and recaptured individual moved 6,500 km from Washington to Japan. Falls prey to other sharks, killer whales and seals. Its dorsal spines can inflict painful wounds, fishermen have to take care; otherwise not dangerous. An important commercial species, pieces of back (fish and chips) and belly (smoked) are sold on the European market for human consumption. But due to its low fecundity and extensive (over-) fishing, especially in the North Atlantic, catch rates have recently declined dramatically.

Akihiko Mishiku   **Shizuoka, Japan**

Squalus mitsukurii
Japanese spiny dogfish

Length: At birth 22-26, max. 110, mature females 72-110, mature males 65-89 cm. Distribution: Common in western North Pacific, but found in most tropical and warm-temperate seas. Depth: 4-740 m. General: Grey above, white below. Fins with white edges. Body rarely with white spots. Litter size 4-9. Preys on bony fishes (hake, conger eels, lanternfish), also on squid, octopus, crustaceans. Caught on longlines and by trawlers, utilised for human consumption.

Squalus asper
Roughskin spiny dogfish
L: At birth 25-28, max. 120, mature females 89-120, mature males 85-90 cm. Di: Gulf of Mexico, southern Africa, Aldabra and Hawaiian Islands. De: 210-600 m. G: Origin of 1st dorsal fin well behind pectoral fins. Eyes and spiracles large, skin rough. Greyish to brownish above, abruptly white below, fin margins mostly white. Preys on bony fish and squid. Filmed from the manned submersible "Jago" in a depth of 344 metres. The dorsal spines distinctive for most family members are clearly visible on the photo below.

Fricke/Schauer   Comoros, Western Indian Ocean

Etmopterus spinax   **Velvet belly**
L: At birth 12-14, max. 60, but rarely over 45, mature at 33-36 cm. Di: Norway and Iceland to South Africa. Also Azores, Cape Verde Islands, Mediterranean, North Sea (rare). De: 70-2,000 m. G: A common lanternshark found on or well above the bottom on the outer continental shelves and upper slopes. Eyes large, retina looks green in reflected light in live (see photos) or freshly caught specimens. Second dorsal fin larger than first, both with a strongly developed sharp spine at their anterior edge. Upper teeth tricuspid, serve in gripping and holding prey; lower teeth unicuspid with strongly oblique cusp and almost horizontal cutting edge, serve in cutting up prey. Dermal denticles with a long median spine, creating the overall appearance of a velvety skin (name). The velvet belly belongs to one of the few genera of luminous sharks. Each genus member has a different colour pattern (or is uniformly dark) the dark areas of which are luminous. Luminescence is generated by symbiotic bacteria in tiny light organs (photophores) which are scattered over the black (mostly ventral) parts of the skin. As the author has witnessed, these bacteria are already present in unborn embryos and thus are passed on from the mother to her young. The resulting pattern of light and dark areas probably serves in species recognition, as these sharks are found swimming together in large schools, segregating by size and sex. Groups of this shark may attack and kill much larger prey such as squid. Preys also on small fishes and crustaceans. Shown are the first uw-photos taken by SCUBA divers at night in the natural habitat (Norwegian fjords) in a depth of 80 m.

Florian Graner   Norway, Eastern Atlantic

81

Etmopterus princeps
Great lanternshark
L: At birth <20, max. 83, mature females 50-83, males 55 cm. Di: Nova Scotia to New Jersey and Iceland to Mauritania. De: 570-2,210 m. G: Uniformly blackish. Like *E. spinax* sometimes infested by the parasitic barnacle *Anelasma squalicola*. Below: *E. lucifer,* 47 cm, luminous, with distinct pattern.

Pedro Niny Duarte / DOP   Azores, Atlantic

Etmopterus pusillus
Slender lanternshark

Length: Max. 48, mature females 38-48, mature males 31-45 cm. Distribution: Widely distributed in Atlantic, western Indian Ocean (South Africa) and Indo-West Pacific (Japan to Australia). Depth: 275-1,000+ m. General: Pale brownish or chocolate brown, belly slightly darker, fins mostly pale. Living on or near the bottom, in the South Atlantic also oceanic at the surface. Denticles blunt, with low crown as opposed to those of other genus members which have high, pointed crowns.

Pedro Niny Duarte / DOP   Azores, Atlantic

Centroscymnus crepidater
Longnose velvet dogfish
Length: At birth 30-35, max. 105, mature females 80-82, mature males 60-68 cm. Distribution: Eastern Atlantic: Iceland to South Africa. Indian Ocean: Aldabra Islands and India. East Pacific: Chile. Western Pacific: Southern Australia and New Zealand. Depth: 270-1,100 m. General: Uniformly dark brown to black. Snout long. Dorsal fin spines short. Common, but poorly known. Ovoviviparous. Litter size 4-8. Females may breed throughout the year. In Australian waters, the flesh is high in mercury.

Pedro Niny Duarte / DOP   Azores, Atlantic

Centroscymnus coelolepis
Portuguese dogfish

Length: At birth <20-30, max. 120, mature females 100, mature males 85 cm. Distribution: Widespread in Atlantic: Iceland to South Africa, including Mediterranean, and Nova Scotia to northeastern USA. Also in western Pacific (Japan, Australia, New Zealand). Depth: 270-3,700 m. General: The velvet dogfishes of the genus *Centroscymnus* are deepwater sharks with tiny dorsal fin spines, rounded rear pectoral fin tips and a pair of abdominal ridges. The upper jaw (clutching-type) teeth are erect and slender, the lower jaw (cutting-type) teeth are larger, broad and have smooth cutting edges. All teeth have only one cusp.

The Portuguese dogfish is a stocky shark with a short snout (shorter than mouth width) smoothly tapering to a rounded tip, with a large mouth and short labial furrows. Coloration is uniformly blackish brown. Upper jaw teeth with erect lanceolate cusps, lower jaw teeth with short, strongly oblique cusps. Adults have large, round, flat, overlapping dermal denticles, those of young specimens have three cusps. This (?) sluggish shark is mainly found in benthic habitats of the continental shelf and upper slope, but has also been caught in a depth of almost 3,700 m, world depth record for sharks! It is found in water of 5-13°C and is known to be abundant in the Rockall Trough area west of the British Isles and off the west coast of southern Africa. Reproduction is ovoviviparous, litter size 13-17. Prey includes bony fishes, sharks, molluscs (squid, octopus, gastropods, ? crustaceans) as well as whale blubber and meat. Cetacean prey is most probably taken as carrion; some speculate on a removal of flesh plugs from live whales in cookie-cutter fashion, but this has to be verified. The author examined the stomach contents of dozens of these sharks from the Rockall Trough area in the northeastern Atlantic and found mammalian remains in at least one stomach, but these were definitely not removed from live whales. Off Japan, females have been caught deeper than males. In Australia as bycatch in deepwater trawls (orange roughy fishery), but not commonly utilised as the flesh is high in mercury. Elsewhere used for its liver oil (which is rich in squalene), for fishmeal or dried and salted for human consumption.

Below: The **Shortnose velvet dogfish** *C. cryptacanthus*, to 104 cm, western S-Atlantic (Uruguay), eastern N-Atlantic (Senegal; photo: Azores).

Pedro Niny Duarte / DOP   **Azores, Atlantic**

Somniosus microcephalus
Greenland shark

L: At birth about 40, max. 730 cm. Di: N-Atlantic and Arctic Ocean. Scattered records from S-Atlantic and Antarctic Ocean. De: Surface to 680 m. G: Largest squaloid. Sluggish, inhabits waters of 0.6 to 12°C. In the Arctic in shallow bays and river mouths in winter. Preys on bony fishes and marine mammals. Often with parasitic luminous copepod attached to cornea of eye (photo) which may serve as lure. Litter size 10. Meat toxic when fresh, dried for human consumption; skin used to make boots, teeth used as knifes by Eskimos. Has only recently been filmed in shallow water at Iceland.

George W. Benz / IV   **Baffin Island, Western Atlantic**

Somniosus rostratus
Longnose sleeper shark
L: At birth 21-28, max. 140, mature females 82-134, mature males 71 cm. Di: Eastern North Atlantic: Madeira, France, western Mediterranean. Western Pacific: Japan. De: 200-1,000 m. G: Biology and habits little known. Ovoviviparous. Probably preys on deepwater benthic fishes and invertebrates. Caught on longlines and with bottom trawls in eastern Atlantic. Utilised as fishmeal and possibly for human consumption. The large *S. pacificus* is found from Japan to California and in S-Pacific. All three genus members lack fin spines.

Isistius brasiliensis
Cookiecutter shark
L: Max. 50, mature females 38-50, mature males 31-39 cm. Di: Oceanic and circumtropical. De: 85-3,500 m, also at surface at night. G: Dorsal fins very small, without spines, set far back on body. Lower surface except fins and a dark collar marking luminous and glowing greenish. This small pelagic predator attacks large animals such as tuna, swordfish, whales, dolphins and megamouth sharks. It is considered to be a weak swimmer and probably waits for its prey to pass by. After sensing a victim, the attacker lunges forward to press its fleshy lips sucker-like against the victim's skin; then it rotates around its long axis and thus gouges out a piece of flesh with its large, triangular lower jaw teeth. This action leaves behind a crater-shaped wound and led to the popular name of this extraordinary shark species. The marks and scars of this specific feeding behaviour can be observed in fish markets where tuna and billfish are sold. The cookiecutter's unique way of feeding was discovered after U.S. Navy inspectors had found inexplicable circular marks on the rubber radomes of their nuclear submarines. The ability to generate such marks was tested with a shark caught alive and a peach as prey dummy! May feed on deepwater squid, teleosts and crustaceans as well. Seems to make extended daily vertical migrations in the water column, rising to the surface at night and descending to several thousand metres during the day. Still, its biology is little known.

Dalatias licha
Kitefin shark

L: At birth 30, max. 180, commonly 150 cm. Di: E-Atlantic (British Isles to Senegal, including oceanic island slopes; also W-Atlantic), W-Indian (South Africa) and W-Pacific (Japan to Australia, New Zealand) Oceans. De: 90-1,000 m. G: Uniformly grey, brown or black. Caudal fin large, indicating a powerful swimmer. Found deep in benthic to mesopelagic habitats on the upper continental slope, primarily from 300-600 m. Preys mainly on bony fishes, but also elasmobranchs, cephalopods, crustaceans. Jaws depicted on p.208. Ovoviviparous, litter size 10-16. Caught with bottom trawls for their liver oil rich in squalene.

Rudie Kuiter   South Australia

Oxynotus bruniensis
Prickly dogfish

L: At birth about 24, max. 72, mature males about 60 cm. Di: New Zealand and South Australia: NSW to Great Australian Bight, including Tasmania. De: 45-650 m. G: Unmistakably hump-backed, cross-section of body nearly triangular, prominent ridges between pectoral and pelvic fins. Dorsal fins tall, with spines. Biology and habits poorly known. Benthic in temperate waters of shelf and slope. Ovoviviparous, litter size up to 7. The depicted specimen was trawled from a depth of 520 m. The genus Oxynotus comprises at least four species and is also placed in a family of its own (Oxynotidae).

Rudie Kuiter   Tasmania, Southern Australia

Centrophorus lusitanicus
Lowfin gulper shark

Length: At birth about 36, max. at least 160, mature females 88-160, mature males 72-128 cm. Distribution: Eastern North Atlantic: Portugal to Nigeria. Western Indian Ocean: South Africa and Mozambique. Western Pacific: Taiwan. Depth: 300-1,400 m. General: Ovoviviparous, litter size 1-6. Preys on bony fishes, squid, small dogfish sharks and lobsters. Caught in eastern Atlantic with bottom trawls, set nets and longlines. Dried and salted for human consumption, processed for fishmeal.

P. Niny Duarte / DOP   Cape Verde Islands, Eastern Atlantic

Pedro Niny Duarte / DOP   Cape Verde Islands, Atlantic

Centrophorus granulosus
Common gulper shark
L: At birth 30-42+, max. 160 cm.
Di: E-Atlantic: Portugal to Senegal, also Mediterranean; probably Gulf of Mexico. De: 350-500 m.
G: Rear pectoral fin tips elongated, pointed (short, rounded in other genus members). Grey above, light below, fins with white margins (juveniles), uniformly greyish (adults). Lives benthic in habitats on the upper continental slope. Ovoviviparous. Preys on bony fishes and squid. All members of this genus are small to medium-sized deepwater dogfishes with large green eyes and a strong spine (with lateral grooves) at the front margin of each dorsal fin.

Rudie Kuiter   Tasmania, Southern Australia

Centrophorus uyato
Little gulper shark
L: At birth , max. 100, mature females 75-89, mature males 81-94 cm. Di: E-Atlantic: Portugal to Namibia, western Mediterranean; also reported from Gulf of Mexico, southern Mozambique, Taiwan, Australia. De: 50-1,400 m. G: Rear pectoral fin tips usually greatly elongated and narrowly pointed. Dark greyish-brown above, lighter grey below, fins darker than upper body surface, but with whitish or transparent tips. Benthic on outer continental shelf and upper slope. Ovoviviparous, usually bears only one young. Prey includes bony fishes and squid. Occasionally caught by bottom trawlers.

Pedro Niny Duarte / DOP   Azores, Atlantic

Centrophorus squamosus
Leafscale gulper shark
Length: At birth 35-40, max. 160, mature females 110-160, mature males 100-103 cm.
Distribution: Eastern Atlantic: Iceland to southern Africa. Indian Ocean: South Africa and Aldabra Islands. Western Pacific: Japan to Australia and New Zealand.
Depth: 200-2,400 m.
General: Greyish brown to dark grey. Ovoviviparous. Litter size 5-8. Preys on cephalopods and deep water bony fishes. The author found a *Stichopus* sea cucumber in the stomach of one specimen from the NE-Atlantic. Meat dried and salted for human consumption. Teeth are shown on p.203.

Deania quadrispinosa
Longsnout dogfish

L: Max. 115, mature females 110, mature males 87 cm. Di: Namibia, South Africa and western Australia to Queensland, New Zealand. De: 150-820 m. G: Snout very long, both dorsal fins about equal in height and anteriorly with strong, grooved spines, the second spine being more exposed than the first one. Greyish, brownish or black, underside paler. Fins sometimes with white edges. The large eyes of living specimens strongly reflect yellow light. Anterior teeth in upper jaw narrow, with erect cusps; in lower jaw with strongly oblique cusps. Dermal denticles high, pitchfork-shaped. On outer continental shelf and upper slope, mainly deeper than 400 m, in dense schools. Preys on fishes.

Right: **Arrowhead dogfish,** *D. profundorum,* to 76 cm, North Atlantic, western Indian Ocean, Western Pacific, 275-1,785 m. Preys on small bony fishes, squid and crustaceans.

Below: **Rough longsnout dogfish,** *D. hystricosa,* to 109 cm, eastern North Atlantic (Madeira), western North Pacific (Japan), 600-1,000 m. Biology and habits of these dogfishes are poorly known.

Rudie Kuiter   Tasmania, Southern Australia

Pedro Niny Duarte / DOP   Azores, Atlantic

Echinorhinus cookei
Prickly shark

L: At birth 40-45, max. about 400, mature females 250-300, mature males 200 cm. Di: Tropical and temperate Pacific. De: 10-425 m. G: This family comprises two little-known sluggish bottom-living species with dorsal fins set far back on the body. This rare photo was recently taken by the German scientist Hans Fricke and his crew during the successful search for coelacanths in Indonesian waters. It is probably the first one of a living specimen in its natural habitat and was taken from a submersible in 382 meters.

Fricke/Schauer   Northern Sulawesi, Indonesia

SHARK GAMES

A sport with shark participation? The history of SCUBA diving holds many examples of divers misusing sharks for fun and the entertainment of public crowds. One of them is shark tagging that originally has a scientific background, but was also conducted just for the fun of it. Today, seen from a distance and after many years of learning how to respect nature, underwater filmer Howard Hall recalls moments of fun and fear off the coast of California.

This blue shark has been tagged by divers from all three nations participating in the competition.

When adjusted for inflation, the First Annual International Shark Tagging Competition cost more than a million dollars to produce. The scope and splendour of this 1978 production was only exceeded by its failure to serve as entertainment. Originally intended as a two-hour special episode of The CBS Sports Spectacular, it eventually aired as a twenty minute segment, which served to fill the gap between segments of legitimate sports such as professional bowling and championship pocket billiards. The shining highlight of this twenty-minute underwater extravaganza was a six-second sequence of the second-place finishers painlessly jabbing a stainless steel scientific tag below the front dorsal fin of a six-foot Caribbean reef shark. This scene was the show's highlight because it was the only tag actually captured on videotape.

Despite the anesthetic qualities of the finished program, I invited several good friends over and we watched the program. We watched all twenty minutes, start to finish, largely because it was the first television program I had ever participated in, and also because in 1978 remote control hand-sets were not common-place and were seldom found associated with the run-down, second-hand televisions like the one that sat on the cinder-block shelf in my apartment. This is to say that we watched the program because my friends and I were too lazy to get up to change channels or leave the apartment (and beer supply) or, in my case, too lazy to get up and go outside to commit suicide.

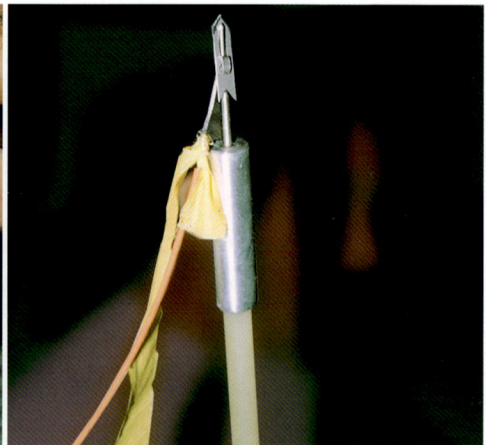

Divers preparing tags. The steel tips of the tags are sharpened and each tag is marked with a coloured piece of cloth.

Participants of the shark tagging competition, from left to right: Organiser Howard Hall, the Australian team Ron and Valerie Taylor, competition judge Paul Tzimoulis, chief editor of *Skin Diver Magazine*, and Helmut Debelius, member of the German team.

My friend and fellow television producer, Dan Walsh, calls television production "network time killing." Our seemingly glamorous professional purpose is to kill the time that separates the television commercials. The trick is to put something on the air that is entertaining enough that the slack-jawed audience is neither painfully bored nor so offended that they throw a brick through the picture tube. The First Annual International Shark Tagging Competition was only slightly less boring than being suspended in a sensory deprivation tank. Such failure to entertain usually destroys the market for any future proposals of similar nature that may be submitted to television executives. I was therefore stunned when, two years later, the show's producer, Stuart Goodman, called me and said CBS wanted to do another one. I suspect this desire to fund a sequel was due less to a careful examination of the first show's ratings and more to a stellar sales presentation by Stu. Also contributing may have been the fact that most television executives fail to actually watch much of the programming their studios release.

When Stu suggested that I participate in the Second Annual International Shark Tagging Competition once again as a competitor, I proposed an alternate idea which I thought would produce more sharks for the show (and a larger paycheck for me). I suggested that Stu employ me as production manager and that we do the show in California waters, rather than in the Virgin Islands where the last show was made. The idea of actually having sharks in the water for competitors to tag appealed to Stu and he agreed.

The Second Annual International Shark Tagging Competition was conducted off the coast of San Diego during the summer of 1980. Ron and Valerie Taylor were invited to defend their championship. The United States challengers would be Chuck Nicklin and Jack McKenney, both famous underwater cameramen with lots of experience filming sharks. From Germany we invited Helmut Debelius and Herwarth Voigtmann to be our European team, also experienced shark divers of international renown.

Just before dawn on a warm summer morning the entire production went to sea aboard the sixty-five foot charter vessel, *Sand Dollar*. By 8 a.m., the boat was adrift thirty miles offshore and bait was in the water. A single shark cage was lowered and trailed thirty feet behind the drifting dive boat. Buoys were attached to twenty-foot lines and tied to the two down-stream corners of the shark cage. The competition area was defined as the area in front of the cage and up to twenty feet down-stream as marked by the buoys. Two cameramen would be positioned on either side of the shark cage, each with a safety diver. We would be using the same gigantic RCA studio cameras used in the first show. Competition judges Paul Tzimoulis and Gerry Murphy would occupy the interior of the cage. The competition was called for 3 p.m.

At 10 a.m. our first blue shark showed up. By 3 p.m. we would have several dozen. I delivered a competition briefing on the stern of the *Sand Dollar*. Each team was issued a tagging spear that was two feet long, had a tagging tip at one end and a rubber sling at the other. We decided to use exceptionally short tagging spears because we knew that getting close to blue sharks was not going to be an issue. During the briefing I suggested that one diver act as the tagger and the other diver act as safety. With clouds of fish blood in the water and dozens of sharks swimming around, safety was a serious concern. The safety diver would concentrate on prodding away sharks that threatened to bite the tagger as he concentrated on jabbing tags into sharks.

Of all the contestants, Herwarth Voigtmann was probably the most famous shark diver. Herwarth conducted what was probably the world's first commercial shark dive in the Maldives Islands. Photographs of Herwarth hand-feeding grey reef sharks and silver-tip sharks began appearing in European dive magazines in the mid-1970s. But Herwarth's program really attracted attention when he began publishing photographs of his daughter Fernande feeding sharks with fish she held between her teeth. These photographs were stunning and seemed completely crazy by our standards at the time. But the popularity of these images was undoubtedly enhanced by Fernande's practice of feeding these "man-eaters" while posing completely nude. Many would-be shark divers collected magazines containing these images and spent hours studying Fernande's mouth feeding technique.

Fernande's underwater performance not only became highly commercial fodder for Herwarth's camera, but also attracted a great number of sport divers to Herwarth's resort. As Herwarth's and Fernande's popularity soared, rumours of accidents began to circulate. Each rumour described a different portion of Fernande's anatomy that had been tragically removed by a shark. It was only recently, while preparing for this story, that I learned from Helmut that Fernande had not been horribly disfigured as I had believed. It turns out that the only accident that ever took place was one involving Herwarth himself. He was bitten on the cheek by a shark that was aiming for the fish he held in his teeth. According to Herwarth, the accident only happened because he was looking the wrong way when the shark approached.

Considering Herwarth's extraordinary shark expertise, Helmut agreed that Herwarth should handle the tagging spear. But despite Herwarth's exceptional diving experience, he had never actually used a pole spear before. When presented with the tagging device, he couldn't make heads or tails of it. To practice up for the competition, Herwarth and Helmut constructed a shark facsimile of cardboard. Helmut would dodge and weave along the decks of the *Sand Dollar* while Herwarth pursued with the spear. Unfortunately, Herwarth's technique improved little in the hours before the competition. He insisted upon jabbing the cardboard shark with the spear before releasing tension on the rubber sling. If the Australian or United States team members noticed Herwarth's flawed technique as he jabbed at the cardboard shark, they gave little notice.

During the Virgin Islands competition two years before, it was obvious to the competitors that the team who went first would have the best chance at victory. There were only five sharks seen during the competition and the first team in the water tagged four. The San Diego competition would be different. I knew that the team that went last would have the most sharks, since blue sharks tend to gather throughout the afternoon and peak in number around sunset.

In the 1970s, divers all over the world marvelled at the daring shark feedings practised by Herwarth and his daughter (facing page).

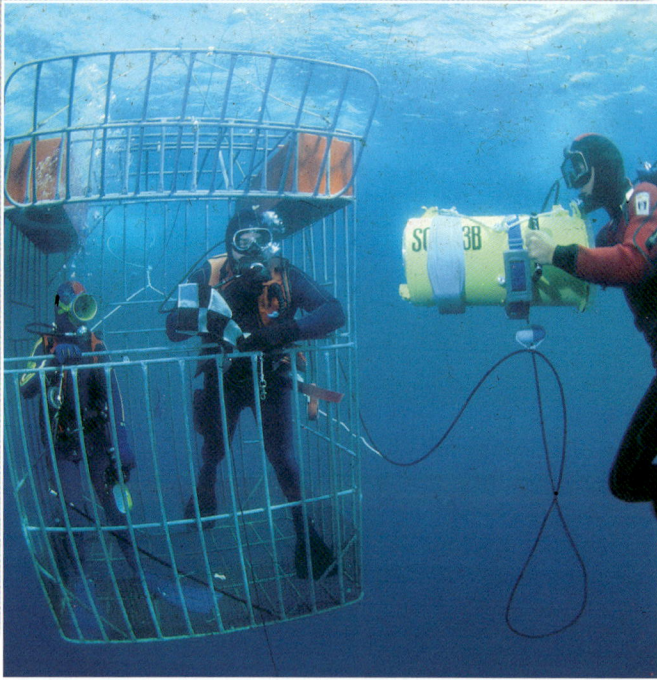

Howard Hall is filming, while Paul Tzimoulis gives the signal to start the competition.

Again, the order of competition was determined by drawing lots. According to the draw, Ron and Valerie would go first, Herwarth and Helmut would go second, and Jack and Chuck would be third. Points would be allotted depending on location of the placed tag. A dorsal fin tag would be worth three points, and a body tag would be worth one.

The stage was set and at 2:45 p.m. the cameramen and judges entered the water. In addition to helping keep score, Gerry Murphy was tasked with continually dispensing chum from a large mesh bag filled with frozen mackerel, thus creating an enormous cloud of bait that drifted down-stream into the competition area. Gordy and I positioned ourselves at the side of the cage with our safety divers behind us. Everything was set for the competition to begin. Water conditions were not great. The combination of plankton and drifting clouds of fish blood produced a greenish hue to the water and visibility of about forty feet. But the shark action was spectacular. Literally dozens of sharks were dashing through the small space between the shark cage and the two buoys twenty feet down-stream. Sharks were constantly banging into the cage. The smaller ones were wedging their heads between the bars and threatening to squirm inside. The judges backed up and pinned themselves at the rear of the cage. Since the chum line started at the down-stream side of the cage where the bait was being released, sharks were not likely to bite Gordy, our safety divers, or me since we were five or six feet up-stream of the chum. Still, I could feel Larry whacking away at sharks that swam past my shoulder and disappeared from my field of view behind me. Suddenly there was a splash and I saw Ron and Valerie appear in the area. The game was on!

A competition participant with tagging spear ready and spare tags fixed under a rubber band around his wrist approaches a shark.

It immediately became evident that Ron and Valerie intended to completely disregard my safety briefing where I suggested one member of the team act as safety diver. Instead, Ron began placing tags while Valerie dispensed additional bait around Ron's head and shoulders from a mesh bag she carried. The sharks were on them like flies on elephant dung. Ron was placing tags as fast as he could, concentrating most on sharks that were in the process of mouthing his legs and arms. I cringed as I watched through the viewfinder. This was not safe diving. In fact, a year later, Valerie was badly mauled by a blue shark in virtually the same area, requiring plastic surgery on her lower leg.

After a few minutes, it became evident that the rate at which Ron could place tags was determined by how quickly he could load the tagging device. Underwater, this became a surprisingly difficult procedure. To do so required removing a tag from a tangle of tags strapped

The German team enters the water: Just where are those sharks?

beneath a rubber armband, focusing on the tip of the spear, inserting the steel dart in a tiny groove, then slipping the spaghetti tag under a tight rubber band. Ron had to do this while sharks constantly bumped into him in response to Valerie pouring blood and guts on his head, a practice that often seemed to disrupt his focus and concentration. But Ron's fingers seldom wavered. After only ten minutes, he had tagged seven sharks and amassed a score of nineteen points.

The German team of Helmut Debelius and Herwarth Voigtmann were next. They moved into the shark infested competition area with grace and confidence. But they suffered two handicaps. First, they took my safety briefing seriously and Helmut took his position as safety diver and did not dispense bait directly on Herwarth's head, which had proved such a successful if seemingly suicidal technique for Ron and Valerie. Second, Herwarth's pole spear technique was almost totally inadequate. He would approach a shark, jab the tag against its hide, and then release the sling. The tag repeatedly failed to penetrate and as the shark flinched away, the tagging spear left Herwarth's hand and plummeted toward the ocean floor 2,000 feet below. Only a diver of Herwarth's extraordinary skill could have caught the rapidly descending spear the first time he dropped it. But he succeeded after racing straight down one hundred feet through the clouds of sharks and chum. His second failed attempt to tag a shark resulted in an improvement on his ability to retrieve the spear. On this second attempt, he caught the spear at a depth of only sixty feet. By the end of Helmut and Herwarth's ten minute competition time, Herwarth had perfected his technique of retrieving the spear requiring a descent of only ten or fifteen feet. Unfortunately, this jab, release, and chase-the-spear technique was inefficient and time consuming. The German team finished having tagged only two sharks.

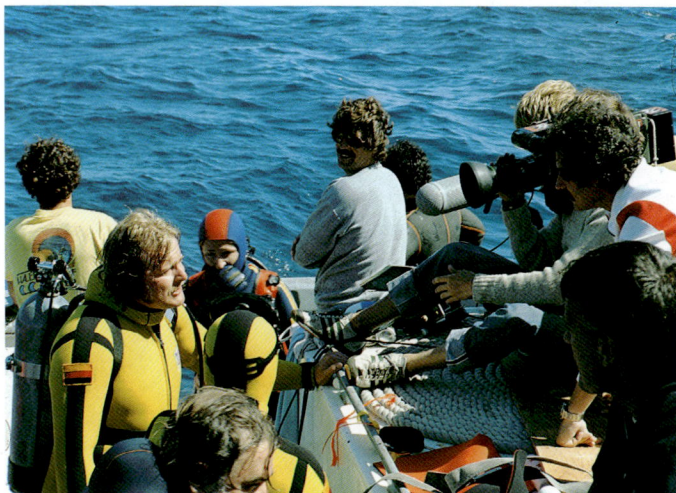
As soon as the Germans have safely returned, they are interviewed by a CBS sports reporter.

By the time Chuck Nicklin and Jack McKenney entered the water, the sharks were everywhere. Both Jack and Chuck were intense competitors and they came to win. Caution was thrown to the wind. Chuck came down with a huge bag of mackerel and immediately began mashing it to create a gigantic cloud. They also began competition with an improved tagging technique. Each time Jack placed a tag, Chuck had another in hand waiting to hand off to Jack. This saved the considerable time required for removing a tag from the rubber armband. Still, getting the tag positioned on the end of the spear while sharks continually bumped into him proved challenging for Jack. But every time Jack got the spear loaded, he was immediately able to spear a shark. He tagged so many sharks so fast that the judges had difficulty keeping track, especially since Chuck had significantly reduced water visibility by banging his bag of bait against the side of the cage.

Things got completely out of hand as sharks began biting Chuck's mesh bag. As Chuck struggled to free the bag from the jaws of three or four large blue sharks, Jack tagged away. The cloud of bait reduced visibility to fifteen feet. Sharks were biting at everything. A small shark squirmed its way into the shark cage distracting the judges from their duties and further hindering their ability to accurately keep score.

Gordy Waterman and I watched in horror. I couldn't bring myself to concentrate on the viewfinder. The sharks were bouncing off the front of the camera and overwhelming Larry's ability to keep them off him and me. I just pointed the camera, let it roll, and prayed we would all survive the last few minutes of competition.

I believe that it was pure dumb luck that no one was bitten during the last five minutes of Chuck's and Jack's competition period. But we all survived with all appendages intact. Jack had tagged twelve sharks in ten minutes and his score was a perfect thirty-six points. The American team had won.

The Second Annual International Shark Tagging Competition again aired as a segment on The CBS Sports Spectacular. It was certainly a more entertaining program than the first. But by the time it aired, the tenor at CBS Sports had changed. Shark tagging was labelled a "Trash Sport." Trash sports were defined as overly sensationalised, unconventional competitions created entirely by broadcasters to make programming. Certainly, shark tagging qualified. But although these manufactured-for-television stunts produced solid entertainment, they also produced some spectacular lawsuits and some really bad reviews. The networks backed off and hence the Second Annual International Shark Tagging Competition was also the last.

The Californians Jack McKenney and Chuck Nicklin knew 'their' resident sharks - and won the competition almost hands down.

Haploblepharus pictus   **False Bay, South Africa**

PETER VERHOOG

The catsharks are the largest shark family and comprise at present about 100 valid species. Their eyes are elongated and have nictitating membranes. The first dorsal fin originates over or behind the pelvic fins. There is no precaudal pit, the spiracles are large. Although many catshark species are deepwater dwellers (e.g. the members of the genus *Apristurus* account for about a third of all catshark species), several are reef inhabitants. These species have elongated bodies and narrow heads that allow hunting and resting in narrow crevices and small caves. Many reef species have attractive colour patterns consisting of stripes, spots, saddles and blotches. These patterns disrupt the shark's outline against the varied background of the reef. Some species are nocturnally active and rest on the ground during the day. The prey of the reef-dwelling species consists of a wide range of food items including invertebrates such as worms, gastropods and crustaceans, but also bony fishes and elasmobranchs. The swell sharks of the genus *Cephaloscyllium* may wedge themselves into reef crevices and inflate their abdomen by swallowing considerable volumes of water so that they are difficult for a large predator to eat. However, one balloon shark (*C. sufflans*) was found inside the stomach of the coelacanth *Latimeria chalumnae*.

While the majority of catsharks (including the shallow water species) are oviparous, a few are ovoviviparous. Copulation has been observed in several species. In many catsharks, males have longer teeth than females, an adaptation for biting and holding a mate. All catsharks are harmless to divers, but may bite if molested. Most family members readily adapt to captivity and some even breed in public aquaria.

Haploblepharus edwardsii   Puffadder shyshark

L: At birth 10, max. 60, mature females 41-60, mature males 42-59 cm. Di: Endemic to south and east coast of South Africa. De: Intertidal to 130 m, more common in shallower depths in the more northern part of its range. G: This common shyshark lives well-camouflaged on sand and rock bottoms. Its dorsal fins are about equal in size, its nasal flaps are broad and extend to the mouth. Two colour morphs are distinguished: The southern Cape form is sandy brown with seven dark reddish-brown saddles bordered by black and with numerous small, dark brown and white spots between the saddles. The rarer Natal form possibly represents a different species and has much darker brown saddles with irregular white spots on a cream background; both forms are white below. Oviparous, one egg per oviduct, two egg cases of about 5 cm in length develop simultaneously. Egg cases have hair-like filaments that anchor them to structures on the ground. Individuals from the Cape Province may mature at a smaller size and occur in deeper, colder water than those from Natal which occur close inshore in warm water from the surfline to a depth of 30 m. Prey includes bony fishes (anchovies, jacks, gobies, gapers), crustaceans (mysids, decapod shrimp, mantis shrimp, crabs, hermit crabs), squid and polychaete worms. Caught by anglers and commercial trawlers. The sympatric brown puffadder shyshark *H. fuscus* is similar, but uniformly brown.

Below: **Dark shyshark,** *H. pictus*, to 60 cm, Namibia to Cape Agulhas, SA. Intertidal to 100+ m. Common, benthic in shallow sandy areas, caught by sport anglers, but little utilised. Seven dark saddles over head and body. See also previous page.

Peter Verhoog   False Bay, South Africa

Poroderma africanum
Pyjama catshark

L: At birth 15, max. 100, mature females 65-93, mature males 58-101 cm. Di: Eastern South Atlantic, western Indian Ocean: Endemic to South Africa. De: 1-100 m. G: Common, found singly or in aggregations in inshore caves, rocky areas and sea grass, where divers easily can observe it. Unmistakable by 7 longitudinal stripes along the body. Nocturnal, mainly preys on crustaceans (mantis shrimp, crabs), but also eats bony fishes (anchovies, gurnards, hake), cephalopods (squid, octopuses, cuttlefish), worms and clams. Two egg cases are laid at a time every three days during the breeding season. Eggs hatch after 5.5 months. A hardy species, can readily be kept in large aquaria.

Christoph Gerigk   Cape of Good Hope, South Africa

Poroderma pantherinum
Leopard catshark

Length: At birth about 12, max. 84, mature females 58-73, mature males 54-84 cm. Distribution: Eastern South Atlantic and western Indian Ocean: Endemic to South Africa.
Depth: Intertidal to 256 m. General: Occurs sympatrically (lives in the same area of distribution) as its close relative, the pyjama catshark *P. africanum* (see above). Differs from the previous species by a variable colour pattern of dark spots or broken rings on a white to grey background. A common nocturnal catshark that inhabits the shallow temperate continental waters of South Africa, particularly around Algoa Bay, but also west of the Cape, ranging north to Natal. Preys mainly on bony fishes, but worms, cephalopods and crustaceans are also eaten. A hardy species, can readily be kept in large aquaria.

Don DeMaria / IV   Port Elizabeth, South Africa

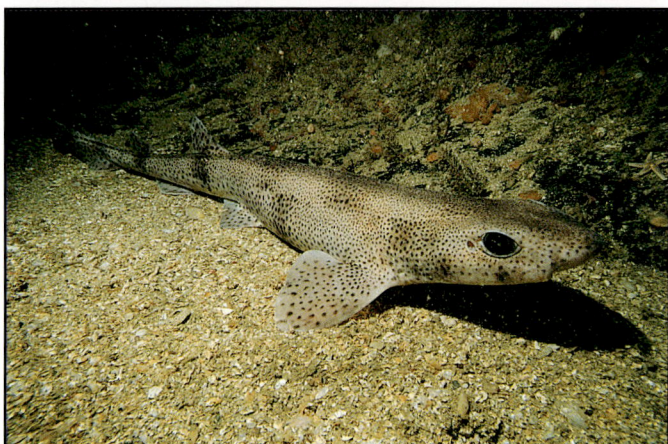

Charles Hood   Cornwall, England

Scyliorhinus canicula
Lesser spotted catshark
L: At birth 9-10, usually 75, max. 100 cm; mature females 44-60, mature males 39-60 cm (in Mediterranean). Di: Eastern Atlantic: Norway to Ivory Coast, and Mediterranean. De: Intertidal to 400 m. G: Many small dark (photo left) and sometimes light spots (centre photo) over body. Sometimes with 8-10 dark saddles on body. A common temperate water shark that most often is found in rocky areas, but also on sand and mud. By day it inactively rests on the ground, at night actively hunting. Preys mainly on molluscs (squid, whelks, scallops, clams) and crustaceans, also on bony fishes (sea horses, gobies, flatfishes), worms, sea cucumbers. Oviparous, two egg cases (4-7 cm) with tendrils are laid at a time, 18-20 are laid in one breeding season. Female swims around algae, or other suitable anchoring sites, until extruding egg case tendrils catch and egg is pulled from cloaca. The young hatch after 5-11 months. Copulation observed in the wild and in captivity. Male coils around female and inserts one clasper. Does well in captivity. Below: Catshark egg case attached to coral.

Helmut Debelius   Helgoland, Germany

Andreas Vilcinskas   Ligurian Sea, Italy

Scyliorhinus stellaris
Greater spotted catshark
L: At birth 16, max. 160 cm. Di: E-Atlantic: Norway to Mauritania, and Mediterranean. De: Intertidal to 125 m. G: In rocky areas, rests on the ground by day, hunts at night. Preys on crustaceans (hermit crabs, swimming crabs), squid, octopuses, bony fishes (mackerel, dragonets, flatfishes, gurnards) and other sharks (S. canicula, above). Oviparous, two egg cases with tendrils are laid at a time among algae in spring and summer. Egg cases are 12 cm long, often found in tide pools and hatch after 9 months. Hardy in captivity, survived for 19 years in aquaria, see OCEANOPOLIS on pp.106-108.

Scyliorhinus comoroensis
Comoros catshark
L: Max. 60 cm. Di: Known only from the Comoros. De: 40-300 m. G: This is the first photo showing the species in its natural habitat.
    Below: **Chain catshark** *S. retifer,* to 47 cm, Western N-Atlantic (New England to Nicaragua), 73-550 m.

Fricke/Schauer   Comoros, Western Indian Ocean

Galeus melastomus
Blackmouth catshark
L: Max. 90, mature females 66-90, mature males 59-68 cm. Di: E-Atlantic: Norway to Senegal, Mediterranean. De: 55-1,200 m. G: Interior of mouth black (popular name). Benthic on shelf and slope. Preys on molluscs, crustaceans, bony fishes. Oviparous, brown egg cases (5 by 1 cm) laid in pairs; each corner with short spike, but no tendril.
    Has been filmed and photographed by SCUBA divers at night in the natural habitat (Norwegian fjord) in a depth of about 80 m.
    See also following species.

Florian Graner   Sognefjord, Norway

Galeus murinus
Mouse catshark
Length: Max. 63 cm (adult male).
Distribution: Eastern North Atlantic: Iceland and Faeroes to Azores.
Depth: 475-1,400 m.
General: Distinctive crest of enlarged dermal denticles on the upper caudal fin margin. Benthic, biology and habits little known. Occurs in the Rockall Trough area west of the British Isles in depths of 880-1,385 m along with other deep sea catsharks (e.g. *Apristurus laurussonii,* 790-1,490 m) but deeper than the previous species (200-800 m).

Pedro Niny Duarte / DOP   Azores, Atlantic

A MATTER OF PATIENCE

Three of us waited on the bottom for the shark to come rushing out. Chip Matheson was to my left kneeling behind a rock. I could see that his hand was trembling violently. It was probably the cold water. The largest and most ferocious sharks dwell in temperate waters. Los Coronados Islands, the most northern islands on the Pacific Coast of Baja California, are definitely temperate. We had been down over an hour, waiting. Howard Hall and his crew were freezing.

Like many catsharks, also the swell shark *Cephaloscyllium ventriosum* lays egg capsules on the ground where they are anchored by threads extending from their corners.

Mark Conlin was to my right. He was steady. Mesmerised by the spot from where he expected the shark to emerge. Mark had much less experience with sharks than Chip or I and perhaps the adrenaline rushing through his veins was keeping him warm. Although I own a *Neptunic* anti-shark suit, I wasn't wearing it. It's effective against all but the largest sharks. But it would do me no good here. Powerheads, the explosive spear points used to kill sharks, would be equally useless.

My 16-mm motion picture camera was set on a heavy tripod. It may have been possible to hand-hold the camera, but the current was strong and I was afraid that when the action got started I wouldn't be able to hold the camera still without the tripod. I didn't want to blow it. We would probably only get one chance. And when the shark began to move, things would happen fast. My movie lights were mounted on the camera and were burning. "Don't let a lamp burst now", I thought to myself. "Or a battery die, or a film break, or any of the dozens of things that can

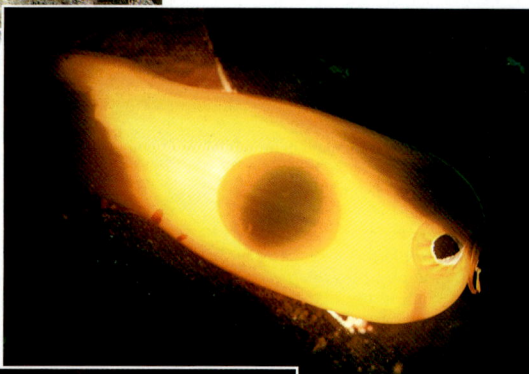

The brownish translucent egg capsule is made of a protein material and contains one egg.

go wrong when making wildlife films underwater. And when the action starts, don't let me do something stupid to blow the shot". I started reviewing all the times I blew a shot by turning the camera off too soon, or zooming the wrong direction, or losing focus. How many times had I actually risked my

The large amount of yolk in each egg is gradually consumed by the developing embryo until hatching.

life to get a shot and then blown it by doing something stupid? And how many times had the subject been sharks?

"This is madness!" I realised. "I'm down here freezing to death, waiting for this shark to come flying out. I'm wasting my time. I should be home. I want to become a couch potato. I should get a real job, 9 a.m. to 5 p.m., indoors where it's warm. I'm freezing. I must be entirely out of my mind! I'm going to spend two years of my life shooting

The young catshark has two rows of enlarged dermal denticles along its back which aid in leaving the slippery interior of the egg capsule.

this film and it's going to be on television for one hour. ONE LOUSY HOUR! And when it's on, people will just page through it with their remote control buttons. Two years of freezing like this. And the film may not be any good. Two miserable years and I could fail, I could fail! WAIT, SOMETHING MOVED!"

Suddenly, Chip and Mark are screaming through their regulators, "shoot, SHOOT, SHOOT!" I pulled the trigger even before looking through the viewfinder. Then I could see it through the lens. The shark was coming out! Its head emerged first. Using a series of modified hook-like denticles along its back the shark caught the top edge of the egg case opening and began pushing itself out. The shark egg was hatching!

The swell shark emerged quickly into the world, 15 cm long, covered with spots, and with a large bronze cat-like eye. A beautiful, jewel-like creature. It hesitated only a moment in the bright pool of incandescent light before dashing off in search of a place to hide and rest. I turned the camera off. "Wonderful, absolutely wonderful". Mark and Chip were still hooting through their regulators: "God, that was wonderful!" I checked my air pressure and decompression computer. "Damn, time to go. Never enough time. Never enough air. I wish I could stay forever! My God, that was wonderful!"

A replica of its parents, the newborn catshark is very agile and immediately able to hunt small prey.

Howard Hall   California, Eastern Pacific

***Cephaloscyllium ventrio-
sum*** **California swell shark**
L: At birth 14, max. 100 cm. Di: E-
Pacific: California to S-Mexico,
Chile. De: 5-460 m. G: Similar to
C. umbratile, but pattern different,
underside without spots. Among
rocks, kelp. Resting in caves by day.
Oviparous. Embryos are eaten by
snails that bore into the egg case.
Young have enlarged dermal denticles along the back that aid in
opening the egg case while hatching; these denticles are lost soon
thereafter. Preys on bony fishes,
crustaceans. Has disappeared -
probably due to global warming -
from many areas at the coast of
California where it once was
common. See also previous pages.

Rudie Kuiter   South Australia

Cephaloscyllium nascione
Whitefin swell shark
L: Max. 100+, mature males >80
cm. Di: Western Pacific: Australia
(NSW, South Australia). De: 146-
220 m. G: Has long been considered a subspecies of the
following sp., but was established
as a valid species in the 1980s.
Colour pattern superficially similar
to that of *C. umbratile,* but has 8-9
prominent greyish-brown saddles
on a light tan or grey-brown
background and numerous flecks.
Ventral surface without spots. Fins
with distinct white margins. Little-
known, benthic on the outer
continental shelf. Most probably
oviparous, occasionally taken by
bottom trawls, but not utilised.

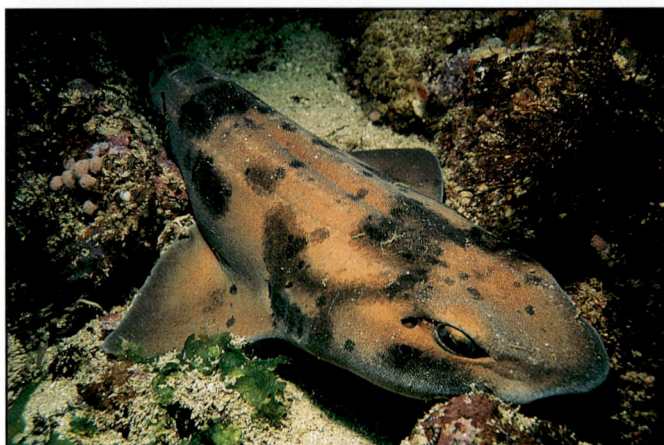

Glenn Edney   Poor Knights, New Zealand

Cephaloscyllium isabellum
Draughtsboard swell shark
L: At birth 16, max. 100 cm. Di:
New Zealand (common around
Stewart Island). De: 18-220 m.
G: Dark saddles and blotches in
a chess-board pattern (name) on
a brown or golden background.
Ventral surface light, without
spots. In caves during the day.
Preys on a wide variety of fishes
(teleosts, spiny dogfishes) and
invertebrates (molluscs, krill,
crabs, lobsters) at night. Sucks
burying worms out of their
burrows. A very unusual component of this shark's diet are tunicates (sea squirts). Oviparous.
Egg case cream in colour, 12 cm
long. Adults segregate by sex.

Cephaloscyllium umbratile
Blotchy swell shark

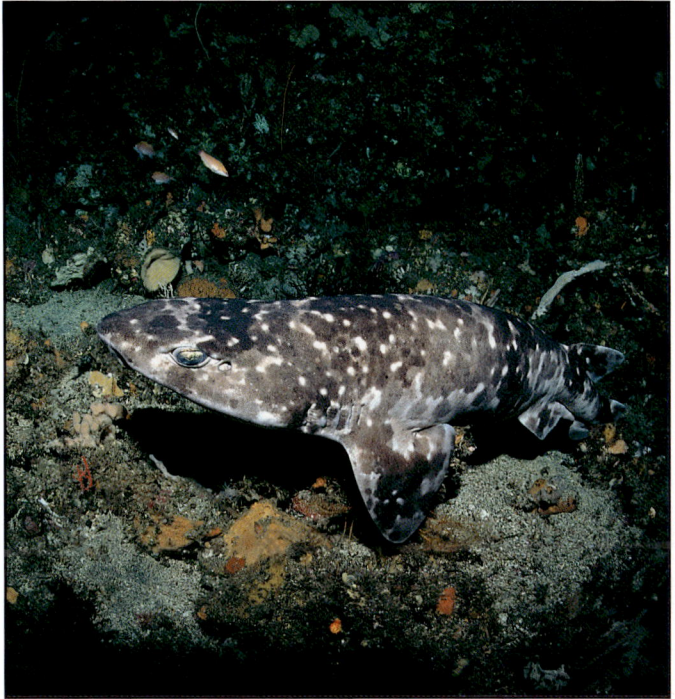

Length: At birth 16-22, max. 120 cm, females are larger than males.
Distribution: Western Pacific: Japan to South China Sea and possibly to New Guinea.
Depth: 20-200 m.
General: Similar to *C. isabellum* (previous species), but with different colour pattern: Background lighter, with irregular saddles and numerous blotches. Ventral surface without spots. An inhabitant of rocky bottoms.

Prey includes a wide variety of smaller bony fishes (mackerel, sardines, filefishes, morays, dragonets, flatfishes, cowfishes), the main component of the diet, and commonly also squid. But also elasmobranchs are eaten, eight species of sharks (including juvenile swell sharks), skates, electric rays and egg cases of a catshark and a skate have been found in their stomachs.

Mode of reproduction is oviparous, two egg cases with tendrils are laid at a time. Eggs hatch after about 1 year. The reproductive season is not well defined. Harmless to divers. This swell shark does well and even reproduces in public aquaria of sufficient size.

The popular name swell sharks was attributed to the members of this genus because most - if not all - are able to inflate themselves by swallowing water or air when threatened by a predator, disturbed or harrassed, e.g. taken from the water, much in the way the porcupine and puffer fishes of the teleost order Tetraodontiformes do. When expanding its stomach while in a confined space such as a rock crevice, a swell shark can wedge itself in and consequently is very difficult to remove from its position.

both photos Kazu Masubuchi   Izu Peninsula, Japan

Fred Bavendam   Edinburgh, South Australia

Asymbolus vincenti
Gulf catshark
Length: Max. 60, mature females 45-53, mature males 51 cm.
Distribution: Australia: Victoria around southern coast to Western Australia.
Depth: 2-200 m, usually 130 m or more.
General: Nasoral grooves are absent, the nasal flaps do not reach the mouth. The anal fin is about the size of the second dorsal fin and positioned far in front of its origin. Coloration dark brown with white spots, belly lighter and without spots. Found in temperate waters on mixed reef and sea grass areas. Oviparous, two egg cases with tendrils are laid at a time.

The deepwater catshark genus *Apristurus* is one of the largest and least known genera of sharks with about 30 named and several unnamed species. Judging from the frequency that new species are discovered, the wide geographic range of the genus and the paucity of knowledge of slope faunas in many areas of the world, this may well become the largest genus of sharks in the future.

During two cruises in the eastern North Atlantic, the author was able to preserve a few specimens of *A. laurussonii* and a similar, yet undescribed species - both caught in deepwater bottom trawls between 800-1,500 m - for further studies by specialists.

Rudie Kuiter   Victoria, Australia

Rudie Kuiter   Port Phillip Bay, Australia

Asymbolus analis
Greyspotted catshark
L: Max. 90, mature females 57, mature males 55 cm. Di: Australia: NSW around southern coast to Western Australia. De: 10-175 m. G: Anal fin about the size of 2nd dorsal fin and positioned far in front of its origin. Coloration light brown with faint, medium brown saddles and scattered dark brown spots, belly lighter and without spots. Found in temperate waters on mixed sand and rock areas, also in bays and estuaries. Oviparous, the brown egg cases with long tendrils are often found in groups, their tendrils tangled around a common holdfast.

Asymbolus rubiginosus
Orangespotted catshark

L: Max. at least 53, mature males 37 cm. Di: E-Australia: Queensland to Tasmania. De: 80-290 m. G: Described only recently. Has been confused with the greyspotted catshark. Overlaps in range with the latter but occurs deeper. Light brown dorsally, with faint greyish-brown saddles and blotches above gill area, beneath dorsal fins and on sides. With dark brown spots with orange-brown borders, each spot larger than spiracle diameter, scattered irregularly and numerous, but variable in number. Underside, pelvic and anal fins pale. Teeth minute, with 3 or 5 cusps. Biology little known.

Saul Gonor   **Jervis Bay, Australia**

Atelomycterus marmoratus
Coral catshark

L: At birth 10, max. 70, mature females 49-57, mature males 47-62 cm. Di: Pakistan to New Guinea, Philippines, China. De: 5-35 m. G: Coloration extremely variable, but dorsal fins always with white tips. Shallow water, coral reef species with elongate body that allows for movement into crevices and among branching corals. Does not 'walk' on paired fins like bamboo sharks. Oviparous. Nocturnal, preys on benthic invertebrates and teleosts.

The banded coral catshark A. *fasciatus* with a pattern of brown bands on lighter background was described in 1993 (W-Australia).

Werner Thiele   **Komodo Island, Indonesia**

Aulohalaelurus labiosus
Blackspotted catshark

L: Max. 67, mature females 67, mature males 54-62 cm. Di: Western Australia. De: Shallow, recorded down to only 4 m. G: Greyish- to yellowish-brown with few faint dusky saddles, many large black spots and scattered smaller white spots. Often tips of both dorsal fins white. Underside pale. Common, but little-known shallow water catshark, only rarely seen due to nocturnal habits. Hides in caves by day. Preys on benthic invertebrates and small teleosts.

A second species of this genus, the Kanakorum catshark A. *kanakorum,* was only recently described from New Caledonia.

Rudie Kuiter   **Victoria, Australia**

OCEANOPOLIS

The shipyard at Brest, on the French Atlantic shore, encloses one of Europe's most modern aquariums. Since its opening in 1990 over two million visitors have discovered the wonders of Oceanopolis. The successful concept behind the hi-tech site is based on a well balanced combination of adventure and information. Here the visitor gains a variety of rich impressions of the Atlantic and its denizens of the deep.

2

ALL PHOTOS: CHRISTOPH GERIGK

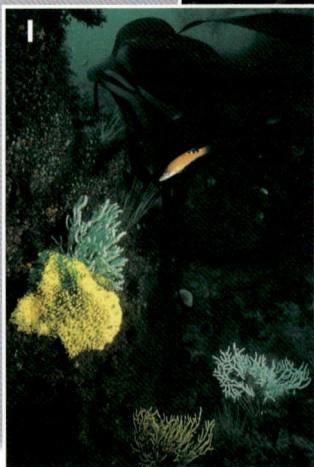

1

1. A dusky ambience greets the visitor at the aquarium. Lighted displays, silvery screens and regularly spot-lighted, life-size models of marine mammals create a mysterious atmosphere. This effect is heightened by the diffuse daylight that streams through the tang forests inside the large cold water aquariums in all shades of green. All important habitats of the North Atlantic are represented in separate tanks, such as the one shown here with its mussel-encrusted wall, laminaria, sponges and sea fans.

2. The sprawling white building resembles a huge crabshell. Large glass fronts partially flood the interior of the stone crabshell with sunlight. There are two levels at Oceanopolis: the exhibit and the aquarium. The latter is especially popular with visitors.

3+4+5. The 500,000 liters of water required for the site are pumped in from the adjacent shipyard. Besides the fourteen primary aquariums, there are an additional seventy tanks located out of sight, in the quarantine area. The main purpose for all this modern technology is to reproduce the shown ecological systems as accurately as possible, that is to say, to guarantee a natural environment.

The daily underwater show is the absolute highlight for many of the visitors who come to Oceanopilis. After the announcement is made, divers gather at the "Grand Tombant," the largest tank on the site with projecting glass panels. Two large screens have been installed on either side of the glass front. Supplied from the surface with compressed air, a diver donning a dry suit begins his descent to the bottom of the tank. On his head he either wears a transparent plastic helmet with an installed microphone, or a full-face mask with a TV camera mounted on top, which transmits images to the two screens.

Spectators watch in fascination as the diver plays with skates, dogfishes or nurse sharks, and take note of how he points out the peculiarities of fishes to the children behind the glass panes, filling in the details

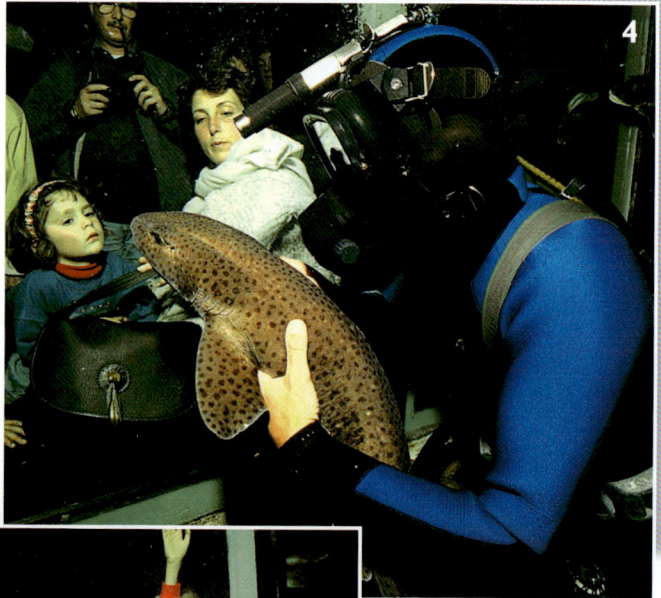

through his microphone. At feeding time, even shy fishes like the John Dory make their appearance. When the show is over, visitors have learned a great deal about the animals of coastal Europe and, perhaps, also gained a new perspective on the "product" fish. Oceanopolis's goal is to familiarize people with the ocean.

This shark family at present comprises 37 species, all of which have well developed dorsal fins, the first dorsal being located well ahead of the pelvic fins. Their eyes are elongated and have nictitating eyelids. A precaudal pit is lacking, the anterior nasal flaps are short, only the whiskery shark *(Furgaleus macki)* has long, barbel-like nasal flaps. The smoothhounds are active swimmers that move most of the time along the ground. Occasionally they rest on the bottom, but only for a short while. Most are drab in coloration, but there are exceptions. The mode of reproduction is not the same for all family members: Most of the species are ovoviviparous, while those of the genus *Mustelus* may be either ovoviviparous or viviparous. Also most species swim in large mating aggregations at certain times of the year. An old report states that captive male dusky smoothhounds tried to copulate with each other in the absence of females. The same report states that both claspers are inserted during copulation. Several family members are commonly eaten by other sharks, e.g. white and sevengill sharks. It is not easy to approach these sharks with SCUBA diving gear, snorkelling is much easier. Many of the species, especially those of the genus *Mustelus,* do well in public aquaria. To avoid losses, however, they should be kept in a tank without larger sharks.

Mustelus antarcticus   **Australian smoothhound, gummy shark**

Length: At birth 30-35, max. 175, females mature at about 85, males mature at about 80 cm.
Distribution: Western South Pacific: Australia: Western Australia (Geraldton) to New South Wales (Port Stephens), including Tasmania. Depth: Shallow waters to about 80 m, but sometimes also down to 350 m.
General: The only genus member in temperate waters of Australia. A shark with a slender body and pavement-like dentition in both jaws. Teeth have low crowns with a single, indistinct cusp (see dentition of *M. mosis* on p.207). Coloration of upper surface bronze to greyish-brown, usually with numerous small white spots, rarely also with some black spots. Underside light in colour. Often swimming in small schools with individuals of similar size and same sex. Tagging experiments conducted off eastern Tasmania and in Bass Strait have shown that this shark is able to make long migrations. Some tagged females were recaptured in South Australia and Western Australia. Males do not seem to migrate. Mode of reproduction ovoviviparous, litter size 1-38, mostly about 14. Young are born in December after a gestation period of 11-12 months. Preys on cephalopods, crustaceans and bony fishes. This species constitutes the major part of the southern Australian shark fishery but has been heavily exploited and over-fished in recent years. Flesh utilised for human consumption, marketed as 'flake', popular in Victoria and Tasmania.

David Fleetham / IV   South Australia

Mark Conlin   California, Eastern Pacific

Mustelus californicus
Grey smoothhound
L: At birth 20-30, max. 124 cm, mature females 70-124, mature males 57-116 cm. Di: Eastern North Pacific: Northern California to Gulf of California. De: 1-25 m. G: Uniformly grey above, underside lighter, no spots or stripes. Common, bottom-dwelling, enters shallow muddy bays. Summer visitor in northern California, resident in warmer southern waters. Placental viviparous, litter size 2-16, gestation period 11 months. Preys primarily on crabs (grapsids, cancrids), also on worms and small fish. Caught and utilised for human consumption. Often seen together with *Triakis semifasciata*.

Christoph Gerigk   Inner Hebrides, Scotland

Mustelus asterias
Starry smoothhound
L: At birth 30, max. 140 cm. Di: Eastern N-Atlantic: North Sea to Morocco and Canary Islands, Mediterranean. De: 1-100 m. G: Snout moderately long, distance between nostrils narrow. Pectoral and pelvic fins small. Coloration grey with white spots on back and sides of body, ventral side lighter in colour. The only spotted family member in European waters. Found on sand, mud and gravel bottoms. Placental viviparous, each embryo is nourished by a placental structure within the uterus of the mother until hatching. Litter size 7-15, gestation period 12 months. Preys on crustaceans.

Helmut Debelius   Galicia, Spain, Eastern Atlantic

Mustelus mustelus
Common smoothhound
L: At birth 39, max. 160 cm. Di: Eastern Atlantic: British Isles to South Africa, and Mediterranean. De: Intertidal to 350 m. G: Brown or grey, some specimens have dark spots. Benthic, inshore. Litter size 4-15, dependent on size of mother. Gestation period 10-11 months. Preys mainly on crustaceans (crabs, hermit crabs, lobsters, mantis shrimp), but cephalopods and bony fishes (snake eels) are also eaten. Harmless, does well in captivity. It is caught for human consumption, but easily overfished due to its low fecundity of only 15 young per year and litter.

Galeorhinus galeus
Tope or school shark

L: At birth 35, max. 190 cm. Di: E-Atlantic: Iceland and Norway to Senegal, Namibia to South Africa; Mediterranean. Western S-Pacific: Australia, New Zealand. W-Atlantic: Brazil to Argentina. E-Pacific: British Columbia to Gulf of California, Peru, Chile. De: 1-471 m. G: Active, schooling species. Sexual segregation known in some populations, the males occur closer inshore than females. Males reach sexual maturity at 8, females at 11 years of age. Litter size 6-52, larger females also have larger litters. Regularly frequent nursing grounds. Ovoviviparous, females give birth to their young in shallow bays and estuaries in spring and early summer. Juveniles may stay in these areas for as long as 2 years before moving offshore in schools. Preys mainly on bony fishes (herring, salmon, cod, croakers, wrasses, damselfishes, gobies, kelpfishes, scorpionfishes), but also eats squid, octopuses, crabs, shrimp, gastropods, worms, echinoderms and other elasmobranchs (small sharks, stingrays, skates). Is in turn prey for larger sharks. Will adapt to captivity if captured and handled carefully. Ram breather, swims constantly in the aquarium. Utilised for human consumption in many countries. Once very common in North Atlantic, but its population has been heavily over-exploited by the fishery industry.

Bottom photo: *Enteroctopus dofleini* eating a school shark.

both photos Doug Perrine   Azores, Atlantic

Fred Bavendam   British Columbia, Canada

Iago omanensis
Arabian houndshark

Length: At birth 15, max. 63, mature females 40-63, mature males 30-37 cm. Distribution: Red Sea and Indian Ocean: Arabian Gulf to southwestern India. Depth: 80-1000+ m. General: This small houndshark is characterised by a relatively small first dorsal fin that originates directly above the pectoral fins, a long gill region and a very small lower lobe of the caudal fin. Coloration grey or brownish above, margins of dorsal fins darker, underside lighter; otherwise without markings. Found living on or near the bottom in tropical deepwater habitats of the continental shelf and slope, in the Red Sea possibly down to 2,195 m. The gills and gill slits of this species are larger than those of other genus members, possibly an adaptation to survive high temperatures (up to 25°C in the Red Sea), low oxygen levels and probably also hypersaline conditions.

Reproduction is placental viviparous (yolk sac viviparity), litter size 2-8, depending on the size of the mother. Young are born after a gestation period of about 9 months. The anatomy of the yolk sac placenta is unique among sharks: After the connection between the embryo's yolk sac and the mother's uterus wall has been established, the latter grows to form a mushroom-shaped bulb around the remnants of the nearly empty yolk sac. From now on and until hatching at term size, the embryo is supplied with nutrients through this connection by its mother. The large mushroom-shaped bulb (each placenta's uterine part) is obviously lost during or soon after birth because after parturition each uterus wall shows a number of scars corresponding to the number of young recently born. There also is a remarkable sexual dimorphism in this shark species: Males are about 33% shorter than females and attain only about a sixth of the weight of females.

Preys on bony fishes (lanternfishes), crustaceans (shrimp, mantis shrimp larvae) and molluscs (cephalopods, bivalves, gastropods). Seagrass, mud and other unusual items have also been found in the stomach contents, including peas and cigarette box remains. The latter appeared in specimens caught off the marina of Elat at the northern end of the Red Sea, an area with a lot of tourist boat traffic; obviously these indigestible morsels have been thrown overboard and subsequently swallowed by the sharks shortly before they reached the ground. Due to its deepwater habitat, this shark is not seen by divers and was once thought to be rare in the northern Red Sea (Gulf of Aqaba) because only a few specimens were known from that area. However, the author investigated about 200 specimens caught during experimental gill netting in 400-600 m over a period of several months in 1986/87. After that the species has also been filmed alive in its natural habitat from aboard a manned submersible operating down to 400 m. Attempts to keep this cute little shark alive in tanks for extended periods of time failed; being the only true deepwater shark of the Red Sea, it has large eyes which often are injured and become infected during catching, subsequently leading to the death of the animal. Additionally, specimens with intact eyes seem to suffer from stress induced by bright daylight.

Another peculiarity seen in this shark are abnormal jaw teeth which occur in the dentitions of many specimens from the Red Sea as well as from the Indian Ocean. These abnormalities are not sex-related and

can only partly be the result of injuries to the dental lamina in the mouth cavity. Injuries usually result in simply de-formed teeth. But many of the dental aberrations seen in this houndshark look as if there have been mistakes in the 'program' generating the den-tition, e.g. tooth cusps pointing not toward the corners of the mouth but toward the sym-physis, teeth with two cusps pointing in either direction, and the like. Because these peculiar teeth are found in individuals from totally differ-ent locations their occurrence might be fixed in the genes of the species.

The species is gill-netted in India and utilised fresh for human consumption, but also very good eating when pre-pared after freshly frozen.

Itamar Grinberg   Gulf of Aqaba, Red Sea

Triakis semifasciata
Leopard shark

L: At birth 20, max. 180 cm. Di: Eastern N-Pacific: Oregon to central Mexico. De: 1-90 m. G: Coloration distinct, with dark saddles and spots on bronze-greyish background. In coastal waters, on rocky ground, among kelp, also in muddy and sandy bays. Often aggregates in shallow water, occasionally mixes with other sharks. Swimming just above the ground, also resting on it. Litter size 4-29, dependent on size of mother. After 12 months, young are born in coastal bays in spring. Preys on crabs, shrimp, worms, octopuses, bony fishes, fish eggs and elasmobranchs (smoothhounds, guitarfishes, bat rays). Shy, approaches are best at night. Harmless, long lived in captivity (over 20 years).

Mark Conlin   California, Eastern Pacific

Triakis scyllium
Banded houndshark

L: At birth 23, max. at least 150, males mature at 99-108 cm. Di: Western North Pacific: Southern Siberia, Japan, Korea, China, Taiwan, possibly also Philippines. De: Shallows to about 150 m. G: Coloration grey with darker saddles and scattered black spots of various sizes, underside lighter. Bands and spots faint to absent in adults. Teeth blade-like with strong cusps and cusplets, not molariform. Common, benthic, often close inshore, in brackish estuaries and bays, on sand flats with algae or seagrass. Not in schools, but resting in groups. Nocturnal, preys on fishes, crustaceans. Often seen in public aquaria in Japan.

Kazu Masubuchi   Izu Peninsula, Japan

REMOTE CONTROL SHARKS

The coconut described a high lazy arc over the tropical reef, framing a small island on the barrier reef before plopping into the shallow cerulean water on the outside of the lagoon. A split-second after impact, a geyser of water and foam erupted from the ocean, as if a grenade had exploded. This was no ordinary coconut. It was loaded with caesium 137. Coconuts like this one had been responsible for completely de-populating Bikini Atoll. Underwater photographer Doug Perrine follows the flight path of the coconut that leads to Shark Pass, one of Bikini's hottest spots...

There are not many spots in the world where well over a hundred grey reef sharks *Carcharhinus amblyrhynchos* gather around a boat shortly after anchoring. One such spot is found at the address Shark Pass, Bikini Atoll, Marshall Islands, Micronesia, Pacific Ocean.

The evacuation of the Bikini Islanders in 1978 due to the hazards of ingesting radioactive coconuts was their second exodus from their traditional homeland. The first occurred in 1946, prior to the 'Operation Crossroads' series of 67 atomic tests, 23 of which were detonated in Bikini Lagoon. The tests started with the 'Able' and 'Baker' blasts, and concluded with the 15-megaton 'Bravo' hydrogen bomb explosion - the most powerful nuclear test ever conducted above ground.

In addition to contaminating the islands of Bikini Atoll with radioactivity, the tests sank 21 ships, including the battleship *Nagato,* from which the attack on Pearl Harbour was launched, and the U.S.S. *Saratoga.* At 264 metres, the *Saratoga* is the largest diveable shipwreck in the world. It was the first aircraft carrier ever built, and is the only aircraft carrier in the world accessible to sport divers.

Head divemaster Fabio Amaral, a big gregarious bear of a man, proudly wears a t-shirt which proclaims "Size Does Matter." On the back is a diagram of the *Saratoga* with a caption that explains "Saratoga is larger than the Titanic." If you have adequate training and experience, and are willing to adhere to his rigid and carefully worked-out procedures, Fabio will take you on dives to wrecks lying as deep as 54 metres, involving up to an hour of decompression. One thing which Fabio will NOT do, however, is take you on a dive at Shark Pass, which is where the coconut landed after I heaved it over the side of our boat.

A caesium-laced coconut could easily kill a man - if he was foolish enough to sit directly underneath a tall palm tree. It would also be possible to contract radiation sickness from consuming a large number of coconuts over a period of many years, drinking well water, or eating anything grown in the soil of Bikini Island. For this reason, all of the food is imported, and water is produced by reverse osmosis from seawater. Background radiation at Bikini has declined to a level lower than that in most major cities during the

Facing page: Think twice before you enter the water! More sharks are coming, and all of them seem to be very, very hungry.

half-century since the tests. The worst radiation hazard is from the tropical sun, which can fry you in short order. The lagoon is essentially free of radiation, even inside the shipwrecks.

Fish are perfectly safe to eat, and in more than 50 years without fishing taking place, populations have returned to healthy levels rarely seen anymore elsewhere in the world. 'Healthy' does not even begin to describe the population of grey reef sharks at Shark Pass, though. Well over a hundred sharks gathered around our boat each time we went there, and many more sharks are scattered along the reef all the way around the lagoon.

Generally, a large concentration of sharks in a particular area is a good indicator of an abundant food resource at that spot. However, while there is certainly no shortage of fish around Bikini, there does not appear to be an especially copious supply at Shark Pass. Furthermore, while not emaciated, these sharks are definitely not 'fat and happy'. A leaf or jellyfish drifting by prompts a mad rush to the surface by up to two dozen sharks competing to reach it first. A slap on the water with a paddle incites a boiling frenzy. And if you toss a coconut into the water, you might be forgiven for thinking the coconut had exploded.

No, in this case it seems evident that there is a shortage of food relative to the number of sharks present. The aggregation probably serves some social purpose - an idea which is bolstered by the fact that all of the sharks we

Above: If you toss a coconut into the water at Shark Pass, you might be forgiven for thinking the coconut had exploded...

Right: A 2.4-metre tiger shark also showed up and nearly swallowed two of our camera housings!

Above: Healthy does not even begin to describe the population of grey reef sharks at Shark Pass, but...
Below: ...these sharks are not 'fat and happy'. A jellyfish drifting by or a slap on the water with a paddle incites a boiling frenzy.

After having experienced Shark Pass, any thought of making a dive there will vanish from your mind. However, it is possible to dive elsewhere on the Bikini barrier reef, where you will be assured of seeing plenty of sharks in a healthy and vibrant coral reef.

saw were female, save for one lucky male. However, it does not appear to be a breeding aggregation. The sharks are present all year round. They did not appear to be pregnant, and we only saw one with a bite mark that could possibly be a mating scar. Because of its position on the shark's body, the scar could just as likely be a wound incurred in a competitive feeding situation.

"You guys are professionals, you've got the boat chartered, do what you want," said Fabio, "but there's no way I'm taking anybody diving there." And so we found ourselves anchored at Shark Pass, gearing up at an extremely measured pace, each of us probably subconsciously attempting to be the last person ready to enter the water. Eddie Maddison, a Bikinian with over ten years experience diving the wrecks, and expat divemaster Brad Watson eyed us warily, wondering if we were really going to go through with our crazed plan. Finally, veteran shark photographer James D. Watt tentatively stepped out to the edge of the platform and dipped his mask to rinse it. Two sharks shot out from beneath the boat and made a grab at the mask. "Let's re-think this idea," said Jim.

An attempt to photograph the sharks by hanging a camera over the edge of the platform produced similar results. Five sharks hit the housing at once, ripping it out of my hand and sending $5,000 worth of photo gear to the bottom below. Eventually the camera was retrieved. When I processed the film I saw that one shark had hit the housing hard enough to take its own picture - and it wasn't just a bump. The mouth is opened most of the way around the 92 dome port, extending almost up to the handle I was holding.

Fortunately we had realised in advance that the best way to photograph these sharks would be by remote control, and had come with the appropriate equipment. However, we still experienced problems as sharks bit through shutter release cables and knocked off video eyepieces. On the last day, a 2.4-metre tiger shark came in and nearly swallowed two camera housings. We also saw a few silvertip sharks offshore, and a number of blacktip reef sharks. Silky sharks are sometimes seen feeding on baitballs out in the blue water.

A normal week of diving at Bikini Atoll includes six days of diving the wrecks, and a half day trip (weather permitting) to Shark Pass to view (from the boat) the feeding frenzy which is incited when a few fish scraps are dropped into the water. Whether this is worth giving up a wreck dive depends, of course, on your priorities as a diver. For 'After viewing this demonstration, any thought of making a giant stride entry at Shark Pass will vanish from your mind. However, it is possible by group consensus to arrange a dive elsewhere on the barrier reef, where you will still be assured of seeing plenty of sharks, not to mention an extremely healthy and vibrant coral reef. wreckies and tekkies' it's hard to compete with intact submarines, battleships, and a carrier with eight decks and a hold full of airplanes. The cargo hold in the *Saratoga* is so large that dogtooth tuna school inside it.

For excellent shark and reef-diving the potential of Bikini Atoll has barely entered the exploratory phase. But if you are a diver, the prospects of diving there may certainly cause your mouth to water and your credit card to start squirming around in your wallet.

Members of this family are familiar to most divers, because several are commonly encountered on tropical coral reefs. The large family at present comprises about 50 species. Many of these, especially the species of the genus *Carcharhinus,* are difficult to distinguish underwater. Tooth morphology and dental counts from both jaws are the best characteristics to determine a *Carcharhinus* species, a method, which is hardly applicable while diving with live animals, however. Field characters helpful in identification include build, snout shape and size, shape, relative position and black or white markings on fins. All family members share circular eyes with nictitating membranes, first dorsal fin well ahead of the pelvic fins, a precaudal pit and a well developed lower lobe of the caudal fin. An interdorsal ridge (a ridge on the back between the first and second dorsal fins) may or may not be present, depending on the species. Most family members are active predators on large invertebrates, bony fishes and other elasmobranchs.

All but one species in this family are placental viviparous, the exception being the tiger shark, which is ovoviviparous. They range in length from 70 cm to over 5 m, several species averaging around 2 m. Many make long, seasonal migrations, moving to warmer water in cooler months. Because of their large average size and their affinity for areas frequented by many people (e.g. tropical, shallow, coastal areas), requiem sharks are probably responsible for more attacks on people (including divers) than any other shark family. Although some species are potentially dangerous, many family members are shy and very difficult to approach underwater and often baiting is the only way to bring them into camera range. Some species (lemon, whitetip reef, bull, oceanic whitetip, Caribbean reef and sandbar shark) do well in large public aquaria, while others (tiger shark) hardly survive for a longer period of time in captivity.

The photo below shows a typical member of this family, the blacktip reef shark *Carcharhinus melanopterus,* a group of which is circling over a reef in the crystal clear waters of French Polynesia in the South Pacific Ocean.

KURT AMSLER

Carcharhinus acronotus
Blacknose shark

Length: At birth 38-50, max. 200, mature females 103-137, mature males 97-106 cm.
Distribution: Western Atlantic: North Carolina to Florida, Bahamas, Gulf of Mexico, Virgin Islands, Puerto Rico, Antilles, Guyana, Venezuela, southern Brazil.
Depth: Intertidal to 83 m.
General: Underside of snout usually with a distinct black blotch, may be absent. Second dorsal and tip of upper lobe of caudal fin with dark markings. A common species in the tropical and warm-temperate waters of continental and insular shelves. Seen over sandy bottoms, in coral reefs, over coral debris and bottom covered with shell fragments.

    Mode of production placental viviparous. Reaches maturity after 2 years. Litter size 3-6, gestation period 10-11 months. Pregnant females are found in Florida waters from the winter months until spring. However, most individuals are caught from March to November, indicating that the species locally migrates. Adults seasonally segregate by sex. Prey includes small bony fishes (pinfish, croakers, porgies, spiny boxfishes, porcupine-fishes) and octopuses. Often associates with schools of small fishes (mullet, anchovies). Regularly falls prey to larger sharks. A possible threat display similar to that described for the grey reef shark has been observed also in captive specimens of this species when confronted with divers or other sharks. This includes arching of back ('hunch'), lowering of tail and raising of snout. Threat displays can be indicative of territoriality and a home range. Nevertheless, this shark is considered harmless to divers, no attacks have been reported.

Marilyn Kazmers / IV

Doug Perrine   both photos Bahamas, Western Atlantic

Carcharhinus albimarginatus
Silvertip shark

Length: At birth 63-68, max. 300, mature females 160-200, mature males 160-180 cm.
Distribution: Wide ranging in most of the tropical Indo-Pacific: Red Sea and South Africa to Australia, Japan to New Guinea. Eastern Pacific: Colombia and Costa Rica. Depth: 1-10 m.
General: Tips and trailing edges of first dorsal, pelvic and caudal fins white; coloration otherwise uniformly grey above, lighter below. Seen singly, in pairs or small aggregations near coral reefs. Especially obvious, when patrolling in front of steep, deep (30+ m) drop-offs. Adults have a home range that may extend for 5 km along the reef face.

Litter size 1-11, often 5-6, gestation period about 12 months. Females with mating wounds and scars have been observed; sometimes even the tip of the first dorsal fin has been bitten off by an 'overacting' male! Young occur in shallow waters including lagoons. Active predator on a great variety of benthic and pelagic bony fishes (scorpionfishes, lanternfishes, flyingfishes, tuna, wrasses, wahoo, parrotfishes, porcupinefishes), other sharks, eagle rays and octopuses. Groups of rainbow runners have been observed swimming beside and chafing against the rough 'sand paper' skin of this shark, a 'scratching' behaviour, that relieves the mackerel of skin parasites and obviously is tolerated by the shark. A shy and wary species in the absence of bait, only rarely approaching divers. Juveniles are reported to be more inquisitive than adults. May make close, persistent approaches in feeding situations. Silvertip sharks have chased divers out of the water and attacks have been reported from Micronesia and New Guinea, however, in baited situations.

Mark Strickland　　　　**Andaman Sea**

Helmut Debelius　　　　**Egypt, Red Sea**

121

Silvertip shark C. *albimarginatus* Mark Strickland, Andaman Sea

First discovery of a cleaning station for silvertips, cleaned by *Thalassoma lucasanum* Peter Kragh, Cocos Island

Carcharhinus amblyrhynchos
Grey reef shark

Length: At birth 45-60, usually up to 185, max. possibly to 255 (one record of a male), females mature at 122-137, males mature at 130-145 cm.
Distribution: Wide ranging in the tropical Indo-Pacific: Red Sea and Madagascar to Easter Island, north to Hawaii, south to Lord Howe Island.
Depth: 1-274 m.
General: Trailing edge of tail with prominent black band. First dorsal fin without markings; coloration otherwise grey or brownish above, lighter below. Interdorsal ridge absent. Some Red Sea and Indian Ocean specimens have a white margin on the trailing edge of the first dorsal fin. This variant has been described as a distinct species (blacktail reef shark, *C. wheeleri*) in 1982, which, however, is considered to be a synonym of *C. amblyrhynchos* by some. The graceful shark *C. amblyrhynchoides* is a valid, more stout-bodied species from the tropical Indo-West Pacific (Gulf of Aden to Philippines and Australia) with a maximum length of 170 cm.

    Grey reef sharks are often seen near reef drop-offs and in reef channels, especially in areas with a strong current. Sometimes they are observed in loose groups of up to about 100 individuals during the day in a limited area. At night, they disperse and move over a larger area, often penetrating into the shallower parts of the reef including lagoons. At ebb tide, individuals may also move into reef passes. The behaviour of this species has been studied in detail since the 1960s. It is a curious shark, which should not be rapidly approached by a diver, as it is territorial (average home range 4.2 square kilometres) and may readily show its well-known threat display: Arching of back, lowering of pectoral fins and swimming ineffectively in an exaggerated manner with the head swinging from side to side. However, there is a marked difference in the behaviour of specimens from the Indian Ocean and those from the Pacific: While mostly the latter have been studied by ethologists (behavioural scientists) and clearly show the aggressive behaviour described above, those from the Indian Ocean region (including the Red Sea) are curious but much less aggressive. Still, the grey reef shark is probably involved in more attacks on humans than any other shark species encountered in the coral reef environment.

    Litter size 1-6, gestation period about 12 months. Prey includes mainly small bony fishes (morays, needlefishes, coronetfishes, soldierfishes, surgeonfishes, butterflyfishes), cephalopods and crustaceans. Groups of grey reef sharks have been observed herding schooling fish up against the reef before attacking.

    The grey reef shark is one of the three most common reef sharks in the Indo-Pacific region, the two other species are the blacktip and whitetip reef sharks. It shows so-called microhabitat separation from the blacktip reef shark; where both species occur together, the blacktip occupies shallow flats while the grey reef shark is usually found deeper. In areas where the blacktip is absent, the grey reef shark also inhabits the reef flats.

    For more information on this species see also SHARK GAMES starting on p.88, REMOTE CONTROL SHARKS starting on p.114 and SHARKS SIGNIFICANT FOR THE MALDIVES starting on p.125.

Herwarth Voigtmann   Maldives, Indian Ocean

C. amblyrhynchos   Helmut Debellus, Sudan, Red Sea

C. amblyrhynchos   Doug Perrine, French Polynesia, Southern Pacific

SHARKS SIGNIFICANT FOR THE MALDIVES

What do divers and maritime tourists expect to find in the Maldives? Big fish like sharks and rays, as well as snappers, sweetlips, mackerels and tunas, all of which can still be encountered there in huge schools. Maldivian fishermen are primarily after the bony fishes mentioned above, but they have also been catching an increasing number of sharks recently. Charles Anderson, a marine biologist living in the Maldives, reports:

Several methods of catching sharks are known in the Maldives. Three important ones are: targeted deepwater angling of spiny dogfishes, specifically Gulper shark (*Centrophorus* spp., family Squalidae), for their valuable liver oil; offshore angling of pelagic shark species like Oceanic whitetip shark (*Carcharhinus longimanus*) and Tiger shark (*Galeocerdo cuvier*) for their fins and meat; and coastal fishing of reef associated species like Silvertip shark (*Carcharhinus albimarginatus*) and Grey reef shark (*Carcharhinus amblyrhynchos*) for the same products using nets and handlines. All shark products are exported. For several years now the activities of the reef shark fisheries have been in conflict with the tourism industry of the Maldives. "Shark watching" is one of the main attractions for divers. According to rough estimates the sharkmania of divers alone earns the government of the Maldives US $ 2.3 million per year. If one accepts further stipulations, a living Grey reef shark is worth a hundred times more than a dead one in a fishing boat.

In the Maldives sharks do not have the false reputation of being maneaters as is common in other countries. Although there are a few cases in which fishermen at work have been bitten by sharks, there is no confirmed unprovoked attack on divers or snorkellers in the Maldives. Thus an unbelievable shark boom has developed there. The guarantee for an underwater shark encounter has become an important argument in the tourism business. Naturally, sharks are only part of the Maldives package in which white sand, palm trees, clear water and colourful coral reef fishes are likewise important. But a questionnaire submitted to 32 experienced diving instructors in the Maldives resulted in a list of 35 diving sites frequented only to meet sharks there. These include Fish Head, Maaya

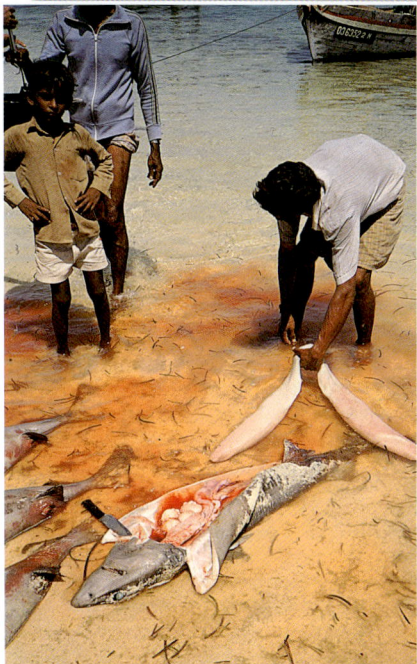

Cutting open a gulper shark. Note huge bilobed liver.

Tila, and Lion's Head, as well as Madivaru and Kuda Faru. It is hard to estimate the number of "sharks only" dives per year. Season, periods of absence of sharks, and the obvious lack of interest in sharks among some divers taken into account, there are still 77,000 pure "shark watching dives," which make up or about 15% of half a million tourist dives in the Maldives per year. If one dive costs US $ 30, then shark watching dives alone yield the yearly amount of US $ 2.3 million mentioned above.

However one choses to interpret these figures, it becomes clear that travel agents must have a great (financial) interest in healthy reefs. So the loud protests following the arrest of Maldivian shark fishermen at Fish Head and Lions's Head were no surprise. When the sharks did not show up at all at Fish Head between February and June 1992 consternation spread among the diving base operators. Today we know that this was only an extraordinarily long period of an annually recurring phenomenon whereby the sharks (mainly Grey reef sharks) simply disappear in the deep for several weeks. After all, who wants to be observed during love games? Yet travel agents have already urged the Maldivian government several times to totally ban shark nets and lines inside the atolls. The following calculation was used to reinforce the demand: about 20 adult sharks are regularly to be seen at Fish Head. If the US $ 670,000 net income from shark observations is divided by the number of sharks, each one at Fish Head is worth US $ 33,500. An analog calculation for all 35 shark diving sites still results in the fictitious shark value of US $ 3,300 per year. Naturally, also the value of a dead Grey reef shark can be calculated: if one adds the proceeds from dried meat and fins, jaws and liver oil the sum is about US $ 32. Altogether, a living Grey reef shark at the diving site is approxi-

Head of a sixgill shark caught from a Maldivian dhoni.

125

A much better way to 'use' them: living sharks, every uw-photographer's delight.

HERWARTH VOIGTMANN

mately a hundred times more valuable than a dead one. This is true for the period of one year, but if one takes into account that a Grey reef shark reaches 18 years of age and is stationary, this sum has to be multiplied several times.

It is not only important to reason how much profit there will be, but also who will benefit from it. The Maldivian fishermen directly depend on the income from their catch. They would not profit from any catching prohibition. This does not mean that they do not profit indirectly from diving tourism as about US $ 8 of the charge for one dive are for the boat's rent, including the dhoni's crew, who otherwise would be fishing. To give an example: between July 1991 and August 1992, seven out of nineteen shark fishing dhoni crews of Dungati accepted the offer to work for a newly opened diving resort nearby. In addition, the diving bases themselves offer work to Maldivians and indirectly increase their well-being as they pay taxes for all imports and guests. The government passes on these taxes for use in health and education programs.

Even today the Maldives still have a big competitive advantage over other diving destinations. Especially in neighbouring areas like South East Asia, there is the threat of overfishing the reefs due to the high food demands of a growing population. If fishing in the Maldives is continued sensibly, the advantages will prevail. Most of the diving base managers, however, are convinced that the numbers of visitors will decrease drastically if there are no more sharks to be seen in the Maldives.

Hit list of Maldives' sharks

Twenty-six shark species have been identified in a scientifically correct way in the Maldives until now. Three species of spiny dogfish from the depths of the Indian Ocean are frequently caught there, but have not yet been correctly identified. Four additional species are represented by dried jaws alone. Consequently, there are *33* species of shark reported from the islands. Furthermore, Maldivian fishermen regard Bowmouth guitarfish *(Rhina ancylostoma)* and White-spotted shovelnose ray *(Rhynchobatus djiddensis)* as sharks! Those five species which every diver - almost guaranteed, depending on the season - can observe in the Maldives, are presented in order of their abundance.

GREY REEF SHARK *Carcharhinus amblyrhynchos*   Maldivian name: thila miyaru
This stout, impressive shark of up to 180 cm in length often appears in groups. An encounter with 10 or more

of these animals is THE Maldivian experience for most divers. Generally this species is quite shy, but some diving instructors - at least in the past - have fed them in order to make them approach people. Those who approach are females, while males and juveniles live in deeper regions. Because they can be seen there throughout the year, the best diving sites for Grey reef sharks are Fish Head (Mushimasmingili Thila) in the Ari Atoll and Lion's Head in the North Male Atoll. At Miyaru Kandu, Guraidhoo and the Embudu Channel they are seen only during the Northeast monsoon (November to April), while around Kuda Boli and Rasfari they occur during the Southwest monsoon. With the monsoons the currents also change. The species can always be found at the channel mouths, which are exposed to the current. Between March and May the female Grey reef sharks also retreat to

Another kind of uw-photography. Even coloured lights do not hurt!

HERWARTH VOIGTMANN

mate. A 144-cm female caught in August in the Ari Atoll is known to have carried two young of a length of 48 cm each. The species is preferred by Maldivian fishermen and caught with standing gillnets, bottom longlines and handlines.

Most divers visit the Maldives mainly to see living sharks. What will be in the future for the sharks and the tourists?

WHITETIP REEF SHARK *Triaenodon obesus*   Maldivian name: faana miyaru

Compared to the Grey reef shark this species is much less impressive even though it grows larger (up to 210 cm). Sighting a Whitetip reef shark is guaranteed in the Maldives as it is not very shy with divers. This is true throughout the year, including both monsoon seasons. During the day the species is often encountered resting in caves or under table corals, frequently also on open sand patches and in water so shallow that even snorkellers can observe it. The nocturnally active species hunts in groups and feeds on parrotfish, goatfish and triggerfish. Each individual is stationary and returns to its resting place. Maldivian fishermen catch the Whitetip reef shark at night using standing gillnets and handlines.

COMMON NURSE SHARK *Nebrius concolor*   Maldivian name: nidhan miyaru

The Maldivian name means "sleeping shark." This is exactly what this nocturnally active shark does during the day in caves along the reef where it can be seen more often by divers than the Leopard shark, which has similar habits. The bottom-dweller reaches a length of 320 cm and specialises in catching octopus and other cephalopods. With its barbels it also probes coral thickets and crevices for crabs and small fish which are sucked into its mouth. One female reportedly carried 8 embryos. Some time ago the fishermen also sold living specimens to the tourist resorts where they were presented to the guests in so-called "shark pools." Fishermen catch this nurse shark at night using standing gillnets and handlines.

SILVERTIP SHARK *Carcharhinus albimarginatus*   Maldivian name: kattafulhi miyaru

This species is widespread in the Maldives and can be seen frequently. The local name refers to the clearly visible white-edged fins. The Silvertip shark swims in a much more majestic way than the often hectic Grey reef shark. This is also due to the former species' greater length, which may reach 300 cm. However, large Silvertip sharks can only be seen along outer reefs, which also is where the fishermen catch individuals up to 230 cm in size. Juveniles of 75 to 125 cm have been caught inside the Maldives atolls. Stomach content analysis revealed that the species feeds on bony fishes and cephalopods. In the Maldives it also occurs singly in shallow water (around 10 m).

WHALE SHARK *Rhincodon typus*   Maldivian name: fehurihi

The largest living fish species (length up to 12 m) is a harmless plankton feeder. The massive fish - in the Maldives mainly 5- to 7-m-long specimens are encountered - often swims in shallow depth and can easily be identified by the white spots on the back. For filter-feeding the huge terminal mouth is opened wide. It has to be pointed out that Whale sharks cannot be observed regularly everywhere in the Maldives. During the northeast monsoon the currents run from east to west, and vice versa during the southwest monsoon. Every time the currents change direction lots of sediment is stirred up in the Maldives. This results in plankton blooms which attract the Whale sharks, especially in the north, where the water around the two atoll chains is more thoroughly mixed and the monsoon change is more pronounced than in the south. It needs to be mentioned that Whale sharks are also caught in the Maldives. According to questionnaires, the catch is about 30 specimens per year, which in the long run is far too many of this relatively rare species, especially if they are meant to attract tourists. Meat and fins are not utilised; they die only for 100 to 200 litres of oil yielded per individual. Thirty dead Whale sharks net no more than US $ 4000!

SHARK FEED IN THE CORAL SEA

My beginning dives in the Bahamas thirty years ago were preceded, usually as I tried to sleep the night before, by a nagging vision of being grabbed by a shark. Heroic stories abounded of miraculous escapes from shark attacks by super-human divers. Dive knives were purchased with their shark fighting qualities of first consideration, which is why they looked rather more like swords than knives, and divers were advised to leave the water if a shark was ever seen. In fact very few ever were, which of course just added to their terrible reputation, and if one was ever seen this became an excuse to huddle close to your buddy, and utter prayers for salvation while backing out of the water.

Grey Reef Sharks *Carcharhinus amblyrhynchos* ignore the divers at this SCUBA ZOO gathering. All photographs on these two pages were taken at Flinder's Reef in the Coral Sea.

I moved on to PNG where, in the early days at least, making a dive without seeing a shark was about as easy as making one without getting wet. I quickly learned that sharks had been slandered and that the risk to a diver from a shark was minuscule. I also found that sharks preferred to stay away from sites regularly dived. This became a problem as better educated divers started to request dives where they would see sharks, and this led to dive operators selecting certain sites and deliberately attracting sharks to them by using baits.

SCUBA ZOO was planned and developed by Mike Ball to give divers close-up experiences with sharks in controlled conditions. A carefully selected sheltered site at Flinder's Reef in the Coral Sea on a flat sandy bottom in 15 metres of water enables the vessel Spoil Sport to anchor nearby. Two huge shark cages have been constructed and anchored to the bottom making an L shape. Access is at the back through large doors. Baits are placed in a metal garbage can with holes in the side to allow the aroma to waft in the water.

Crew members of the Spoil Sport use lines to pull the baits closer to the divers.

All divers enter together and descend to the cages, where they may enter or just lie on top. Once everyone is in position the bait bin is connected to a series of lines which enable it to be pulled close to all the divers. Sharks follow the bin and approach the cages. The water is wonderfully clear and the white sand bottom makes everything bright even if the sun is not shining - we actually had a rain squall while on our dive, and never realised

The sharks gather around the bait bucket just before it is opened. Apparently the baits are much more attractive than divers well wrapped in neoprene.

until later. This is the best time for photographers as the sharks are slowly cruising and come as close as you want them to. They are obviously interested in the bait bin and not the divers.

Grey Reef Sharks were abundant, but there were also Whitetips and some magnificent Silvertips. It was impossible to count them all but I estimate between thirty and fifty sharks coming and going. If you were concerned that you could not watch your back - there were always sharks behind me - you can go into the cage, but the sharks were so well behaved that I never felt any threat at all.

After about twenty minutes of this everyone was instructed to enter the cages, the top of the bait bin was released and the string of baits floated out starting a frenzy of shark feeding. Now the pace was quite frantic but everyone could get a close up view with the absolute, but probably unnecessary, security of the cages. After only a couple of minutes the baits were consumed, sharks returned to their slow cruising and a frenzy of divers left the cages to look over the sand below where the baits had been to see if they could find any shark teeth.

Those low on air just followed the direct line from the cages to the stern of the Spoil Sport where later in the day sharks were given another feed direct from the platform.

I rate this dive as truly excellent for both new and the most experienced divers. If you are a diver who has some nagging concern about a shark taking a nibble one day then you should definitely make the dive. The reality of shark diving is far more wondrous than any fantasies you may have, and far less fearsome.

On occasion a Silvertip Shark Carcharhinus albimarginatus joins the feed. Usually this species is more shy towards divers and patrols steep reef walls.

Carcharhinus falciformis
Silky shark

Length: At birth 70-87, max. 330, mature females 213-305, mature males 187-300 cm.
Distribution: Circumtropical.
Depth: 1-500 m.
General: Appearance sleek, snout relatively long. Interdorsal ridge present. Tip of dorsal fin rounded, fin originates behind pectoral fin. Second dorsal with long free rear tip, pectoral fins long. Underside of pectoral and
pelvic fins may be dark. Coloration dark brown to bluish-grey. Oceanic species, seen at drop-offs near deep
water, around offshore islands (e.g. Cocos Island, Costa Rica) and seamounts. Silky sharks associate with
schools of tuna or scalloped hammerhead sharks (e.g. in the Sea of Cortez). Segregates by size: Young specimens live, sometimes in loose groups, in shallower areas than the adults. Litter size 2-14. Prey mainly comprises bony fishes (scad, chubs, tuna, jacks, mackerel, mullets, porcupine fishes), but also includes squid and pelagic
crabs. A possible threat display similar to that described for the grey reef shark has been observed also in this
species, including arching of back, raising of snout and lowering of tail. Divers should treat larger specimens
respectfully, especially in baited situations. Silky sharks have chased divers out of the water. Small adults and
subadults are reported to be curious and to make slow, close, inquisitive passes in non-baited situations.

This is one of the three abundant sharks of the open ocean realm, only number two behind *Carcharhinus
longimanus*, but superior to *Prionace glauca* within the hierarchy of species when considering boldness of similar-sized specimens at a major food source like a whale carcass.

For more information on this species see also THE ETHICS OF SHARK FEEDING on pp.150-155.

Hans-Michael Hackenberg   Egypt, Red Sea

Peter Kragh   Cocos Is., Eastern Pacific

both photos Mark Conlin   Cocos Island, Eastern Pacific

Carcharhinus altimus
Bignose shark

Length: At birth about 70-90, max. 282, mature females 226-282, mature males 216-267 cm.
Distribution: Western Atlantic: Florida, Bahamas, Cuba, Nicaragua, Costa Rica, Venezuela. Eastern North
Atlantic: Senegal, Gambia, Sierra Leone, Ivory Coast, Ghana. Mediterranean. Western Indian Ocean: Red Sea,
South Africa, Madagascar, India. Possibly western Pacific: China. Central Pacific: Hawaii. Eastern Pacific: Gulf of
California, southern Mexico, Colombia, Ecuador, Revilla Gigedo Islands.
Depth: 90-400 m.
General: A large and relatively slender-bodied requiem shark. Snout large, long and broad, nasal flaps long,
teeth of upper jaw high, triangular, saw-edged. Interdorsal ridge high and prominent. Pectoral and dorsal fins
large. Coloration brown to greyish without conspicuous markings, but sometimes with a bronzy sheen, fin tips
dusky, underside white. Inhabits tropical and subtropical waters of all three oceans. in North American
Atlantic waters, common around the Bahamas and the West Indies, but an unusual catch off southern Florida.

A bottom-dwelling shark, usually found deep on the outer shelf and uppermost slope, sometimes also at
the surface or closer inshore, especially at night. Young specimens may occur in shallower water of up to 25 m
in depth. For a well-defined species with such a wide distribution (and some utilisation in different regions of
the world), only relatively few precise data on its biology exist. Prey includes benthic fishes (croakers, lizard-
fish, batfish, sole, dogfish, catsharks, stingrays and chimaeras), cuttlefish, squid and octopus. Mode of reproduction is placental viviparous. Maturity is reached at a length of approximately 200 cm. Litter size 3-15, usually 7.
Gestation period about 10 months. Gives birth in summer (August and September) in the Mediterranean, in
September and October in Madagascar waters. Caught by offshore trawlers and, rarely, by anglers. Has been
extensively taken in the Caribbean region (especially off Cuba, also southern Florida) on deep-set longlines;
utilised there for fishmeal, as food for animals, for oil and shagreen. Despite its large size not dangerous to
humans because of its deepwater habitat.

The author examined two subadult, yet immature specimens, a male and a female, each about 160 cm in
length. Both had been caught in a gill net set overnight for benthic deep sea sharks at a depth of about 300 m
a few kilometres south of Elat, Gulf of Aqaba, northern Red Sea. The widely distributed bignose shark is only
very rarely reported from this northernmost Red Sea location. Dorsal coloration of the freshly caught sharks
was brown with a bronzy sheen. The stomach of one specimen contained partly digested bony fishes. Part of
the dentition of one of the specimens is shown in SAWBLADES AND MILLSTONES on p.206.

Pedro Niny Duarte / DOP   Cape Verde Islands, Eastern Atlantic

Carcharhinus brachyurus
Copper shark

Length: At birth 64, max. 292, mature females <240-292, mature males 200-230 cm.
Distribution: Wide ranging in warm temperate and subtropical waters of both hemispheres. Especially common around southern Australia, northern New Zealand, Japan, South Africa, southern California and Peru.
Depth: 1-100 m.
General: Fins plain in colour, without obvious markings, but may have darker edges or slightly dusky tips. Dorsal fins relatively small, pectoral fins long. Snout long and sharply rounded. Interdorsal ridge absent. Coloration bronze to grey above, lighter below; sides with indistinct light band. Found in rocky areas, close inshore in shallow bays, harbours and also around offshore islands. Migrates into warmer waters during the winter. Mode of reproduction is placental viviparous. Litter size is 13-20, the gestation period about 10 months. Prey includes bony fishes (sardines, kahawai, sea catfishes, hake, sole), sharks, rays (torpedo rays, stingrays, sawfish), sea snakes and squid. Follows and feeds on schooling fishes, sometimes pursuing them into shallow water.

This large, powerful shark is usually not considered dangerous, but can become aggressive in the presence of bait, freshly caught or wounded fish. Attacks on bathers and surfers have been reported. Not suited for a life in aquaria, as it abrades its snout on the walls of the tank. Such abrasion wounds always lead to secondary infections with fungi and bacteria (e.g. *Vibrio* spp.) and become lethal for the shark if not treated correctly and as soon as possible.

In Australia, some large species of the genus *Carcharhinus* are called whaler sharks because they regularly feed on whale carrion. The copper shark, for instance, is called bronze whaler in Australia. In the high times of whaling, those sharks were a nuisance to the crews of the whaling ships because they were tearing considerable amounts of blubber and meat from the whales which were transported tied outboard alongside the ships.

In the USA, this species is called narrowtooth shark. Small specimens have upper jaw teeth with erect symmetrical cusps with finely serrated edges. Larger individuals have narrowly triangular to scythe-shaped, finely serrated cusps which become increasingly oblique towards the corners of the mouth. The lower jaw teeth have erect or slightly oblique cusps with very finely serrated edges. Tooth counts are 14-16 (right upper jaw), 1-3 at upper jaw symphysis, 14-16 (left upper jaw); 14-15 (right lower jaw), 1-3 at lower jaw symphysis, 14-15 (left lower jaw).

Howard Hall   **Southern Australia**

Carcharhinus galapagensis
Galapagos shark

Length: At birth 57-80, max. 300, mature females 235-300+, mature males 170-292 cm. Distribution: Circumtropical. Depth: 1-180 m. General: Snout long and rounded, interdorsal ridge present. First dorsal fin high and with sharply rounded tip, originates over pectoral axil. Pectoral fins large, their underside often with dark tips. Oceanic species, usually seen around offshore islands and seamounts (the similar dusky shark rather prefers continental coastlines), may be common at drop-offs and steep slopes of coral and rocky reefs. A very curious shark, often aggregates around anchored vessels. Young specimens live in shallower waters than the adults. Litter size 6-16. Prey comprises bony fishes (morays, parrotfishes, surgeonfishes, squirrelfishes, porcupinefishes), elasmobranchs (sharks, rays), octopuses, squid, crustaceans and whales (carrion). Obviously eats more sharks, rays, porcupinefishes and whale meat than smaller family members. Also preys on pinnipeds (sea lion), e.g. off the Galapagos Islands.

A threat display similar to that described for other requiem sharks has been observed in this species; it includes swimming ineffectively in an exaggerated manner and swinging the head from side to side. Considered dangerous, especially young individuals persistently approach and even charge divers. At least one attack on a swimmer has been reported. Adults should be treated respectfully. Galapagos sharks do well in captivity. This is only one of about twenty species present around the Galapagos Islands and one of eight requiem and two hammerhead shark species encountered there. Unfortunately, Galapagos has been raided by overseas fishing fleets in recent years that literally wiped out most of the resident sharks (finning!).

See also facing page, photo from **St. Paul's Rocks, Brazil, Central Atlantic**.

Paul Humann   Galapagos Islands, Eastern Pacific

Steve Drogin   Galapagos Islands, Eastern Pacific

OUT FOR BULL SHARKS

The Bahamian islands are a favourite holiday destination of American divers. And one of these islands is very famous for its sharks: Walker's Cay is known for its 'shark rodeo', a chumsicle experience which attracts over a hundred Caribbean reef sharks. But it is also a real hot spot for meeting bull sharks which were our primary goal this time. Georgina Wiersma takes the plunge...

PHOTOS: PETER VERHOOG

Dozens of Caribbean reef sharks and black-tipped spinner sharks are milling around the chumsicle, a big frozen lump of fish remains.

When the seaplane lands on the lagoon of Walker's Cay, there is nothing that indicates that this is the home of many sharks. Walker's Cay is primarily a destination for enthusiastic fishermen. Every day these big-game fishermen return to the small marina, and visit the fish cleaning station to clean their catch. The waste of this procedure is put in barrels, and frozen with a wire hawser. The result is a so-called 'chumsicle': A large fishy, icy lollipop, a delicious treat for all sharks. Captain Snoopy takes us out to the place where it's all going to happen, a sandy patch, surrounded by reefs, close to the open ocean, at a depth of 8-10 metres. The engines of the boat are switched to full power, and the boat makes a few swift circles, attracting the sharks. Waiting at the sea bottom, we soon see sharks swimming in from all directions. Even before the chumsicle is in the water, there must be dozens of them. The captain slowly lowers the frozen lump, float attached, into the blue water. By now, there are more than 100 Caribbean reef and spinner sharks, circling around the bait and the divers. Occasionally, a nurse shark will enter the crowd, approaching the bait from below, pushing away the reef sharks to get a good grip. The chumsicle defrosts slowly, while shark after shark takes a bite. Every now and then, one of the sharks will tear off a large piece, and swim to the reef at an incredible speed, chased by some of the other predators.

But Walker's Cay has another great experience to offer, although it's only for experienced shark lovers and photographers. The scraps from the fish cleaning station are not all used for the 'shark rodeo'. The staff of Walker's Cay dump the remains on the other side of the island, where there are no swimmers or divers. But there sure are bull sharks...

Considered one of the most dangerous shark species, bull sharks grow to a length of 2.5 metres, but make up for that by developing an impressive girth. They are very ponderous and fast, and tend to be very curious. They prefer the murky waters of river mouths and harbours, and are known to swim up rivers

Bull sharks are powerful predators and said to eat almost everything. But they sure are not menacing eating machines as it has been believed for a long time. This rare shot shows two of them close together in the warm waters of a Bahamian coral reef.

over long distances. At Walker's Cay, their home is indeed murky. It's quite amazing to lay down in the waist deep water without dive gear, but with a snorkel, being surprised by a large shark, investigating the divers with his piggy-eyes from a very close range.

A divemaster throws in bait, while we lie in the water, motionless, holding our breath until the sharks are very close. And we take pictures, of course. These bull sharks are very different from other sharks we've encountered: They are very inquisitive, and not at all afraid of us. We can clearly imagine that bull sharks are famous for swimming up rivers, attacking bathing people in shallow waters. When approaching us, they will come as close as up to thirty centimetres from our dive mask. As the water is quite murky, and sharks stir up the silt from the bottom, it's quite an exciting experience. Peter leaves the water every thirty minutes to get a new roll of film. There are hardly any bull shark pictures available, and he's delighted to get an opportunity to take shots of these impressive creatures; some of his photos even show more than one spec-

imen. When I am standing on the shore, I watch Peter in the water, surrounded by not less than nine adult bull sharks, all circling him. They're even touching his fins with their bellies, swimming. We stay in the water for over four hours. By then, we're ready for a bite ourselves, but can tell a truly different story about the not-so-infamous bull shark.

A group of at least four of the stoutly built bull sharks can be seen in this spectacular photo from extremely shallow water.

Carcharhinus leucas
Bull shark

Length: At birth 56-81, max. 340, mature females 180-324, mature males 157-226 cm. Distribution: All tropical and subtropical seas, most abundant near continental coastlines. Eastern Atlantic: Mauritania to South Africa. Also in freshwater systems, see below. Depth: Intertidal to 150 m. General: A large, stocky species. Snout very broad and short, eyes small. Interdorsal ridge absent. First dorsal fin large, triangular, with pointed tip. Coloration dusky above and pale below. Fin tips and trailing edge of caudal fin may be dark, especially in juvenile specimens. Found in coastal areas from deep flat sand and rubble areas near reefs. Usually seen near the bottom, often lying torpid ('sleeping') there.

Litter size 1-13. Gestation period 10-11 months. The only wide-ranging shark species, that moves into estuaries, up rivers and large streams and even into lakes to give birth. Well-known for its ability to enter and live for extended periods of time in freshwater. May not be able to maintain its entire life cycle in freshwater, has no landlocked populations there. The bull sharks in Lake Nicaragua (connected to the Caribbean via the San Juan river) were once thought to be landlocked, but are known today to move freely between the lake and the sea; occasionally breeds in the lake, but only to a minor extent compared to the usual birthing habitats such as estuaries and other brackish coastal waters. Bull sharks have been reported from the following rivers and lakes (among others): Mississippi and Atchafalaya rivers (southern USA); Lake Nicaragua and San Juan river (Nicaragua), Lake Isabel (Guatemala), Patula river (Honduras), Panama Canal (Panama); Amazon river (to Peru, where recorded 3,700 kilometres from the sea!); Gambia river (Gambia), Ogooue river (Gabon), Ruenwa river (Zimbabwe), Zambezi and Limpopo rivers (South Africa); Tigris river and Shatt-el-Arab (Iraq); Hooghly river (Ganges system), India (where occurring sympatrically with the Ganges shark *Glyphis gangeticus);* Lake Jamoer (New Guinea), Lake Macquaire and Swan, Brisbane, Herbert and East Alligator rivers (Australia).

Segregates by size, juveniles frequent shallower waters than the adults. The bull shark is a schoolbook example of an opportunistic feeder. Prey is very varied and comprises many species of bony fishes, sharks and rays, sea turtles, marine and terrestrial mammals, crustaceans, squid and sea urchins. While young individuals focus on teleosts as prey, adults are notorious for their predilection for other elasmobranchs. The species is an important predator on bramble sharks *(Echinorhinus* spp.), spiny dogfish sharks, requiem sharks (including young bull sharks), sharpnose sharks and hammerhead sharks, guitarfish, sawfish, skates, stingrays, butterfly rays, eagle rays and devil rays *(Mobula* spp.). Adults are particularly fond of young sharks in inshore nursery grounds and stingrays.

Because of its size and varied diet, the bull shark has to be considered dangerous. Said to be one of the three or four shark species most dangerous to people, several attacks on humans (and dogs) have been reported. It is responsible for numerous attacks in the tropics, where it has attacked bathers, spearfishermen and aquarists underwater. There are, however, no confirmed reports of attacks on 'regular' SCUBA divers.

See also OUT FOR BULL SHARKS on the previous two pages.

Peter Verhoog   Bahamas

Carcharhinus limbatus
Blacktip shark

Length : At birth 38-72, max. 260, mature females 120-255, mature males 135-226 cm.
Distribution: All tropical and subtropical continental waters, also around the island groups in the Pacific.
Depth: Intertidal to 30 m.
General: Snout long, pointed. Interdorsal ridge absent. First dorsal fin high and pointed. Black tips on pectoral fin, second dorsal fin and lower lobe of caudal fin. Tips of other fins may also be black, often so in specimens from the Indo-Pacific region (compare photos on following page). Side of body with conspicuous light band. Found in turbid inshore waters like lagoons, estuaries and mangrove areas, also in reef channels, near drop-offs and seamounts. Occurs in the lagoon of Rangiroa Atoll and moves out of it through a pass during ebb tide. More active during the day than at night. Litter size 1-10. Gestation period 11-12 months. Females give birth in shallow, often turbid, coastal waters. Prey mainly includes bony fishes (sardines, menhaden, anchovies, sea cat-fishes, jacks, croakers, snook, boxfishes), sharks and rays, but also gastropods, squid and crustaceans. One of the few requiem sharks that jumps out of the water like the spinner shark *C. brevipinna*. Spins around its body axis while falling back, hence also called spinner shark. Behaviour towards divers varies, sometimes bold and aggressive, otherwise shy and circling at a distance. Aggressive in baited situations, especially in turbid water.

In October 2000, almost 50 young blacktip sharks and about a dozen Atlantic sharpnose sharks *(Rhizo-prionodon terraenovae)* were found dead on the beach of Panama City, Florida. More dead sharks - up to 150 - were estimated to drift in the waters of St. Andrews Bay off Panama City. Shark specialists could not deter-mine the cause of their death after a first superficial examination on the beach. The sharks showed no exter-nal signs of bleeding as had been reported at first, also no wounds from hooks or other catching gear. Bleeding from gills and nostrils would have pointed towards a viral infection which, however, could not be excluded from the possible causes of death. The scientists took tissue samples from the sharks' gills, livers, intestines and muscles in order to investigate the case further, but their examination in the laboratory could not unravel the mystery. At that time, a red tide was reported from the area. A red tide is a population explosion of red-coloured dinoflagellates, unicellular algae exuding a potent toxin affecting the nervous system of vertebrates. Red tides, however, always kill animals other than sharks first, but no dead bony fishes had been found on the Florida beach. Low oxygen levels in the sea water in the aftermath of the red tide cannot be ruled out entirely as a possible cause of the sharks' death. Mass dyings of sharks are very unusual and only rarely reported, con-trary to the relatively frequent mass strandings of whales and dolphins. The causes of such events involving either elasmobranchs or mammalians obviously are different. Shortly after the Florida incident, a similar mass dying of sharks was reported from Fortaleza at the Atlantic coast of Brazil. At least 40 sharks were beached while still alive on a 100-km-long strip of coast. Species involved were shortfin makos *(Isurus oxyrinchus)*, black-tip reef sharks *(Carcharhinus melanopterus)*, smalltail sharks *(C. porosus)* and other, unidentified species. As in the Florida case, the cause of their death was not obvious. See also following page.

Helmut Debelius   Cuba

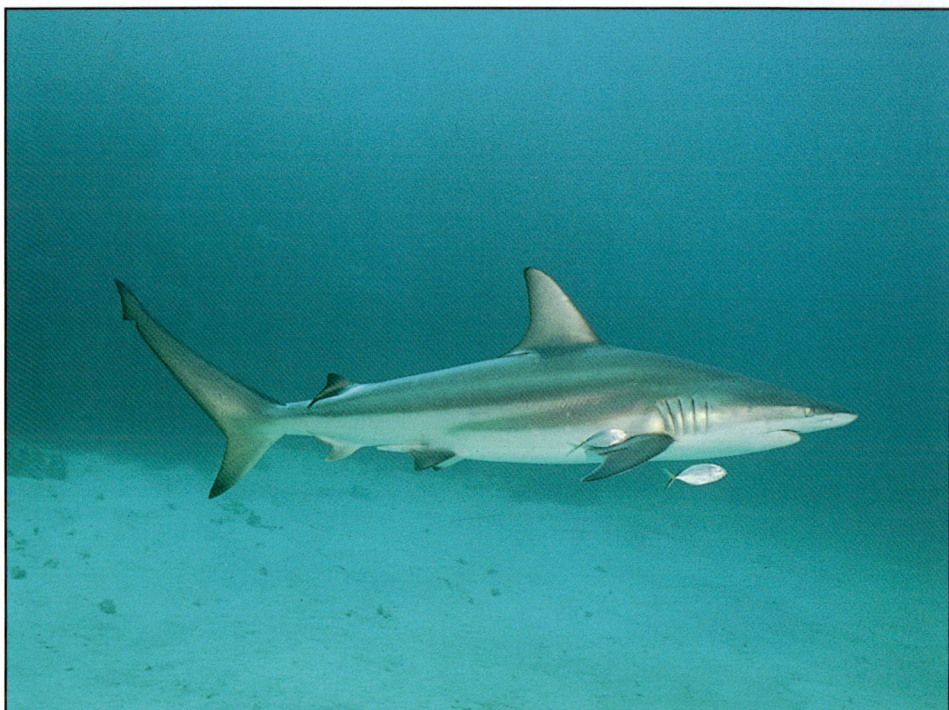

Blacktip sharks *Carcharhinus limbatus*   Doug Perrine, Caribbean

Carcharhinus limbatus   Thomas Reich, Ras Muhammad, Red Sea

Carcharhinus longimanus   **Red Sea**

Carcharhinus longimanus
Oceanic whitetip shark

Length: At birth 60-65, max. possibly to 395, most <300, mature females 180-270, mature males 175-245 cm.
Distribution: Circumglobal in tropical and warm temperate seas. Depth: 1-150 m.
General: This impressive species is easily recognised by its large dorsal and pectoral fins, which are broadly rounded and conspicuously tipped with white. Dorsal coloration of Pacific specimens grey, of Red Sea specimens brown. Underside light in colour. Newborns with dusky rather than white fins tips.

This primarily oceanic shark is usually encountered in water at least 180 m deep, also close to drop-offs near deep water, but only occasionally close to the reef itself. Often seen singly or in loose groups from boats as it swims slowly at the surface, usually accompanied by groups of pilot fish or other sharks. Migrates to warmer waters in winter. Mode of reproduction is placental viviparous. Litter size 1-15. Gestation period about 12 months. Females give birth in shallow, often turbid, coastal waters. In the western North Atlantic, this species gives birth in early summer. The same is true for southwestern Indian Ocean specimens. In the Central Pacific, however, pregnant females have been found to contain small embryos all year round which suggests that giving birth (and probably also mating) is not confined to a certain season there.

This magnificent predator generally moves quite slowly but is equally active during the day and night hours. By day, it is often seen cruising at or just below the surface, with its large, conspicuous pectoral fins spread like wings. It preys on a wide variety of vertebrates including bony fishes (lancetfishes, marlin, oarfishes, threadfins, barracuda, jacks, dolphinfishes, tuna, skipjack and other scombrids), marine turtles, sea birds, but also squid, gastropods, crustaceans, mammalian carrion and garbage. Occasionally also feeds on pelagic stingrays, most probably the unique *Dasyatis violacea*. Sometimes several individuals aggregate around a major food source, e.g. a whale carcass. Known to associate with groups of pilot whales in order to feed on the remains resulting from the whales' attacks on schooling fish.

A bold shark that persistently (sometimes for hours) circles divers and makes occasional slow, close passes. But in spite of its large size, great power and stubborn aggressiveness, the oceanic whitetip shark is not as dangerous as generally assumed, with only a few attacks on swimmers and boats on record (one such case has been reported by former professional diver Bret Gilliam: In a depth of 60 m, two Oceanic Whitetips attacked a group of three during a decompression dive on St. Croix's Cane Bay wall, one diver was killed and Gilliam survived severely bent). Its behaviour is much more deliberate than that of its smaller, fast-moving and sometimes even frenetic reef shark cousins. A group of oceanic whitetips are shown in the shark film 'Blue water, white death' while feeding on a sperm whale carcass off Durban, South Africa. A report of the Oceanographic Research Institute in Durban suggests, that sharks of this species are responsible for the deaths of many people after the ship 'Nova Scotia' was torpedoed and sunk by a German submarine off Natal, South Africa, during World War II.

One of the three abundant sharks of the open ocean realm, dominating silky sharks (*Carcharhinus falciformis*) and blue sharks (*Prionace glauca*) within the hierarchy of species when considering boldness of similar-sized specimens gathered around a major food source. See also previous page, photo from the Cape Verde Islands, Eastern Atlantic.

Werner Thiele   Egypt, Red Sea

Chris Newbert   Hawaii, Central Pacific

Walti Guggenbühl   Egypt, Red Sea

Carcharhinus melanopterus
Blacktip reef shark

Length: At birth 33-52, max. 180, mature females 96-131, mature males 90-180 cm.
Distribution: Wide ranging in the tropical Indo-Pacific: Red Sea to Society Islands, north to Japan, south to New Caledonia and Australia. Also found in the eastern Mediterranean (see below).
Depth: Intertidal to usually no deeper than 10 m.
General: First dorsal and lower lobe of caudal fin always with black tips, other fins may also have black on their tips and/or trailing edges. Snout blunt, interdorsal ridge absent.

Mainly encountered on reef flats and in lagoons, often in very shallow water, with the distinctly black-tipped dorsal frequently showing above the surface. Will aggregate in reef channels at low tide and move on to the reef flat at high tide. Has a home range of a few square kilometres.

Litter size 3-4. Gestation period 8-9 months. The copulation has been observed to occur in shallow water, with a belly-to-belly posture. Prey includes bony fishes (surgeonfishes, wrasses, triggerfishes, parrotfishes, goatfishes), stingrays, squid, octopuses, crustaceans and sea snakes (preferred food in some areas). Feeds during darkness. Has been filmed during nighttime hunting ventures in the restricted space of coral reef habitats, where fast swimming packs of this agile shark hunt down moving prey or 'extract' sleeping fishes from their retreats in cracks and crevices. Has been observed to attack spawning aggregations of surgeonfish above the reef. Falls prey to other sharks and large groupers (Serranidae).

Not considered dangerous to divers and rather difficult to approach. However, larger individuals are possibly aggressive when fish are speared, bleed and die in the water. Small specimens are notoriously reported to have bitten peoples' legs while the victims were wading in the shallows on top of the reef flat. One source advises to lie down in the water if attacked in such a way!

A truly Indo-Pacific species, the blacktip reef shark entered the Mediterranean from the Red Sea as a so-called Lessepsian migrant. Since the opening of the Suez Canal in 1862, many organisms migrated in this or the other direction (anti-Lessepsian migration, from the Mediterranean into the Red Sea). On their way, these migrants have to cross the hypersaline water of the Bitter Lakes. Animals able to do so, are termed euryhaline, which means they can adapt to changing water salinity relatively quickly. In the case of the sharks this adaptation is achieved within a few days, as studies under experimental conditions have shown.

Generally a hardy species in marine aquaria of suitable size. However, the author had the opportunity to regularly observe five blacktip reef sharks held together with larger sharks (several sandbar sharks *C. plumbeus*, one sicklefin lemon shark *Negaprion acutidens*) during the period of about 12 months. While the smaller sharks gave way to their larger cousins and none of them was seriously injured or even killed, all of them refused to eat, lost weight, became infected and finally died. Stress induced by the presence of potential predators should be avoided by carefully selecting the tank mates. The large lemon shark was definitely not the correct choice.

Kurt Amsler   Maldives, Indian Ocean

Carcharhinus obscurus
Dusky shark
Length: At birth 70-100, max. 420, mature females 260-365, mature males 280-340 cm.
Distribution: Cosmopolitan (but patchy) in tropical and warm temperate waters. Western Atlantic: Southern Massachusetts to Florida, Bahamas, Cuba, northern Gulf of Mexico, Nicaragua, Brazil. Eastern North Atlantic: Canary Islands, Cape Verde Islands, Senegal, Sierra Leone, possibly also Portugal, Spain, Morocco, Madeira, western Mediterranean. Western Indian Ocean: South Africa, Mozambique, Madagascar, possibly Red Sea. Western Pacific: Japan, China, Viet Nam, all around Australia, New Caledonia. Eastern Pacific: Southern California to Gulf of California, Revilla Gigedo Islands, possibly Chile. Depth: Intertidal to 400 m.
General: Snout broadly rounded. Interdorsal ridge low. First dorsal fin originates behind pectoral axil. First dorsal tip pointed. Pectoral fins sickle-shaped, these and pelvic fins often with dark tips and trailing edges. Similar to Galapagos shark, but dorsal fins of the dusky shark are lower, also the latter species is more often found around continental coastlines. Migrates in temperate and subtropical waters of eastern North Pacific and western North Atlantic, moves northward during warm summer months and back in winter. Litter size 3-14. Adults mature at age 6, may live at least 18 years. Females move to shallow nursery grounds to give birth (e.g. off Natal, South Africa), after parturition they leave the area. Young sharks move into deeper water as they grow; fall prey to other sharks. Prey mainly comprises bony fishes (eels, lizardfishes, cuskeels, spadefishes, mackerel, jacks, tuna, flatfishes, gurnards), sharks (angel sharks, sawsharks, spiny dogfish sharks, catsharks, smoothhound sharks, other requiem sharks), skates and rays, but also crustaceans, squid, octopuses and sea snakes. Feeds by day in the entire water column. Fished in many regions for meat, fins, oil and shagreen.

Rudie Kuiter   New South Wales, Australia

Helmut Debelius   Mauritius, Indian Ocean

Carcharhinus plumbeus
Sandbar shark

Length: At birth 56-75, max. 240, mature females 144-183 (max. 243), mature males 131-178 (max. 224) cm.
Distribution: Worldwide in subtropical and tropical seas. Depth: Intertidal to 280 m.
General: First dorsal fin very large and high. Pectoral fins large and broad. Interdorsal ridge low. Seen singly or in large schools (adults segregate by sex). Most often encountered swimming close to the ground over sand and mud not far from shore, but also near reefs and at the surface in water several hundred metres deep. This is the first shark species the author saw in the notorious pose of swimming with its first dorsal fin protruding from the water (witnessed on an unusually windless early morning in the Gulf of Aqaba, northern Red Sea, just off the marina of Elat). Litter size 1-14. Gestation period 8-12 months. Nursery areas in shallow bays. Juveniles in coastal bays move with tidal currents and remain in a relatively small area during ebb tide. Apparently the males follow and bite the female until she swims upside down, then they copulate with one male inserting both claspers.Prey mainly comprises benthic bony fishes (flounders, porcupinefishes, eels, parrotfishes, surgeonfishes, goatfishes), sharks (sharpnose, spiny dogfish, bonnethead sharks), rays (guitarfish, skates and their egg cases, stingrays, cownose rays), crustaceans (mantis shrimp, crabs, shrimp), squid and octopuses. Feeding activity is highest in the early morning hours. Not considered aggressive, but makes close inquisitive passes. Sandbar sharks have been photographed on the Mediterranean coast of Turkey near Bonçuk, where they are the attraction of the local diving school. A group of 1.5-2 m long sharks is resident in a remote bay there. However, the bold sharks may be approached only without SCUBA gear! Encountering these animals in the Mediterranean is a rare opportunity, despite the fact that the species has been described from there.

David Fleetham / IV   Hawaii, Central Pacific

Helmut Debelius   Turkey, Mediterranean

Carcharhinus perezi
Caribbean reef shark

Length: At birth about 70, max. 300, mature at about 152-168 cm.
Distribution: Western Atlantic: Most common large reef shark in the Caribbean and the Bahamas. Uncommon in the Florida region. Also in northeastern Gulf of Mexico and south to Brazil. Depth: 1-30 m.
General: Snout broadly rounded, short. Interdorsal ridge low. First dorsal fin relatively small, origin behind pectoral fin, its tip acutely pointed. Pectoral fins moderately long and narrow. Tips of all fins dusky except those of first dorsal and upper lobe of caudal fin. Often misidentified as bull shark, but more similar to dusky shark which can be distinguished by a more slender body and larger gill slits. The similar blacktip shark has a pale to white anal fin, the similar spinner shark C. brevipinna has solid black tips on its fins as if dipped in ink, not just dusky. Gill slits relatively small, first gill slit located over plane of pectoral fin. Body stout. Coloration silvery-grey, fading to white on the underside. Side of body with white streak reaching into grey coloration. Usually observed on the seaward side of coral reefs, but also in the shallows. Commonly seen under ledges and in caves where lying on the ground in apparent torpor as if sleeping. Often with dark leeches (parasitic worms) attached to skin. Litter size 4-6. Prey comprises bony fishes and large, motile invertebrates. Most common participant in the shark feeding spectacles organised for tourist divers in the Bahamas; usually not aggressive there, but makes close inquisitive passes. May become aggressive in the vicinity of spearfishing activities. See also the ETHICS OF SHARK FEEDING on the following pages and A SOFT SPOT FOR SHARKS p.236.

Mark Conlin   Caribbean

Douglas D. Seifert   Caribbean

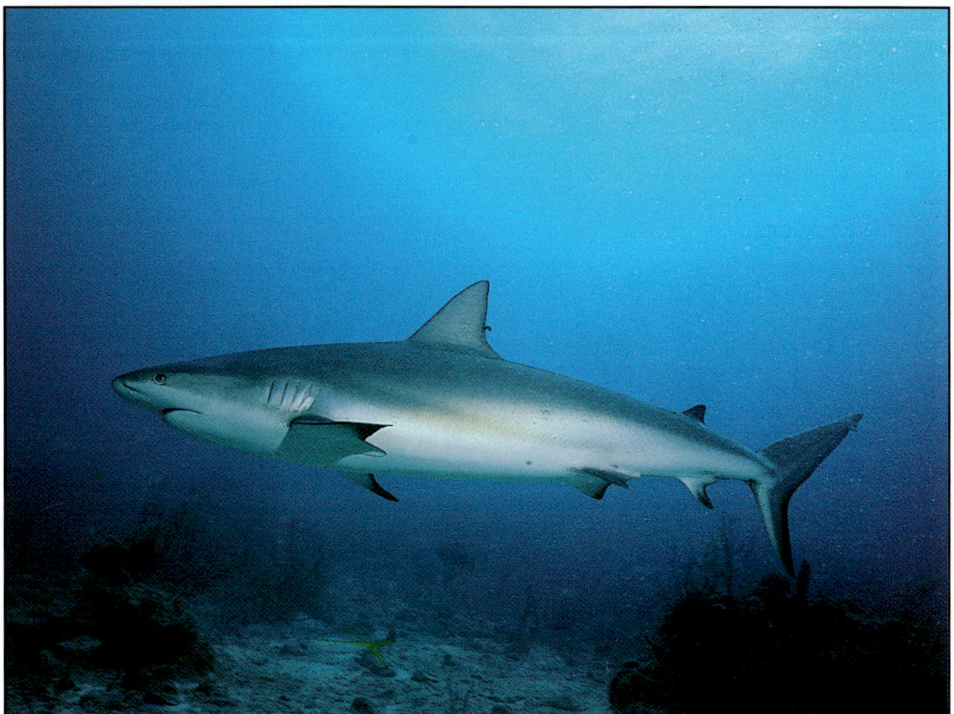

Paul Humann   Bahamas, Western Atlantic

THE ETHICS OF SHARK FEEDING

Ethics is that branch of philosophy that seeks to balance and analyse which actions are right and which ends are good. The application of seemingly-abstract moral judgements to the practice of feeding sharks as an entertainment for scuba diver-tourists stimulates an intellectual feeding-frenzy of emotion and reason among divers, scientists, tour operators, televised documentary program couch potatoes, government officials and deep-thinking journalists like Douglas David Seifert.

PHOTOS: DOUGLAS D. SEIFERT

Shark feeding at Walker's Cay, Bahamas.

There can be no question that diving with sharks is the hottest ticket in the dive travel market today. Public fascination with sharks by people in general and by divers in particular has never been greater and - compared to the Jaws hysteria of the seventies - never more enlightened. But only to a degree.

One needn't be a rocket scientist (or marine biologist, for that matter) to recognise that there has never been a more urgent need than at present, for the mass of humanity to take an interest in sharks. The prime factor is man, his indifference to the shark's place in the ocean ecosystem, and the exploitative value of shark products. Humans are destroying millions each year with a terrifying industrial efficiency.

The reason why the great white shark (named culprit in a majority of human fatalities) was granted protection in Australia, California and South Africa occurred because public opinion was manipulated to pro-shark. This would probably not have happened if divers and filmmakers had not been experiencing the animal for two decades and bringing back images and stories that swayed public perception. Scientists publishing research papers obviously played a valuable part in the process, but the swing in opinion was largely achieved via media professionals and laymen preaching to the congregation - not the choir - in print, on television and in a thousand small public forums.

There are places where seeing sharks going about their routine business naturally is common - for example: The islands of the Eastern Pacific (Cocos, Galapagos, Malpelo, Socorro) or the islands and atolls of the Pacific and Indian Oceans (Maldives, Tahiti, Papua New Guinea) - but the human preoccupation, some might call it a lurid fascination, with the business end of the shark - its terrible jaws and teeth - are what the punters are after. Sharks au naturel are, with some exceptions, cautious and wary by nature. For the most part, they will keep a good distance from divers, especially if the water visibility is good (and the diver has a camera filled with unexposed film in his hands). Hence, the business of shark feeding: The best way to get close to a shark is to dangle some food in front of its snout. Guaranteed.

Commercial dive operators in different parts of the world have been feeding sharks for its entertainment value for more than twenty years. The strategies employed for bringing sharks and divers into close contact can be conveniently divided into two types of operations - albeit with variations on a theme: The first is the equipment and crew intensive anti-shark, cage diving set-up; the second the more user-friendly shark feeding stations with some variety of shark feeder professional, be it an experienced divemaster or charter operator.

Anti-shark cage diving is conducted in South Australia and South Africa for the express purpose of experiencing great white sharks. It is also conducted off California and New England for blue and mako sharks and, in Australia, a dive boat operator keeps a permanent cage moored at a site on the Great Barrier Reef for grey reef and whaler sharks. White shark trips are an expensive way to experience sharks (US $ 4,000 plus airfare and up per week). Blue and mako shark dive trips are a quarter the price but again the subjects are decidedly less 'shark'.

For the average recreational diver with an interest in seeing sharks up close and personal, many live-aboard dive boats in the tropics and land-based dive operations created shark feeds for the sake of variety or as a bonus to their traditional diving programs. Probably the first commercial operation to feature reef-situated, non-cage shark feeds was Herwarth Voigtmann's pioneering grey reef shark feeds in the Maldives in the early Seventies. They soon became popular with European diving tourists and reached a level of absurdity when, reportedly, Voigtmann's daughter began feeding the sharks by a new and interesting method. She would take the regulator out of her mouth, hold her breath, place the tail of a dead fish in her mouth and lean out to let the sharks take the bait, all while diving topless (see SHARK GAMES, page xxx). But shark feeding's crowd-pleasing popularity with divers was noted and as word of this activity spread, bookings increased. The economic value of feeding sharks could not be overlooked and the practice spread, first slowly, then like wildfire, to other points on the global diving map.

Nowhere is this better observed than in the commercial diving operations in the islands of the Bahamas. One can not open a dive magazine without being bombarded: Bahamas diving equals shark diving is the message of the advertising campaigns of the Bahamian dive operators association. If you went to the Bahamas and didn't see sharks, you didn't dive! Currently, over 20,000 divers experience sharks via shark feeding operations in Bahamian waters yearly and the figure is increasing by 20% per annum.

Organised shark feeds in the Bahamas probably originated concurrently at Stella Maris on Long Island and by Stuart Cove and Frazier Nevins on New Providence. Stella Maris catered to a German clientele familiar with the Maldives operations and began reef shark feeds to draw divers to their remote operation. Cove and Nevins, both avid fishermen as well as divers, fished for dorado, wahoo, tuna and billfish at the AUTEC buoy (owned and operated by the US Navy to calibrate its submarine fleet) which is anchored in 2,000 metres of water, in the Tongue of the Ocean, twenty-two kilometres off New Providence. They had long noticed an abundance of silky sharks resident under the buoy and one day began to chum them to the boat and diving with them. Over time, they began hand feeding the sharks and bringing out paying guests.

Silky sharks are evil-looking little sharks, quick and quirky, with pointed snouts and long, lean bodies. They appear more fearsome than they actually are, none the less the sudden appearance of a group of them can be intimidating to divers. But the silkies under the AUTEC buoy are small, in the one to one and a half-metre range, and after their initial buzzing of divers, they tend to settle down. Diving with sharks loses its excitement value after long exposure to them, so Stuart Cove began to 'play' with them by grabbing onto their tails. He made a remarkable discovery. When the tail of a small silky shark is grabbed and slightly twisted, the shark stalls, ceases moving and becomes paralysed. It is a phenomenon called 'tonic immobility', which is loosely defined as 'a state of prolonged muscular contraction.'

Apart from the bravado value of catching a shark by the tail and holding it helpless, Stuart discovered tonic immobility to be a tremendous vehicle by which to remove stray fishing hooks from the sharks' jaws. It is

When a diver twists the tail of a small silky shark, it becomes paralysed for a short while. This phenomenon is called 'tonic immobility'.

After being immobilised near the AUTEC buoy, also this shark was released of the fishing tackle.

more common than not to see lost hooks and lures in the mouths of the silkies since the buoy acts as a fish attraction device and there isn't a day that goes by that an angler doesn't stop by to try and catch what's under it. To the delight of paying guests, Stuart performed shark-catching, a display of tonic immobility and the release of fishing tackle from the sharks' mouths. Whether the sharks had any definite feelings about underwater dentistry or being put into a state of paralysis is unclear.

The popularity of diving with silky sharks - and the premium divers would pay to do so - led Stuart Cove to experiment with other programs. He began taking a box filled with filleted fish carcasses to a shallow spot on the reefs (15 metres depth) where Caribbean reef sharks had frequently been seen, not far from a drop-off into deep water. By trial and error, he created a shark feeding routine: Routine for the sharks and reliable enough to bring divers into contact with sharks in a relatively safe manner. The shark dive occurs nearly every afternoon although not necessarily at the same site or with the same sharks.

Regardless of venue, the ritual is the same. The dive boat reaches the dive site and ties up to the mooring. The sound of the boat's engine brings the sharks in from their dispersed areas. The divemaster/shark feeder swims a plastic crate down to a clearing amid coral heads on the reef and waits. Divers enter the water and make their way to the bottom where they sit in a semi-circle, backed against the coral heads, about six metres from the baitbox. A dozen or more Caribbean reef sharks mill around, swimming in from down current, attracted to the scent of the fish carcasses carried downstream from the crate. By keeping the bait enclosed, the shark feed can be drawn out for an extended period. The panorama is awash in bubbles from the exhalations of excited tourist divers, which is somewhat off-putting for the sharks, but their instinct to feed overrides a natural wariness and they will come close to the divers, often with less than a metre's distance.

When the mood strikes, the shark feeder, who is wearing a chainmail gauntlet that runs from fingertips to shoulder on each arm, thrusts a metre-long metal pole into the box and spears out a fish carcass. He offers it to the most dominant shark, although the excitement produced by feeding stimuli causes several sharks to swarm energetically around the feeder. By selectively offering food at delayed intervals, the shark feed can be dragged on for up to an hour. Even when the supply of food is gone, the sharks continue to mill around, although at a more cautious distance from the divers.

Of all the organised shark feeds in the Bahamas, the most spectacular and the most unique

Right: Sharks circle the 'chumsicle' and wait until it has partly defrosted so they can feed on the fish pieces falling off. **Facing page:** Underwater filmers and photographers get excited while masses of sharks mill around.

is conducted on an irregular basis at Walker's Cay, northernmost island of the Bahamas at the top of the Abacos chain. Walker's Cay is a private, intimate, self-contained island (100 acres in total) owned by an American industrialist, and was originally created solely as a sportfishing resort. During the high season, with each boat slip occupied, its horseshoe-shaped marina is filled with more than 50 million dollars worth of fibreglass, aluminum, teak, marine electronics and fishing tackle. Diving was added as an optional activity for guests who did not fish or who sought variety. All that changed in 1991 when Gary Adkison and Barry Albury began experimenting with shark feeding at a dive site where they'd often seen sharks. In just six years, Walker's Cay's main drawing ratio has changed from fishing 90%, diving 10% to fishing 10%, diving 90% of its bookings.

What started as a 'typical' shark feed with thirty or more sharks evolved into what is now called 'Shark Rodeo' which regularly features one hundred(!) or more sharks. The star of the Rodeo - at least as far as the sharks are concerned - is Gary Adkison's own invention, the 'chumsicle.' This is a mass of fish carcasses, weighting approximately seventy-five kilos, packed around an iron frame, which is placed into a large garbage can and frozen solid. The chumsicle is brought out to the site and lowered into the water. A line runs from the chumsicle to a light anchor and a line runs from the top to a large float. Thus, the mass of chum is suspended in midwater, four metres from the sandy bottom, the centre of feeding activity with one hundred sharks circling, as if orbiting, the food source. Additional swirling 'satellites' that find the free meal irresistible are an abundance of yellowtail snappers, groupers, jacks and other reef fish. The dive site is a natural amphitheatre of upon a sandy plateau, fifteen metres deep, surrounded by coral walls. But before the chumsicle is put into the water, there are a few procedures. Gary Adkison briefs the divers on exactly what they are going to witness and how the operation will occur.

The activity is monitored by Adkison and his divemasters, who signal if there is a sudden change in behaviour, most notably, if a chunk of fish breaks off from the chumsicle. When this occurs, the sharks begin to compete for the food item and the result is shark rodeo.

"When a piece of food has broken free from the chumsicle and is free in the water column, the sharks will do shark rodeo and start competing with each other for the fish. That's when we'll ask you to move back. It's not that they're going to bite you, they're pretty smart, they know you're not food, they think you're another predator, and you are - and we tell everyone this - you're much worse, you're the worst predator in the ocean. What we're concerned about is that they're going to accidentally bump into you and they can really, really hurt you if they whack you with their tail."

Sharks are present as far as the eye can see. They mill around, circling, at all levels, from the bottom to the surface. Two species are prevalent: Caribbean reef sharks and blacktip sharks. The actual number is said to be eighty to one hundred, but at certain times of the year the number can be twice that amount. The sharks keep a distance of four metres from the divers, who are exhaling at a rate bordering on hyperventilation. The divers' masks contain eyes staring wide-open which threaten to pop out of their sockets. When all divers are settled in, a divemaster surfaces and signals the boat to lower the chumsicle.

Activity changes to an electric excitement with the addition of a food source. The most vicious predators are a swarm of yellowtail snappers; it is difficult to make out the chumsicle through their cloud. But the more dominant sharks make passes, parting the snappers, bumping the chumsicle and gnawing at it. The bait mass is too large and too solidly-frozen for them to get a mouthful, at first. Over the next thirty minutes to an hour, as the chumsicle thaws and is worked upon by fish and shark, filleted carcasses separate from the frozen form and bursts of feeding activity are witnessed. It becomes apparent that of the one hundred sharks, only a small percentage actually tries to feed on the bait. There is a social hierarchy at work, called 'bite or flight' and the dominant sharks keep subordinates in check.

And then, ultimately, one shark will dislodge the remaining mass from the frame. It is too large to swallow but the shark will swim off with his prize, carrying it like a soccerball in his jaws as he swims away, and like a soccer game, all the other 'players' suddenly rush after him, trying to take possession of the 'ball'. The spectacle of fifty or more sharks swarming close on the leader as he attempts to swim away with his prize is the highlight of shark rodeo.

Critics of the business of shark feeding for entertainment are many and their arguments seem reasonable upon face value. At the core of all arguments is the underlying concern that a human being may be injured. This fear, which is not backed up but the weight of statistics, is usually annexed to the additional thought: "And there could be a backlash against sharks." There have been accidents, though not well publicised and certainly no one in the Bahamas wanted to go on record about them. Primarily, those bitten were the ones feeding the sharks. Most injuries were minor. And as far as backlash against sharks is concerned, that argument is null and void. Any backlash in this day and age is settled in a court of law and by insurance companies, not justice meted-out upon unidentified animals in the wild.

This unreasoning fear of sharks comes from a lack of field experience. Statistically, more people will be injured and will die as result of putting on scuba gear itself (or driving an automobile to their diving destination) than by experiencing sharks, fed for entertainment or otherwise. You can almost hear the plaintive wail of the armchair ocean specialist: "Sharks are wild animals, sharks are unpredictable." Unpredictable! When the subject is sharks, always this word: Every time I read the word or hear it uttered by experts on television I feel a wave of nausea. Sharks are very predictable! Given a stimulus to feed, they feed. It is the product of the evolutionary process that placed them high on the food chain. Get in between their food and their teeth and you can predict some misfortune.

There is an argument that feeding sharks on a regular basis and at a designated site alters their behaviour, conditions them to contact with humans (associating human activity with food) and impacts the environment. It's all true and as unimportant in the grand scheme of things in the ocean as the environmental impact of spitting in the ocean.

Shark feeds can only occur where sharks occur naturally. If there are no sharks in the area, then all the blood, fish parts and offal in the world will not draw them in. This is one reason why shark feeds are not advertised in Bonaire or in the Virgin Islands. There are sharks in all waters, but if they do not regularly visit a particular area then they are not going to start just because a dive operator has a bucket of fish guts.

Reef sharks do become conditioned to being fed. No question about it. But regular shark feeds do not take the place of a shark's continual hunt to survive. And so long as the food offered is of a variety which the animals would encounter naturally - no land mammal products, please - and the animals suffer no trauma in getting the baits, then there is little cause for alarm. There are added benefits to organised shark feeds for scientists and researchers. The ocean is a difficult laboratory to achieve the necessary standards to replicate results required for scientific theory. By bringing the same animals into the same setting on a regular basis, many experiments can be performed and new insights can be learned. For example, shark tagging has been an ongoing part

Shark tagging has been part of Walker's Cay's shark programs for many years. Here a tagged Caribbean reef shark *Carcharhinus perezi.*

of white shark operations and a part of Walker's Cay's programs for many years and without tagging, population dynamics studies would be impossible. Stuart Cove's interplay with juvenile silky sharks at the AUTEC buoy has given the world a venue to study the phenomenon of tonic immobility.

The only true ethical concern over the business of conditioning sharks to a human presence via feeding comes from the shark's association of a boat's arrival with feeding cues. This is inherently dangerous, not for human swimmers, divers or snorkelers, but for the sharks themselves.

Unscrupulous fishermen (an oxymoron, I know) have capitalised on this aspect of shark learned behaviour to catch and kill sharks for sport, for financial gain, or out of boredom. Commercial long-liners and other fishing types have staked out or targeted the shark feeding areas in the past and will continue to do so in the future as long as economic gain can be realised for shark products. In this area of ethics there can be no compromise. If commercial dive operators are going to make a business partnership with sharks, then they must be ultimately responsible for the stewardship of those animals. Fishing-free zones or reserves must be enacted and resources diverted to patrol them against poaching.

It is the moral imperative for man who alters nature for his own ends to protect nature. There can be no deviation from this course of action. The waters of the Bahamas are still fair game to shark fishermen, even though there are laws against taking sharks and protected zones, but the laws have no teeth and violators go largely unprosecuted. This has to change for the sake of the animals. The officials of the Bahamian government must understand that, if for no other reason than simple economics, living sharks are a renewable resource, an asset bringing millions of dollars in tourist dollars into the country and worth far, far less dead.

Are the shark feeds for tourist entertainment a circus? Unabashedly so. The main difference is instead of the circus coming to town, one must go to the circus. But the animals are not caged after the performance and the only clowns are those back at the dock who go on criticising something they haven't experienced - and will have ultimately missed-out on once the show is over.

Galeocerdo cuvier
Tiger shark

Length: At birth 51-76, max. 550, mature females 250-550, mature males 226-370 cm. Distribution: Circum-tropical. Depth: Intertidal to 305 m. General: Snout broad, bluntly rounded. With well developed caudal keel. Coloration: Juveniles and young specimens have a distinct vertical pattern of dark bands on a light background similar to that of a tiger (hence the popular name), which becomes more and more inconspicuous in older individuals. Found in a wide range of habitats, from clear waters around coral reefs to turbid estuaries and bays. Swims slowly, described as sluggish, but capable of fast attacks on prey. Mainly active during the night, moving into shallower waters at dusk to hunt. Usually solitary, but will aggregate around a rich food source.

Litter size 10-82, gestation period 12-13 months. The only family member with an ovoviviparous mode of reproduction, without nourishment of embryos via a placental structure inside the mother's uterus.

Prey is very varied and includes horseshoe crabs, slipper lobster, conch, squid, benthic bony fishes (goose-fishes, walking batfishes, sea robins, lizardfishes, flounders, porcupinefishes, pufferfishes), sharks, skates, rays and marine mammals (including the endangered Hawaiian monk seal and large whale carcasses, see centre photo on p.158, feeding on dead sperm whale *Physeter catodon*). Preferred prey in certain geographical regions are sea turtles (swallowed whole or cut to pieces; the tiger shark's biserrated sawblade-like teeth - see SAW-BLADES AND MILLSTONES on p.208 - are able to cut through both the horny carapace and the bony skele-ton), sea snakes and birds (land and sea species). Catches young, inexperienced albatross in very shallow water at the Hawaiian Islands during the daylight hours, when the birds try to take off from the water's surface; these dramatic scenes have been captured on film and are shown here in ALBATROSS SNACK on pp.160-161.

Without doubt the tiger shark is one of the most dangerous shark species. Its large size and undifferenti-ating opportunistic feeding habits make underwater encounters with this animal a serious treat. Numerous verified attacks on spearfishermen, pearl divers, swimmers, SCUBA divers, surfers and boats are on record. Additionally there are cases reported for the tiger shark where a single individual attacked several people, which is very unusual when considering shark attacks in general. The number of survivors of tiger shark attacks is much smaller than that of survivors of white shark attacks. This may be due to the fact that (adult) white shark attacks on humans are considered to be prey misidentifications (human = seal) on behalf of the shark; a white shark at first bites only once to check its find, while tiger sharks will attack, 'saw' to pieces and devour almost anything they can handle. Generally this impressive species is rarely seen by divers, but direct encounters with large specimens should best be avoided by retreating from the water.

In 1992, a 'task force' against tiger sharks was established in Hawaii 'to protect tourists'. However, it is obvious that local fishermen simply wanted to get their share in the booming shark fin trade. After killing approximately 70 tiger sharks, the operation was officially terminated, but most of the local fishermen contin-ued to slaughter the magnificent predator which is easily caught at night on baited hook and line.

Continued on page 158.

Helmut Debelius

Kihei, Hawaii, Central Pacific

Marshall Islands, Central Pacific

Meanwhile, a resident scientist, Kim Holland, started to study the local tiger shark population. He caught many tiger sharks and released them alive after tagging them. He also demonstrated tonic immobility in tiger sharks - a state of torpor which is induced by turning a shark on its back and lasts for a short period of time - and attached sonic transmitters to the skin of several specimens to study their migration paths.

In June 1999, Hawaii still allowed the needless slaughter of sharks. Fishermen could still cut the fins off of sharks and throw the sharks back into the sea, dead or alive. 99% of the sharks caught were finned while 86% of the sharks were alive when brought on board the fishing boat. These sharks are considered 'bycatch' and are caught accidentally while fishing for tuna or swordfish. Shark finning played a large part in wiping out a large numbers of sharks not only in the United States, but worldwide. Attempts are under way, to prohibit this practice. The Federal U.S. agency that manages fisheries in the Pacific was asked to take action to end this practice. The reason for such local efforts is obvious. The shark is part of a balanced ecosystem and without sharks the balance is altered which threatens the viability of other species as well. Check the website address of West Pacific Fisheries listed in the bibliography section to get further information.

Even before, the Hawaii-based protection organisation EnviroWatch already tried to make way for legislation to ban the shark product trade in the islands. Their proposal was rejected in April 1999 after the lobby of the WPRFMC (Western Pacific Regional Fisheries Management Council) strongly opposed their idea. In order to get further information on this topic, check the website address of EnviroWatch listed in the bibliography section following the systematic part of this book.

In August 2000, however, the protection organisations finally were successful and shark finning was banned in Hawaiian waters. In continental U.S. waters finning is banned already since 1993!

Tiger shark feeding on sperm whale carcass   James D. Watt / IV, Hawaii

Tiger shark caught on hook and line   Gert de Couet, Hawaii

Juvenile tiger shark   D. Perrine, Bahamas, Western Atlantic

Adult tiger shark   D. Seifert, French Polynesia, Southern Pacific

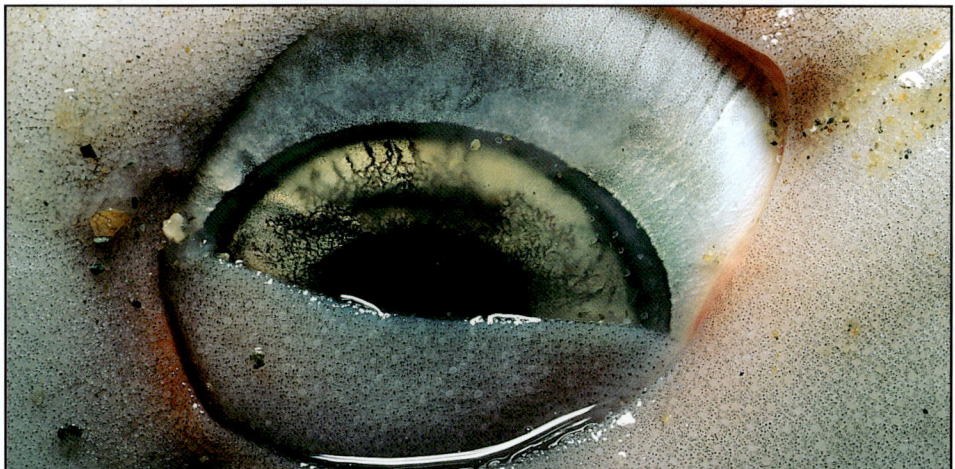

Eye, partly covered by nictitating membrane   D. Perrine, Marshall Islands, Central Pacific

ALBATROSS SNACK

Every year in the Northwestern Hawaiian Islands, thousands of Laysan and blackfooted albatrosses land on the small sand islands to mate and to lay their eggs. Albatrosses form permanent pair bonds, and their courtship dance is an amazing display of preening, bill slapping, and vocalisations. After mating, their softball-sized eggs hatch and quickly grow into 50 cm down covered chicks. Tim Clark tells us about their fate.

The islands are not very large, some only some hundred metres long, making inland flight impossible. Albatross chicks have to wait till their wings are developed enough to take them out to sea. As the down changes into feathers, the chicks start to test their aerodynamic wings by migrating to the windy edge of the island. Hundreds of chicks will line up along the berm each morning from late June to early August, hopping up and down trying to catch the drafts coming off the water. Finally they leave the island for the first time on their wobbly wings, heading straight out to sea. Many young albatross take off and won't set foot on land again until they are ready to mate in seven or eight years. Others have to land in the water, their wings not fully developed. Most chicks are blown back to their island by the gusty trade winds. The others are eaten by the tiger sharks that patrol the waters every year during albatross fledgling season.

Tiger sharks are seen in the deeper waters outside the atolls throughout the year, but during the summer they navigate through the reefs and into the shallow waters of the inner lagoon. Five to eight tiger sharks can be seen each morning, often patrolling waters less than 1 metre deep, waiting for a fledgling to

Top: A tiger shark slowly approaches a young albatross and attacks it. Above: The bird is grabbed and disappears from the surface.

land in the water. Most of the birds fledge in the morning when the winds are gusty, so the tiger sharks are more abundant in the morning as well. They can be seen circling the lagoons, along the fringing reefs, and even next to shore along the water's edge with their dorsal fin sticking out. When a bird lands in the water, any shark near-by will rush to get it before it takes off again. If a shark isn't near, the bird will sit on the water until it is ready to fly again. While this gives the sharks time to see the albatross, the bird may also fly off before any shark sees it.

Tiger sharks don't always get the bird they are after. The shark has to generate enough thrust to get its head out of the water and drag the albatross under. If the shark does not get its head out far enough, it will simply push the albatross away. It may take three or four passes until the shark can finally get its teeth on the albatross. The albatross can also take off again and try to get safety back to shore. If the bird flies close to the water, the shark can see the albatross and will follow and can get the bird if it lands in the water again. If it makes it to shore, the albatross will shake its wings dry and make its way back up the beach.

The same procedure as seen underwater. The tiger sharks gather in very shallow water to feast on the fledglings.

Negaprion acutidens
Sicklefin lemon shark

Length: At birth 45-80, max. 310, males mature at 234 cm.
Distribution: Wide ranging in central and western Pacific and Indian Ocean: South Africa, Mauritius, Seychelles, Madagascar, Red Sea east to Pakistan, India, Thailand, Viet Nam, Malaysia, Indonesia, New Guinea, Australia (Queensland, Western Australia, northern Australia), New Caledonia, Philippines, Palau, Marshall Islands, Tahiti. Depth: Intertidal to 30 m.
General: All fins but caudal more or less sickle-shaped, dorsal fins about equal in size. Head broad, coloration uniformly yellowish-brown. Prefers bays, lagoons, estuaries, sandy flats, outer reef faces, often in turbid, still water. Swims close to the substrate, but can rest on the ground. May also venture offshore into deeper water as one appears at the dead sperm whale shown in the shark film 'Blue water, white death'. Litter size 1-18, gestation period 10 months. Juveniles move onto tidal flats at high tide to feed and seek refuge from larger sharks; bare their dorsal fins in the shallow water. Tagging experiments revealed that these sharks have a relatively small home range: Tagged individuals have been recaptured in the same area after 4 years. Prey includes mainly bony fishes (triggerfish, parrotfish, mullet, batfish, jacks, porcupinefish), but crabs, shrimp, octopuses and stingrays are also eaten. Generally shy, even aggregations of juvenile and adolescent sharks are difficult to approach underwater. There are reports of smaller specimens attempting to bite waders and of larger individuals furiously attacking divers after being molested! Advice: Never provoke or even harass this shark and always leave enough room for it to move about.

Helmut Debelius   **Saudi Arabia, Red Sea**

Doug Perrine   **French Polynesia, Southern Pacific**

Negaprion brevirostris
Lemon shark

Length: At birth 60-65, max. 340, mature females 240-285, mature males 224-280 cm.
Distribution: Western Atlantic: New Jersey to southern Brazil including Gulf of Mexico, Bahamas and Caribbean. Eastern Atlantic: Senegal, Ivory Coast, probably widely distributed off West Africa. Eastern Pacific: Southern Baja California and Gulf of California to Ecuador. Depth: Intertidal to 92 m.
General: Fins less sickle-shaped and teeth of larger individuals more deeply serrated than in *N. acutidens* (previous species). Adults frequent lagoons as well as outer reef faces near deeper water. Juveniles are common on sand flats, in mangrove areas and in lagoons. This shark is solitary or aggregates. Commonly associates with schools of jacks. Adults have a home range of 18-93 square kilometres, younger (smaller) individuals have smaller home ranges. Most active during the twilight hours. Mating involves side by side swimming as has been observed in captive specimens. Females have tooth marks, indicating that biting by males occurs during courtship and/or mating. Litter size 4-17. Gestation period 10-12 months. Reaches maturity at 6.5 years of age. Pregnant females enter shallow nursery areas where they give birth (see also following pages). Prey of juveniles includes bony fishes (toadfish, pinfish) and shrimp. Adults prey mainly on fishes (catfish, jacks, mullet, croaker, spiny dogfish, guitarfish, stingrays, other sharks), but also on crabs, sea birds and large shelled gastropods (conchs). It is not only exploited by fisheries, but also the subject of investigation in many studies on shark behaviour and physiology (e.g sense of vision, breathing) because it readily adapts to captivity. Has been shown to feed until satiation and then stop eating. Not aggressive toward divers, but should be treated with respect.

Doug Perrine   **Bahamas, Western Atlantic**

Helmut Debelius   **Cape Verde Islands, Eastern Atlantic**

GIVING BIRTH

Researchers in the western Atlantic had set longlines for lemon sharks to take blood and urine samples for hormone analysis, as well as for measurement and injection of a growth marker. In order to see how the hormone profile changed during the reproductive cycle, the longlines were set outside of the breeding lagoon during the pupping season. So it was no surprise that a pregnant female was captured. The decision was made to deliver the pups by hand before bringing the mother to the boat for sampling to avoid those stressful procedures for the young lemon sharks.

1. A 60-cm lemon shark pup swims away just after delivery. The exhausted mother rests on the bottom between births.

2. The tail of the next pup to be born protrudes from the birth canal of the mother still resting on the bottom. If she had not been captured, the mother would probably have given birth while swimming.

3. A newborn pup swims through the lagoon, trailing its afterbirth still attached by the umbilical cord. The afterbirth is often consumed by remoras, which gather on the pregnant female and sometimes assist the birth by biting off the umbilical cord!

4. A volunteer gently pulls a pup from the mother's birth canal. All the lemon shark pups were examined to determine their sex as they were released.

5. When it became obvious that the exhausted mother could not give birth to the last pups, a scientist buried his arm in the birth canal trying to reach their tails.

6. The last pup swims across the grass flats to seek shelter among the roots of the mangroves on the fringe of the lagoon. Until it is much larger it will be highly vulnerable to predation by other sharks. After the sampling, the mother was released unharmed. After a year's rest she will return to the lagoon to give birth again.

Glyphis gangeticus
Ganges shark

Length: At birth 56-61, max. about 200 cm.
Distribution: Poorly known due to misidentification of other carcharhinids (e.g. pigeye shark *C. amboinensis* and bull shark *C. leucas,* the latter frequently penetrating into brackish and freshwater systems) which occur sympatrically with the Ganges shark. Definitely found in the Indian Ocean region in the Hooghly River, Ganges system, West Bengal, India; probably also inhabiting rivers in the vicinity of Karachi, Pakistan.
Depth: In rivers, may also occur in marine inshore waters and estuaries, exact data unknown.
General: The sharks of this genus are characterised by a short, broadly rounded snout, extremely small eyes, the lack of spiracles, upper teeth with broad cusps and lower teeth with long, slender cusps conspicuously protruding from the mouth. Coloration is uniformly grey or brownish without a colour pattern. At least two species are recognised, but they are poorly known with only a few genuinely identified specimens in museum collections. Genus members are commonly called river sharks because they inhabit freshwater systems, also brackish estuaries and marine inshore waters in the Indian Ocean and western Pacific region.

Whenever they can afford it, hinduist Indians burn their dead at specific sites on the banks of their sacred river, the Ganges ('Mother Ganga') and throw the ashes into the water. However, not always are the corpses completely burned up and there is enough flesh left for scavengers. Sharks are known to occur in areas where offal is thrown into the water on a regular basis, for example near slaughterhouses situated close to the sea; moreover, they follow ships and eat the left-overs from meals thrown overboard. In the lower reaches of the Ganges, at times quite a lot of carrion is drifting downstream, e.g. dead animals like drowned cattle and - of course - partly burned human bodies that escaped the flames a bit too early. Obviously, at least part of this unusual food source is utilised by sharks. As the water of the river is very murky and underwater visibility absolutely zero, the identity of the sharks involved in 'post-mortem man-eating' can only rarely be ascertained. Hence it is no surprise that the Ganges shark has a very bad reputation as a maneater in India. Nevertheless, it can be assumed with a considerable degree of certainty that the correct identity of the scavenger is the bull shark rather than the Ganges shark. The dentition - especially the slender teeth of the lower jaw - of the latter species, its apparent rarity and the sympatric occurrence of the bull shark with its notorious reputation of penetrating far into streams and eating mammalian remains are sufficient evidence that the Ganges shark must be relieved from the accusation of being a maneater. Although no data are available, it is much more likely that the river sharks of the genus *Glyphis* rather eat fish than mammalian (and human) carcasses when considering the shape of their teeth and comparing it to that of other fish-eating shark species.

The related speartooth shark *G. glyphis* reaches about 1 m in length and has been recorded from inshore waters and rivers in Borneo, New Guinea and Queensland, Australia. It too is rare and a lot of confusion exists on how many species there are and the true identity of specimens reported and/or collected. During recent years, new records of these sharks from Borneo (see photo below) and the Alligator River in Australia's Kakadu National Park were made and several specimens have been investigated by specialists. Hopefully, more information on these mysterious sharks will become available in the near future.

Mabel Manjaji   Sungai Kinabatangan, Borneo

Prionace glauca
Blue shark

Length: At birth 35-44, max. 380, mature females 220-320, mature males 180-310 cm.
Distribution: Probably the widest-ranging cartilaginous fish: Circumglobal in tropical and temperate seas. West-
ern Atlantic: Newfoundland to Argentina. Central Atlantic Islands: Azores, St. Paul's Rocks. Eastern Atlantic:
Norway to South Africa, Mediterranean (into northern Adriatic, but absent from Black Sea). Indian Ocean:
South Africa and southern Arabian Sea (absent from Red Sea and Arabian Gulf) to Indonesia, Japan, Australia,
New Caledonia and New Zealand. Central Pacific Islands. East Pacific: Gulf of Alaska to Chile. Depth: 1-400 m.
General: An unmistakable species with very slender body and wing-like pectorals, large eyes and deep blue
dorsal coloration, lighter below. The oceanic shark sometimes moves closer to shore near offshore islands. In
more tropical regions blue sharks are found at greater depths, e.g. in the Indian Ocean they are most common
between 80 and 220 m. Around the Channel Islands off California, blue sharks move into shallow inshore
areas at night, possibly in response to changes in prey distribution. In the North Atlantic they make regular
dives, from the surface to depths greater than 400 m. These vertical movements usually occur several times
each day and occur most often during the daylight hours.
    Litter size 4-135, with larger females carrying more young. Gestation period 9-12 months. Males bite
females during courtship and possibly also during copulation. Prey includes squid (see THE NIGHT OF LOVE
AND DEATH on pp.169-172), schooling bony fishes and krill (see photo on following page). An adaptation to
economically catch small and slippery crustaceans like krill are the elongate papillae on the gill rakers that prevent prey from leaving the mouth cavity via the gills slits. The nocturnal feeding of blue sharks on mating and
spawning squid off California is one of the earlier highlights of documenting shark behaviour on film. Also eat-
en are sharks, sea birds and mammalian carrion. One of the three abundant sharks of the open ocean realm,
number three behind *Carcharhinus longimanus* and *C. falciformis* within the hierarchy of species when considering boldness of similar-sized specimens at a food source.
    While swimming slowly, usually at speeds of less than 1 knot (52 cm/sec), blue sharks make use of ocean
currents. In the North Atlantic, tagging and recapturing of blue sharks revealed that they migrate in a clock-
wise pattern along a route located in the prevalent local currents. Individuals tagged off the east coast of the
USA have been recaptured off Spain, in the Strait of Gibraltar and in the northern central Atlantic near the
equator. In turn, sharks tagged in the Canary Islands were recaptured off Cuba. The combined results of the
tagging/recapturing experiments show that blue sharks swim in the Gulf Stream from America to Europe, continue southward along the coasts of Europe and Africa and return to America in the North Atlantic Equatorial
current. Also in the Pacific, blue sharks travel very long distances as one specimen tagged off Tasmania, southern Australia, has been recaptured off Java, Indonesia.
    Blue sharks are curious of divers, but not aggressive. They have much more to fear from people than vice
versa because they are caught for human consumption and by 'sport' anglers (e.g. at the British Isles, Madeira
and the Canary Islands, also in the Adriatic Sea). Once one of the most numerous sharks in the open sea, the
species today is endangered by overfishing. See also SHARK GAMES on pp.88-94.

Helmut Debelius   **California, Eastern Pacific**

Diver provokes blue shark to test protective chain mail suit made of steel

Blue shark eating krill   Both photos Mark Conlin, California, Eastern Pacific

THE NIGHT OF LOVE AND DEATH

The ocean was perfectly still on this cold, moonless night in February. As our small boat rounded the west end of California's Catalina Island we could see the squid fishing fleet at anchor two miles away; their powerful fishing lights pushing out against the almost total darkness. So calm and dark was the night that the island and the sea could be distinguished from the sky only by a lack of stars. The squid fishing fleet looked like fourteen brilliant fireflies suspended in space. Howard Hall remembers:

Enormous swarms of squid are attracted to the lights of fishermen. After only a short time blue sharks join the scene.

We approached one of the fishing boats and could see that it was surrounded by a huge glowing mass of squid nearly fifty yards in diameter. Squid are often attracted to artificial lights, and during the breeding season the lights not only attract them, but seem to enhance their courtship activity. The fishermen were busy dipping their six foot diameter brail net into the writhing mass and winching aboard incredible quantities of the seven to ten inch long cephalopods. In two hours they would take over ten tons. At the edge of the fishermen's pool of light and passing through the periphery of the school, were numerous pilot whales and sea lions. And knifing directly through the school, right beneath the brail net were twenty or thirty large blue sharks.

The sharks were the reason we had come. Photographer, Marty Snyderman and I were working on a television special about sharks for National Geographic. But as I watched the scene surrounding the fishing boat, I realised that this was more than just a scene of predation. We were witnessing the culmination of an entire population of squid. They had come up from the dark depths of the sea to mate, lay their eggs, and to die. It was a major event in the marine environment and one which did not go unnoticed by a large variety of the ocean's inhabitants.

The scene directly below the squid boat looked rather intense. The sharks were so engrossed in eating squid that I feared they might accidentally bite a diver in the process. So we tied our boat off a hundred yards behind the fishermen where the density of squid and sharks was greatly reduced. We dropped over the side of our boat and the ocean exploded in a brilliant flash of green bioluminescence. Then we descended twenty feet and swam through the almost total darkness toward our underwater movie lights which were hanging over the side by a power cable. Suddenly the darkness was penetrated by a blinding flash as the movie lights came to life. Surrounding us were thousands of pulsating squid; their

Blinded by love the squids swim about.

169

colours fluctuating in flashes from creamy white to reddish brown. They entirely ignored our presence as individuals frantically searched for mates. Below us the dark water would flash with green light as thousands of arrow shaped squid shot up toward our movie lamps. Then, after entering our tiny pool of light, couples would grasp each other and begin frenzied courtship.

During mating, the squid entirely ignore their surroundings. This is their one chance in life to find a mate and insure the perpetuation of their

Blue sharks snapping out blindly for the inevitable mouthfuls of squid.

genetic code. I watched in rather morbid fascination as an occasional shark passed through the school devouring the mating pairs. Only two feet in front of my face an especially large blue shark grasped the tail of one member of a mating pair. With its ten sucker arms the male squid continued to embrace its mate as the shark casually drew the pair into its maw. Instead of releasing its dying mate and fleeing, the male continued to hold fast until, finally, it too was consumed. But this desperate behaviour makes sense in the world of the squid. They live only one year. And during that time they roam the deep sea growing, developing, and maturing in preparation for this one night. It's the most important night of their lives, and it's also their last. For after the males mate and the females lay their eggs, they all die. So it's understandable that on this night the squid have no fear of death whether it be in the slashing jaws of the sharks or alone on the sea floor waiting for their rapidly degenerating bodies to cease function.

The night is also special for the sharks. If ever sharks dream, this must be a dream come true. So numerous are the squid and so unconcerned are they with their surroundings that the sharks can effortlessly consume mouthful after mouthful. One blue shark passed me with an enormously distended belly. It grabbed six or seven squid as it passed then suddenly stopped, and while shaking its head back and forth violently, vomited out about twenty pounds of the dead cephalopods. After the spasm passed, the shark swam forward and began again refilling its belly.

As we descended to a depth of one hundred and ten feet, what seemed to be a huge white cloud became visible in our lights below. We had reached the bottom. But the expansive plain of sand on the ocean floor was invisible. It was entirely covered by the eggs of the squid. The female squid produces an egg case that is nearly half the size of her body. The first squid to lay her eggs anchors the egg case to the sandy plain by digging head first More than enough for every predator.

From the surface to the bottom, there were mating squids everywhere.

deep into the sand until only her undulating tail fins protrude. After the anchor has been placed, the female squid digs herself out and leaves the creamy white and translucent egg case suspended above the sand. Soon the entire sand plain is covered with eggs and squid begin anchoring their egg cases to the anchors of earlier arrivals. At times the bottom is covered with squid eggs several feet deep and as far as a diver can swim in any direction. Each egg case holds about two hundred individual eggs. After a week, the maturing embryos with their bright red eyes can be seen clearly by a diver if he holds his hand light behind the translucent case. After two weeks the eggs hatch and the tiny squid venture out into deeper water to begin the cycle anew.

The egg covered plain is a flurry of activity. Here the predation on the squid is even easier than below the squid fishing boat. Not only are squid mating and laying eggs here with complete disregard for the numerous animals that are preying on them, but after mating and laying eggs, the squid die. The process is not unlike the rapid aging and dying of salmon when they spawn. After mating and laying their eggs, male and female squid cease to be the quick, darting creatures that instantly flash from one colour to another. Their bodies quickly begin to degenerate; their long sleek tentacles become twisted and disfigured, and their colour becomes a constant ghostly white.

The dying squid offer little resistance to the predators. Numerous species of fish, crab, lobster, and several non-pelagic species of sharks such as angel sharks and

A female squid fixes her egg case on the bottom among thousands of others.

horn sharks, find capture of these failing bodies effortless. But even with the mass gathering of these predators for this event, not to mention the heavy predation on the squid by man, they die faster than they can be taken. Soon the bottom is littered with their tiny corpses. In some places the dead squid lie in fields three and four feet deep.

As soon as the work is done the female dies.

Like the blue sharks, the smaller predators have eaten beyond their capacity. I watched as a comically swollen angel shark stared at a squid that was dying right in front of its face. I think the shark just wanted to sleep and rest its stomach, but the squid kept rolling up to its nose in the mild ocean surge. Finally, perhaps out of frustration or simply lack of self control, the shark snapped up the squid. It chewed on the squid for a while alternating it from one side of its mouth to the other like a cigar. But the shark couldn't swallow it. Finally the shark stopped moving and seemed to go back to sleep; the back half of the squid still sticking out of its mouth.

All the time Marty and I were down, our bright movie lights were attracting more and more squid. A sphere of squid was accumulating around our lights like the large one that had developed around the squid boat. At times this school of squid became so dense that we couldn't see more than two feet and then we would have to swim hard for a few moments to get clear of the school. Blue sharks were also beginning to notice the high density of squid around the lights and soon several of them were passing right before our cameras preying on the school and providing excellent opportunities for

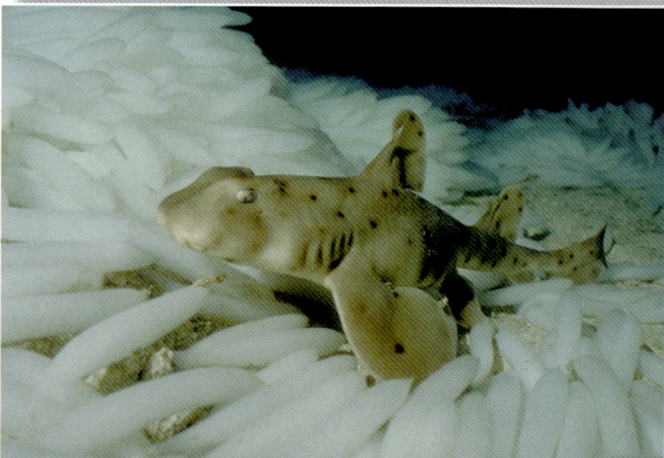

The abandoned egg cases are food for many predators, among them horn sharks...

filming. Soon, however, the situation got out of control.

The number of blue sharks feeding on the squid surrounding our lights was rapidly increasing. They would dash through the flashing mass, their nictitating membranes covering their eyes, snapping out blindly for the inevitable mouthfuls of squid. Marty and I began to worry that the sharks might inadvertently bite one of us, and when the shark action and density of squid became too great we would hastily swim out of the school. But soon the school was too large for us to find our way clear. The squid became so unbelievably dense that they were getting into my

...and angel sharks, both bottom-dwelling species.

mouthpiece and I could hardly see the light coming from the powerful movie lamps that I held in my hand. I felt something strike my left side and flinched. It may have been Marty accidentally kicking me but I wasn't sure. Then I was struck hard in the head and I saw the face of a shark, its teeth lunging as it gobbled squid just inches from my face mask. I didn't like the idea of swimming back to our boat in total darkness, but the alternative of staying with the lights and being the centre of attention had become intolerable.

I dropped the lights and they immediately disappeared into the school. Marty had his hand on my tank, so he instantly (and thankfully) realised what I had done and in the darkness we raced skyward. We were bumped several more times before we cleared the school, but soon we were in the open and we could slow our ascent rate. It was like being in a cold black closet. At first I couldn't even see the rim of my face mask. But soon, when my eyes began to adjust, I could see Marty's firey green outline in the darkness as his body movements disturbed the bioluminescent plankton. Then, as we ascended to our boat, I began to see thousands of streaks of green fire in the water. It was like being in the upper atmosphere on the edge of space during a great meteor shower. All around us brilliant arrows fired down from above. I paused for a moment during my ascent for a last look at this ethereal scene and watched as thousands of squid rained down from above toward the ocean floor to mate, to lay their eggs, and to end their one short year of life.

Many square metres are covered in egg cases. This overfed but friendly angel shark departs to let a hungry diver get its share.

Triaenodon obesus
Whitetip reef shark

Length: At birth 56, max. 210, mature females 105-158, mature males 104-168 cm.

Distribution: Wide ranging in most of the tropical Indo-Pacific. Indo-West and central Pacific: South Africa and Red Sea to Pakistan, India, Sri Lanka, Burma, Indonesia, Viet Nam, Taiwan, Riu Kiu Islands, Philippines, Australia (Queensland, northern and Western Australia), New Guinea. Widespread in the islands of Oceania (Polynesia, Melanesia, Micronesia), northward to Hawaii and southwest to the Pitcairn Island group. Eastern Pacific: Cocos Islands, Galapagos Islands, Islas de Revilla Gigedo, Panama to Costa Rica. Depth: Intertidal to 40 m.

General: Snout very broad. Pectoral and dorsal fins with pointed tips. First dorsal and upper lobe of caudal fin with white tips. Coloration grey to brown above, lighter below, some individuals additionally with darker blotches. One of the three most abundant sharks in the coral reefs of the tropical Indo-Pacific region. Found resting (often in groups) in caves, under ledges or in narrow crevices (see photo below and on following page) during the day, also (more frequently) in lagoons, on reef flats and near reef faces (where this was the second shark species the author encountered underwater in the northern Red Sea; the more obvious, actively swimming grey reef shark was the first one). Resting caves are known to be frequented by the same individuals regularly every day, sometimes for years. Activity is highest during the night, when the sharks frantically hunt singly or in small groups in the reef where they prey on bony fishes (eels, morays, parrotfishes, goatfishes, snappers, damselfishes, surgeonfishes, triggerfishes), crabs, lobsters and octopuses. While poking into the reef crevices, audible noises from the rough, denticle-clad shark skin scratching against the corals and rocks can be heard. Often, pieces of coral are broken off during such action.

Copulation has been observed in the wild: The male grasps one pectoral fin of the female with his mouth and inserts a single clasper. The male also bites and holds the female's gill region during courtship. Both maintain a parallel orientation during copulation. Copulation lasts 1.5-3 minutes. Litter size 1-5, commonly 2-3. Gestation period 13 months. Maturity is reached after at least 5 years, maximum age is estimated to be at least 25 years.

Often the first shark to be seen in a tropical coral reef. Shyness towards divers and snorkelers depends on the experience of the individual shark. Usually swims up from a resting position on the ground when approached during the day, only to return to nearly the same position after circling out of sight for a few minutes. May get aggressive when molested or in the presence of freshly caught fish (spearfishing).

Unique in being the only shark causing ciguatera food poisoning after its flesh or liver have been consumed. The toxic agent (ciguateratoxin) originates from unicellular algae accumulated via the food chain and induces severe gastrointestinal and neurological symptoms. More frequently, it occurs after consumption of large predatory bony fishes such as groupers and snappers, its occurrence strongly depending on geographical region and season of the year.

Mark Strickland   Andaman Sea

Whitetip reef shark actively hunting at night   Great Barrier Reef, Australia

Whitetips resting in a cave during the day, Hawaii   Both photos Helmut Debelius

A group of whitetip reef sharks hunting morays among overgrown rock boulders. Their slender body

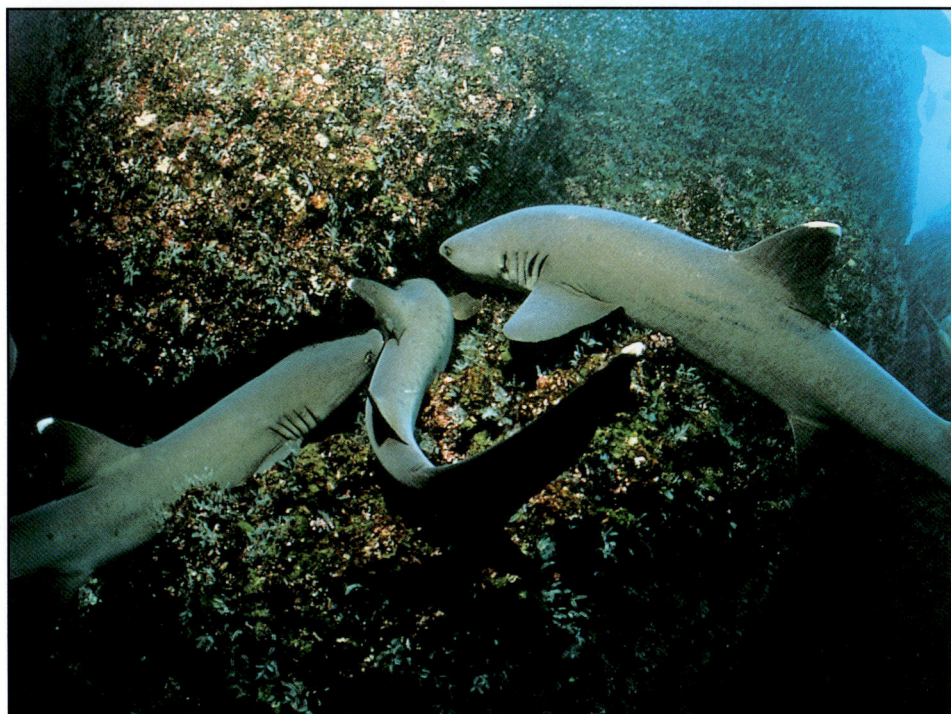

enables them to penetrate into the narrowest of crevices. Photos Dietmar Seifert, Cocos Isl., E-Pacific.

Rhizoprionodon acutus
Milk shark

Length: At birth 35-40, max. 178 (a record from off the Natal coast, South Africa), usually about 100, mature females 70-165, mature males 68-178 cm.

Distribution: Tropical waters of eastern Atlantic: Madeira and Mauritania to Angola. Indo-West Pacific: South Africa and Red Sea to Pakistan, India, Malaysia, Indonesia, Thailand, China, Taiwan, Japan, Philippines, Australia (Shark Bay, Western Australia, to Fraser Island, Queensland).

Depth: Shallow intertidal to 200 m.

General: The milk shark is - like all members of the genus Rhizoprionodon - a relatively small, slender charcharinid. Characteristically, its second dorsal fin originates well behind the origin of the anal fin. In other genera of the requiem shark family the second dorsal fin originates above or in front of the origin of the anal fin. The milk shark is the only species in its area of distribution that has long upper and lower labial furrows. Its eyes are relatively large. The teeth in both jaws are similar and have narrowly triangular cusps with finely serrated cutting edges in adults (in juveniles, the edges are smooth). An interdorsal ridge is usually absent. Tooth counts are 25 (right upper jaw), 23-24 (left upper jaw); 23-27 (right lower jaw), 22-26 (left lower jaw). Coloration of the dorsal surface is bronze to greyish, the underside is lighter. Most fins are plain in colour, the dorsal and upper caudal fin tips are dark in juveniles and sometimes have a dark edge in adults. This abundant species is found in inshore as well as in offshore waters on the continental and insular shelves. It ranges through the entire water column but usually lives near the bottom. It is often seen off sandy beaches, sometimes also in estuaries, tolerates very low salinities, but does not range into pure freshwater of streams. Milk sharks from different areas of the species' wide range vary considerably in size. While members of the Australian population reach a maximum length of about one metre, those from Africa may reach almost twice that size. Litter size is 1-8, usually 2-5, gestation period about 12 months. Off Natal, South Africa, mating occurs in summer, the sharks mature at age 2 and reach an age of at least 8 years. Off Bombay, India, young are born in winter. In Australian waters females give birth each year; birthing, however, occurs throughout the year and not in a specific season. Milk sharks mainly prey on bony fishes, to a lesser extent also on cephalopods and crustaceans. Off Natal their numbers fluctuate throughout the year, being highest during the summer months. There, total numbers have increased during the recent years. This increase is probably the result of the local shark meshing program (long stretches of inshore waters off popular bathing beaches are protected against large sharks by fixed gill nets) which has reduced the numbers of large coastal sharks that usually prey on their smaller cousins. This species is one of the most common sharks that is incidentally taken in northern Australian waters by bottom trawlers operated from Taiwan and Thailand. Larger specimens are processed for human consumption. Elsewhere, it is also an important species utilised for food and fish meal production.

Pedro Niny Duarte / DOP   Cape Verde Islands, Eastern Atlantic

Rhizoprionodon terraenovae
Atlantic sharpnose shark

Length: At birth 29-37, max. at least 110, mature females 85-90 (max. 110), mature males 65-80 (max. 103) cm.

Distribution: Western North Atlantic: Continuously distributed along the Atlantic coast of North America from Southern Newfoundland to Florida and Gulf of Mexico.

Depth: Intertidal to possibly 280 m.

General: Snout long, parabolic, upper labial furrows long. Anal fin originates in front of second dorsal fin origin. Posterior margin of anal fin slightly concave or almost straight. Pectoral fins small. Coloration grey or greyish-brown, fading to white below. Large specimens may have light spots on flanks, white pectoral fin margins and dusky dorsal fin tips.

This abundant small coastal species can be observed close to the surf zone off sandy beaches, in enclosed bays, harbours and marine to brackish estuaries. It readily tolerates reduced salinities, but does not penetrate far into freshwater. Differs from other, similar species of the genus, e.g. the allopatric Caribbean sharpnose shark *R. porosus*, mainly in the number of precaudal vertebrae. Hence, the identification of species of this genus based on external features alone is not always possible.

Litter size 1-7, with larger females carrying more young. Gestation period 10-11 months. Sex ratio of near term embryos is 1:1. Off Texas, Florida and North Carolina young are born from June to August. In the Gulf of Mexico mating occurs from May to July and young are born almost one year later. Gravid females move into inshore waters to give birth and then outnumber males by nearly 3:1 (northern Gulf of Mexico).

Prey includes small bony fishes (herring, snake eels, silversides, wrasses, small jacks, croakers, mojarras, toadfish, filefish), squid, shrimp, crabs, segmented worms (annelids) and gastropods (particularly their feet). A small, harmless shark. Shy, difficult to approach underwater. This common inshore species is fished in Mexico for human consumption.

In recent years, the spiral intestines of several sharpnose shark species (among other carcharhiniform sharks) which are generally not easily distinguishable were investigated for parasites. It could be demonstrated that certain tapeworms (tetraphyllideans, family Onchobothriidae) are restricted to elasmobranch hosts and often are highly host-specific. Some of the tapeworm species parasitise just one host species and thus can aid in identifying the host. The results of these investigations suggest that each carcharhiniform species may possess a unique assembly of parasitic tapeworms that could be used for species identification (Healy, C.J. & J.N. Caira, 1998).

Doug Perrine   **Bahamas, Western Atlantic**

The eight or nine species in this unique family of sharks are placed in two genera, one of them *(Eushpyra)* is monotypic. They range from 1-5 m in length. Together with the requiem sharks, which they resemble in external anatomy of body and fins, hammerheads are considered to be the most highly evolved (advanced) of all sharks. All species are immediately recognised by the more or less pronounced lateral expansions of the head. The head shape ranges from simply broadened and flattened (in *Sphyrna tiburo*) to T-shaped with long and narrow, wing-like extensions on the sides (in *Eusphyra blochii)*. Much is speculated about the function of this unusual head. It may aid in locomotion, serving as a bow plane to increase lift and manœuverability, especially in extremely narrow turns when closing in on prey. Therefore this type of head is also termed cephalofoil, meaning head plane. An analogous technical structure is the so-called canard wing of some types of aeroplane. It consists of two moveable rudders at the aeroplane's nose, which provide lift (or descent) also at the craft's front and not only in the usual position of the elevators at the stern. The special head may also enhance electroreceptive, olfactory and visual perception by spreading the sense organs over a broader area. It is also used to pin stingrays (a favourite prey item of the larger hammerhead sharks) to the ground. The two recognised genera are clearly separated by the width of the nostrils (short in *Sphyrna* spp., enlarged in *E. blochii)*.

Hammerhead sharks are found in many tropical and temperate coastal habitats, with several species occurring on or near reefs where they take refuge, hunt and are cleaned by resident cleaner fishes. Their mode of reproduction is placental viviparous; the near term embryos have flexible 'hammers' that are bent back toward the tail in order to facilitate birth. Some species are known to move from deeper waters into the shallows to give birth. Prey includes squid, crustaceans, other elasmobranchs (two species definitely favour rays) and bony fishes. Several species form large schools and make long seasonal, north-south migrations. The larger species (great hammerhead) can be bold and even dangerous to divers. The smaller species (bonnethead shark) are shy and difficult to approach underwater. The bonnethead shark does well in larger aquaria; the larger species suffer from capture and handling stress and only rarely adapt to a life in captivity. See also A SHIVER OF SHARKS on pp.182-188.

Howard Hall   Sea of Cortez, Mexico

Sphyrna lewini

Scalloped hammerhead shark

Length: At birth 42-55, max. 420, mature females 212-309, mature males 140-295 cm. Distribution: In most warm temperate and tropical seas. Depth: 1-275 m.

General: Hammer-shaped head, with an indentation in the middle of its front and the broad extensions arched backward. Pectoral fins with dark tips. Pelvic fins not falcate. Most abundant family member. Adults are pelagic, but also occur in inshore waters, sometimes near coral reef drop-offs and often around seamounts.

Litter size 15-31. Has nursery areas in shallow, turbid coastal waters.

Prey mainly comprises bony fishes (mackerel, dolphinfish, squirrelfish, scorpionfish, snake eels, groupers), other sharks, rays, squid, crustaceans and octopuses, rarely also sea snakes. Juveniles feed more on benthic fishes and crustaceans, while adults prefer squid. Newborn scalloped hammerheads are eaten by smalleye hammerhead sharks.

Usually indifferent and inoffensive toward divers, but may become aggressive in baited situations. Occasionally makes close passes to inspect divers. Mating has been observed in the Sea of Cortez. The male grasped the female's pectoral fin and curled the posterior part of his body around her. With a parallel orientation of their bodies, the pair sank until they hit the reef below, at which time they separated.

The Sea of Cortez is famous for being visited seasonally by immense schools of scalloped hammerhead sharks. Unfortunately the number of individuals involved in these placid non-feeding, (?) mating aggregations dwindles from year to year due to extreme overfishing at all costs by Mexican and foreign fishing fleets.

See also photo left.

Helmut Debelius — Aldabra, Western Indian Ocean

Douglas D. Seifert — Malpelo Island, Eastern Pacific

Avi Klapfer — Cocos Island, Eastern Pacific

Doug Perrine   Azores, Atlantic

Rudie Kuiter   Great Barrier Reef, Australia

Sphyrna zygaena
Smooth hammerhead shark

Length: At birth 50-61, max.
400, mature females 210-304,
mature males 210-256 cm.
Distribution: Circumglobal in
tropical and warm temperate
seas.
Depth: 1-60+ m.
General: Front margin of hammer-shaped head convexly
rounded, without indentation.
Free tip of first dorsal fin does
not reach pelvic fin origin.
Oceanic species, usually seen
over deeper water of the continental shelf, but also found in
shallow coastal waters around
reefs. Solitary or migrating in
large schools. Moves to
warmer waters in the winter.
   Litter size 29-32, gestation
period about 8 months.
   Prey mainly comprises
squid and bony fishes (mackerel, jacks, herring, mullet, barracuda, needlefishes, filefishes,
groupers), but elasmobranchs
are preferred in certain
regions: Eats small sharks
(including its own species),
skates (in northern areas),
stingrays (in southern areas;
whiptail stingrays and spotted
eagle rays among others).
   Usually indifferent and
inoffensive toward divers, but
may become aggressive in baited situations. Occasionally
makes close passes to inspect
divers.

Sphyrna tudes
Smalleye hammerhead shark
L: At birth about 30, max. about
150, mature females 120-148,
mature males 110-134 cm. Di: W-
Atlantic: Venezuela to Uruguay. De:
1-12 m. G: Head with central
indentation in its convexly rounded
front and almost straight posterior
margins. Greyish-brown above, light
below, fins without markings. A little
known inshore species that lives on
the continental shelf. Placental viviparous, with yolk-sac placenta. Litter
size 6-9. Prey includes small bony
fishes (sea catfish, grunts), but also
newborn scalloped hammerhead
sharks, swimming crabs, squid and
shrimp. The locally abundant species
is taken by coastal fisheries.

Doug Perrine   Trinidad, Western Atlantic

Sphyrna mokarran
Great hammerhead shark

L: At birth 50-70, max. 600, mature females 250-550+, mature males 234-340 cm. Di: In most tropical seas. De: Intertidal to 80 m. Ge: Front of head nearly straight, with central indentation. Fins plain. First dorsal fin very high, sickle-shaped. Free tip of first dorsal fin does not reach pelvic fin origin. Pelvic fins falcate. Largest family member. Litter size 13-42, with larger females having larger litters, gestation period 7+ months. Gives birth in late spring and summer. Prey includes other sharks, bony fishes (catfishes, toadfishes, jacks, herring, boxfishes, groupers), squid and benthic crustaceans. Preferred prey are, however, rays (guitarfishes, whiptail stingrays, cownose rays, eagle rays) and skates. Specimens have been reported with up to 50 stingray and/or catfish spines broken off and stuck in the jaws, throat and sides of the head. Most feeding occurs at dusk. The large, solitary species obviously is avoided by other sharks. Although usually not aggressive, it will make close passes toward divers, and should always be treated with respect. Has been reported to attack bathers. Small photo below from the Caribbean.

Douglas D. Seifert   French Polynesia, Southern Pacific

Bob Halstead   Papua New Guinea

Sphyrna tiburo
Bonnethead shark

L: At birth 37, max. 150 cm. Di: W-Atlantic: North Carolina to Brazil. Eastern Pacific: Southern California to Ecuador. De: Intertidal to 80 m. G: Head distinctively shovel-shaped, without indentation at convex front margin. The diurnally active species is one of the smaller hammerhead sharks. Litter size 11-14. Segregates by sex, with females moving into shallower water before giving birth, often in estuaries. Prey mainly comprises swimming crabs and shrimp, but also mantis shrimp, mussels, barnacles, isopods, cephalopods and small fishes. Harmless, but shy and difficult to approach underwater.

Helmut Debelius   Bahamas, Western Atlantic

181

A SHIVER OF SHARKS

In remote areas of the Eastern Pacific Ocean, diving on seamounts or rocky pinnacles surrounded by the deep waters of the open sea is the ideal setting for introspective, innerspace pondering. Because of their geographic isolation, pinnacles and seamounts attract and densely concentrate numerous varieties of fish and other animal life. They are the underwater equivalent of an oasis in an ocean desert. Most divers are familiar with 'a school of fish,' or 'a shoal of fish,' but when it comes to the subject of sharks, the proper plurality is defined as 'a shiver of sharks.' And, although the terminology has fallen out of use, upon finding oneself in the midst of hundreds of schooling scalloped hammerhead sharks, there is no more appropriate phrase, as Douglas David Seifert admits.

The most beguiling thing about sharks - as lone individuals in general and with reference to schooling scalloped hammerheads in particular - is the way they just appear, seemingly from nowhere. One may enter the water, at a dive site reputed for shark encounters, and, all anticipation to the contrary, detect nothing; nothing but empty sea. Of course, the sea is not really empty. Schooling fish, such as jacks or grunts or Creole fish, swirl about in dizzying patterns of fishy flight. But, to a diver intent on seeing big animals, the beauty of schooling fish dynamics may not be appreciated and the schools may even be likened to an annoying swarm of flies cluttering up the scenery.

Suddenly, a hammerhead shark appears. It moves forward slowly, almost lazily, its head swinging gently in a side-to-side arc. It's body writhes in a sideways, sinusoidal rhythm, propelled forward by sturdy beats of its tail. The skin coloration is brown, but at a distance, it appears grey-blue, remarkably blending in with the open water surrounding. It's size is perhaps two and a third metres in length, exclusive of tail. The head is strange, with eyes situated at the end of flattened stalks, which project outward from the body, ahead of the gill slits. The distance, from eyeball to eyeball, measured across the bridge joining the twin eyestalks into a singular, blunt, scallop-fringed snout is two-thirds of a metre in width. The scallop-fringe pattern along the snout identifies it as a scalloped hammerhead shark (Sphyrna lewini), classified as a shark dangerous to man but without any verifiable record of attack upon humans. It's appearance is the essence of the word supernatural. The head undulates back and forth with the steady measure of a metronome. At all times, one eye regards the entirety of the scene before it, above it, below it and around it, while the other observes the scene in converse, receding behind. The combination of pattern of head movement and exaggerated anatomy ensure that the hammerhead is the master of visual acuity.

The shark moves effortlessly against the current, gently gliding with the grace reminiscent of a bird of prey riding thermal updrafts. As the hammerhead angles upward, it reveals a small mouth, filled with tiny but sharp teeth, jaw hanging loose, almost slack, closing occasionally with rapid twitches and returning to an open gape. One can not help think of a shark without thinking of that infamous mouth filled with those terrible teeth. A shark and its bite - or its potential to bite - are an inseparable paired image in the subconscious and yet the hammerhead's mouth appears undersized, at least in comparison with other shark species of the same, or even lesser, lengths.

While concentrating on the lone, close shark, an awareness takes over that others are present. Next, three hammerheads appear, again, seemingly from nowhere, then twenty. And, as one is focused upon the direction from which they are appearing, it becomes overwhelmingly evident that more and more are swimming forward in greater numbers. The density of individuals increases, filling the surrounding waters with sharks, from barely below the surface to just above the bottom: 35 metres in depth's worth of sharks, spread across another hundred metres in width and perhaps further in breadth. Quickly estimating the total number of individuals in view brings about a mind-boggling total: More than two hundred, perhaps twice that. Sharks everywhere. A sensation of rising goosebumps prickles beneath one's wetsuit but the water is not chilly. Then comes the realisation: A Shiver of Sharks.

As they pass overhead, and are caught in silhouette against the pool of sunlight at the surface, they resemble nothing so much as a mass of human spermatozoa as seen under a microscope. They move together in a common direction but not uniformly or in unison. Each individual has its own relative pace and exhibits a distinct reactive movement all its own in the context of its relationship to its nearest neighbour. There is a certain individualistic variation to the choreography in their parade. They pass beyond and depart into the scrim of translucent visibility just as suddenly as they had appeared. They vanish one by one until no sign of hammerhead remains. Moments later, pause is given to ponder whether one had witnessed the phenomenon at all...

Taxonomists recognise nine species of hammerheads, which range in size from the great hammerhead (Sphyrna mokarran), which may reach six metres in length, to the bonnethead (Sphyrna tiburo), which matures at just over a metre, and place them collectively in the family Sphyrnidae, part of the order Carcharhiniformes. Although hammerheads are most often (and most erroneously) characterised as 'primitive' - due, no doubt, to their unique appearance evocative of 'something prehistoric' to the layman - hammerheads

Facing page: Not a school, but a shiver of sharks: Scalloped hammerhead sharks congregating at Cocos Island.

The largest hammerhead is the great hammerhead shark *(Sphyrna mokarran)*. This one was photographed in French Polynesia.

first made their appearance on Earth during the Pliocene Era, 25 million years ago - or 55 million years after the appearance and 'standardisation' of the shark-body-shape-we-think-of-when-we-say-shark evolved - and are thus the living representatives of the most recent design in shark evolution.

Hammerhead sharks could only be (and are) so named for their distinctive sledgehammer-shaped head, unique among all animals in nature. Because of this strange and fascinating physiology, they are one - often the first choice - of the fewer-than-a-dozen species which can be readily named by anyone when asked to cite examples of the 350 or so shark family members. And why not? Humans have long been fascinated by the hammerhead's odd appearance. Ever since the shark was first encountered, people have been pondering about what function the odd-shaped head serves. This fascinating hammer has been theorised to function as a bow plane, such as is found on submarines - or as an airplane wing functions - to provide additional lift as the animal propels itself through the water. But why should one species of shark require more lift than others? The typical shark body form is the model of hydrodynamic efficiency. A powerful tail pro-vides continuous forward thrust (no shark swims backwards) and the pectoral fin's shape creates an effect where there is decreased pressure on the upper surface of the fin and increased pressure on the lower surface of the fin. By rotating the angle of the fin, the shark directs the current flow of water below, there is decreased resistance above and the result is ease of ascent, called lift. The

Male scalloped hammerhead shark. Males of this species are outnumbered by females by up to 6:1.

hammerhead's hammer appears to function much like a second set of pectoral fins, probably giving the hammerhead not only greater manœuverability, but requiring less forward thrust from the tail to maintain its position against strong currents. The hammer is also utilised for trapping and holding prey against the bottom. One of the shark's preferred food items is the stingray and it has been witnessed to pin the ray to the sand as it bites the wings. Many captured hammerheads bear scars and have imbedded shards of broken-off stingray barbs in and around their mouths.

Scalloped hammerheads are always observed swinging their heads to and fro in an arc as they swim, continuously analysing sensory information between their widespread receptors. By contrast, the great hammerhead moves with the straight-forward linear intensity of an on-rushing truck: All deliberate, unhesitant movement, nothing relaxed or rhythmic about which to wax poetic.

Certainly, the width of the broad snout provides greater area coverage - and presumably enhanced efficiency - for both the shark's olfactory sense and the shark's electrical receptor system, the ampullae of Lorenzini. These jelly-filled ducts, which are open to contact with the surrounding waters via pores in the skin, house sensory organs sensitive enough to detect voltage gradients of lower than a millionth of a volt. All living creatures generate a mild electrical field in water and the shark's electrical sensory system enables the shark to identify - and home-in on - prey. The wide separation between the nostrils allows the hammerhead to continuously compare smell intensities and follow the strongest scent back to its source; the broad hammer provides greater surface area coverage for the ampullae of Lorenzini than in other sharks. The increased width facilitates this functioning in the same manner by constant comparison of fluctuations in electrical intensities to aid in prey detection and in underwater navigation.

Tracking experiments indicate that hammerheads may be able to follow 'highways' in the ocean. The sea floor itself is magnetised along a north-south axis and volcanic features such as contours and the paths of lava flows create magnetic anomalies and produce alternating bands of strong and weak magnetisation along the bottom. These magnetic fields are the result of volcanic activity which formed the geography of the sea floor as well as the seamounts where hammerheads gather. By comparing and analysing the differential between strength and weakness of these magnetic fields, the hammerheads can follow a consistent route, routinely tracing the paths of ridges and valleys which radiate outward from seamounts and pinnacles, to and from deep water feeding grounds and their seamount resting stops.

The scalloped hammerhead shark is the most common species of hammerhead to be found in the tropical waters of the world and is the only known large shark species to regularly engage in schooling behaviour. Further, among all the hammerhead species, only the scalloped hammerhead consistently forms such schools. Why is schooling the exclusive province of scalloped hammerheads to the exclusion of their cousins? No one has an answer. The size of the schools varies by location but the composition of the schools remains consistent. The smaller schools of twenty to thirty individuals may be seen in: San Salvador in the Bahamas Islands; the Southern Red Sea; Rangiroa, French Polynesia; Osprey Reef, Australia's Coral Sea; Papua New Guinea; the Solomons Islands; Taiwan; Sipadan and Layang Layang, Malaysia; Mozambique; and off the coast of Natal in South Africa. The large schools of two hundred or more are usually only encountered - albeit encountered regularly, predictably and as resident - in the Eastern Pacific Ocean. Currently, at the remote islands and seamounts of Wolf and Darwin, Northern Galapagos Islands; Malpelo Island, Columbia; and Cocos Island, Costa Rica, the largest schools of scalloped hammerhead sharks maintain a constant presence, due to the geographic isolation of the sites, the geological composition of those sites - they are all the products of volcanic activity - and the natural conditions that have led to this unique life history adaptation. Originally, however, the world's most famous sites for schooling scalloped hammerhead sharks were, without question, the El Bajo Espiritu Santo and Los Animas seamounts in the Sea of Cortez in Baja California, Mexico. Or, rather, they were once and in the not too distant past...

In the 1970's, Dr. A. Peter Klimley from Scripps Oceanographic Institute began searching for an actual site of hammerhead shark congregations which had been, for a number of years, reliably-rumoured to exist. After much trial and error, large numbers of hammerheads were located at the remote seamount, El Bajo Espiritu Santo, some twenty one kilometres from the peninsular mainland of Baja. Klimley and associates spent the next several years observing scalloped hammerhead schooling behaviour, tracking their movements and developing the theories which form the core of current scientific knowledge about this unique phenomenon. Recreational divers had also 'discovered' the phenomenon and it was at El Bajo that Stan Waterman and Howard Hall first captured this particular, peculiar marvel on film for ABC Television's American Sportsman. The dramatic, never-before-seen footage of hundreds of hammerhead sharks swimming in a unified school, oblivious to the presence of the underwater photographers captured the imagination of the world. Unfortunately, it captured the interest of one segment of the viewing audience a bit too well, however, and it was not long before the seamount was targeted by both the impoverished, local fishermen and wealthy, foreign, long-line fishing fleets. The intense, worldwide demand for shark fins and the large concentration of sharks in one area - identified and publicised - coupled with misguided fishing policies permitted by the Mexican government allowed the shark populations to be decimated to such an extent that the hammerhead schools are now, for all practical purposes, regionally non-existent in the Sea of Cortez. It is an even greater tragedy in light of the fact that virtually all of what is understood about the dynamics of hammerhead shark schooling behaviour was gained in this natural laboratory, now, reportedly, an ocean desert.

The findings of Dr. Klimley and associates serve to answer more than the simple question: "Why do scalloped hammerhead sharks school at remote seamounts?" Instead, the observations reveal several enigmas that can only be explained by the complexity of the scalloped hammerhead shark's natural history. At the heart of these enigmas is the clearly observed fact that schooling in scalloped hammerheads is radically different from schooling in bony fishes. Typically, schooling in fish is understood to be a defensive strategy against predators: Similar-sized fish of the same species have an attraction towards each other; they group together with a consistent spacing between individuals; maintain a commonness of direction as well as change direction concurrently: The result being that a would-be predator has difficulty in targeting an individual and is, instead, faced with a chaos of synchronised similar movement of like animals - a veritable fish fireworks.

Mature and near-mature hammerhead sharks have little need for defensive strategies against predators; their only predators (besides man) are orcas and a few other large sharks: The tiger, white, bull, dusky and silky sharks - the majority of which inhabit inshore areas far from the offshore seamounts (newborn pups and juveniles remain close to the bottom in the shallows of inshore environments, dispersed as a strategy against predation). Nor is the shark school's function a cooperative feeding strategy as is practiced by large schooling fishes such as tuna or yellowtail. There is no evidence that the sharks work cooperatively to drive prey fish together into tightly-controlled densities and take turns feeding upon the confusion wrought of

Such a group of female scalloped hammerhead sharks is always a rewarding sight for an underwater photographer.

multiple predators. The hammerheads as a group have never been reliably witnessed feeding during the day, although the odd individual has been observed to take a fish on occasion. Tracking experiments indicate the schools break apart at night, presumably to feed. The individuals travel great distances away and dispersed from the schools. Their diet is comprised of teleost fishes, such as trevallies (jacks), batoids (rays), and cephalopods (octopus and squid). Much of this prey lives in ocean habitats both distant and at depths greater than the hammerheads inhabit by day. Thus, they must actively hunt out in the open ocean for a specialised, high protein diet rather than in the relative plenty of the immediate seamount environment.

Exhausting other possibilities, the remaining cause for aggregations of fishes is for reproduction: A strategy utilised by many bony fish species which school infrequently but expressly for this purpose. But broadcast spawners such as groupers, wrasses and jacks are often hermaphroditic, and respond to a call and need of nature, not the opportunities necessitated by single-sexed attraction as in sharks. Further, and most importantly, the scalloped hammerhead schools are composed primarily of females - with a sex ratio ranging anywhere from 3:1, females to males, to 6:1, females to males, depending upon the year and the group of the observed studies. But again, the scalloped hammerhead sharks schools are enigmatic in that no reproductive behaviour - at least explicitly, as in penetrative copulation - has been witnessed. Instead, evidence seems to indicate that schooling behaviour in hammerhead sharks fulfills a daytime socialisation function. These schools are comprised of distinctly mature adult and near-mature adolescent females, apparently banding together as a reproductive pool for the occasional sexually mature males, believed to be transient loners, whom happen upon the school, either opportunistically or according to some unknown pattern, possibly by following the magnetic highways in the ocean. The infrequency of these occurrences dictates the males must mate with the most ripe and suitable females to effectively propagate with the greatest benefit to the species. The success of species reproduction, in this case, is the scenario requiring the mature male to impregnate the largest mature female(s).

In the East Pacific, an interesting cleaning association has developed: Angel- and butterflyfishes have specialised in cleaning large elasmobranchs (hammerheads and mantas).
Above: King angelfish.
Left: Barberfish.

Female scalloped hammerhead sharks produce a litter of pups in quantities directly proportionate to the size of the mother. The larger the female, the greater number of pups. Once sexual maturity is attained by the female, fecundity is more a function of size than of age. This is because females grow at a faster rate than males - and necessarily must do so - in order to produce the largest number of pups as guarantors of the survival of each generation. As in other arenas of reproductive biology among a majority of species, the male contribution to the survival of the species is limited to the donation of spermatozoa during a brief union. Once mature, the amount of spermatozoa produced by a male, regardless of volume, does not significantly increase the effect his contribution has in the creation of offspring. After consummation of intercourse, the male swims off without further interest in the process. But for the female, it is the number of eggs and the space within her allocated for embryonic development, not the volume of fertilisation that ensures a successful and productive yield. The greater size of the female corresponds with a greater capacity to gestate and deliver an increasing number of pups on a yearly basis: A litter of twelve pups, her first year, to a maximum of thirty eight pups per year, as she becomes fully grown in size. This is a function of the shark's anatomy. Within a sexually-mature female, the embryos account for as much as one third the female's weight and with larger size, the mother has more cavity space for the eggs to gestate ova into embryo and has a larger liver for energy storage, to be allocated to the nutrition and the development of the embryos.

In order to alert a visiting male to the location of the most productive females, the largest sexually-mature females dominate the core of the school and remain central to the females in waiting. Upon encountering the school, a mature male instinctively heads to its centre to engage the larger females. The male selects a female and initiates shark foreplay, by chasing her and biting her flanks, then the two depart from the school and copulate out in open water away from the school. For this strategy to succeed in the propagation of numerous offspring, it is essential that the males do not copulate with adolescent or small females. To this end, the females actively maintain a social structure with an unmistakable hierarchy. The largest, like-sized, fully mature females compete for dominance at the centre of the school among themselves and defend their spatial territory - their positioning - against all the smaller and would-be dominant females in positioning displays. This manifests itself in physical displays of dominance-submissiveness roles as the members of the schools continuously interact. Dr. Klimley identified these behaviour patterns as torso-thrust, head-shake, cork-screw, acceleration and hit.

Male scalloped hammerheads, especially sexually mature males, perform a torso-thrust when challenged for spatial position. Upon receiving a challenge from a dominant female's body language, the male tilts to one side and beats his tail with great emphasis, showing his underside, possibly emphasising his claspers. It is thought to be a way of providing visual communication that he is a male and therefore to some degree exempt from the dominant-subordinate relationship among females. It is also utilised as a male hammerhead joins or departs from the school. Head-shake and acceleration are withdrawal postures displayed by female subordinates vacating their positions. As a subordinate violates the personal space of a dominant female, a conflict occurs between the impulse to repel the attack or to take flight. The dominant's body language asserts its priority and the subordinate swings its head back and forth over an exaggerated arc several times in quick succession, then propels itself upwards and away with several strong beats of its tail or exhibits a rapid burst of swimming with exaggerated head-shaking, terminating in a long glide away from the spatial intrusion. Should the subordinate fail to heed the body language of a dominant and ignore the command to take flight, the dominant female exhibits the cork-screw and hit behaviours. The cork-screw is the dominant's last warning to the subordinate. She swims in a rapid burst of speed while rotating her torso 360 degrees and resuming her direction. The rotation causes sunlight to reflect off her undersides sending a large flash of bright white to the trespasser. If the subordinate does not back off, the dominant dive bombs the subordinate, bringing the underside of her jaw into contact with the intruder's body ahead of the dorsal fin and causes contusions to the subordinate's body. The large number of scalloped hammerhead sharks seen with these fresh scars initially led scientists to believe that they were mating scars until the hit behaviour was witnessed as a normal and regular part of the shark's pecking order.

These contusions provide a site for infestation of ectoparasites, such as amphipods, which feed on the exposed tissue and irritate the hammerheads. The sharks seek relief from these parasitic infestations by utilising the cleaning services of small fish, such as the king angelfish *(Holacanthus passer)* and the barberfish *(Johnrandallia nigrirostris)*, a bright yellow member of the butterflyfish family. Infested hammerheads swim to cleaner stations and linger, barely moving, as small swarms or individual cleaner fish swim out like the mechanics of a Formula One pit crew to perform their services. In the spirit of mutualistic symbiosis, both the cleaner fish and the shark benefit.

The cycle of life in hammerhead schools is only a recently discovered phenomenon, although it may have been on-going for 25 million years, and the places to witness it are few and remote. Tragically, just as insights into their life history are becoming understood, so too comes the realisation that we may be witnessing the countdown to extinction of the species. The worldwide demand for shark fins has set the fishing fleets of every nation on an all-out hunt for these magnificent animals. Their populations are geographically vulnerable and often within the economic zones of Third World countries all too willing to trade pesos for tiburones. While it is a fact that a living shark has a significant dollar value to international scuba diving tourism, the voices of the few operators for whom the hammerheads are the foundation of their business are often drowned out in the outcry from the thousands of impoverished fishermen seeking a commodity to sell at market. At the time of this writing, only Costa Rica has enforced protection of its population of scalloped hammerhead sharks at Cocos Island. There, the park service rangers are limited to what they can protect by proximity to the island. Any fishing boat ten miles offshore is exempt from their jurisdiction. Park, refuge, and reserve are noble ideals and must be encouraged and commended, but they are only applicable to animal life within a proscribed boundary; the sea recognises no boundaries nor do the animals living within. The hammerheads of the Sea of Cortez were the first victims of the fishing-until-there-are-no-more-fish-to-fish attitude prevalent in the mentality of those for whom the ocean is a place of resources to be exploited, not treasured; now the shark populations of the Galapagos Islands are under the severest threat. There are laws banning long-line fishing within the national park boundaries, but sharks do not adhere to delineations on human charts and even the enforcement of these laws has a very low priority for the Ecuadorian government. Unless actions are forthcoming to halt this slaughter of hammerhead sharks - and all sharks, for that matter - within the next decade, school's out forever.

Squatina japonica   **Izu Park, Japan**

About 15 species of these dorsoventrally flattened sharks have been identified to date, a few of them are only poorly known or yet undescribed. Their flattened bodies have a terminal mouth, no anal fin, pectoral fins that are not attached to the sides of the head (like in rays) and no spines in front of the dorsal fins. Although these sharks are not usually found in the coral reef environment, they often occupy soft substrates near them. Family members are benthic, spending most of their day resting, usually buried just under the surface of sand or mud bottoms. If a prey item moves too close to a motionless angel shark, it rapidly lifts its head and throws its jaws forward to snatch its prey. This feeding mechanism is extremely fast, rivalling the speed of bony fish predators that ambush their prey. After the strike, the angel shark settles back down into the sediment and waits for its next victim. A famous high-speed film made in California waters shows an angel shark demonstrating this behaviour: It swallows an approaching small horn shark, only to spit it out again after a second. The horn shark had coiled itself up and let its defensive dorsal fin spines do the rest!

Angel sharks are found singly or (more rarely) in aggregations. They move over the bottom at night and hunt sleeping and dead fish and nocturnally active invertebrates (squid, octopuses, crustaceans). All family members are ovoviviparous (yolksac viviparity), but not much is known about their mating behaviour. One species from the Sea of Cortez, which has yet to be scientifically named and described, has spines on the edges of the pectoral fins like the wing (alar) thorns of male rajid skates. In skates, these thorns help anchor the male to the female during copulation. Angel sharks will bite if provoked or if a diver places a hand or foot near their head. They readily adapt to captivity in larger aquaria.

Squatina tergocellata
Ornate angel shark
Length: Max. to at least 100 cm.
Distribution: Southern Australia: Western Australia (Geraldton) to Southern Australia (Port Lincoln).
Depth: 120-400 m. Apparently most common around a depth of 300 m.
General: A little-known shark of temperate waters. Upper surface yellowish-brown, with three pairs of dark granular ocelli and a dense pattern of bluish-grey to white spots. The ocelli resemble cells in the process of mitotic division when viewed through a microscope; they are unique within the family of angel sharks and thus distinctive. Sides of head at level of eyes slightly concave. Several large thorns above eyes, but no thorns on back. Underside uniformly light in colour. Lives on or near the bottom on continental shelf and upper slope. Frequently caught in small numbers by trawlers in the Great Australian Bight (fishing for flathead *Platycephalus* spp. and morwong *Nemadactylus macropterus*). The flesh is tasty but obviously not marketed as regularly as the Australian angel shark (p.192). Angel shark teeth have a distinctive rhomboid base (root) with one sharp cusp.

Mark Strickland   Andaman Sea

Squatina squatina
Common angel shark

Length: At birth 24-30, max. possibly 244, females mature at 126-167, mature males to 183 cm.
Distribution: Eastern North Atlantic: Southern Norway, Sweden and Shetland Islands to Morocco and West Sahara, Canary Islands, also in most of Mediterranean.
Depth: 2-150+ m.
General: This angel shark lives on or near the bottom in the temperate waters of the continental shelves of Europe and Northwest Africa. It is found on mud, sand or gravel bottoms. During the day, it lies buried with hardly more than its eyes protruding from the substrate. Its pectoral fins are very broad and high and have broadly rounded rear tips. The trunk is very broad. The anterior nasal barbels are simple (not furcated) and have spatulate tips; they are smooth or only weakly fringed. The midline of back and tail is with or without a row of small thorns from head to dorsal fins and between the fin bases; patches of small thorns present on snout and above eyes. The denticles on the sides of the body have very narrow crowns with sharp cusps. This angel shark is grey, yellowish or light brown in coloration, often with a darker mottling of small and very small spots, young have also white lines. There are no ocelli. Litter size is 7-25, with larger females having larger litters. Birth occurs in summer in northern Europe and in winter in the Mediterranean. The nocturnal species swims strongly and preys primarily on bony fishes (hake, argentine, flatfishes), but also on skates, crabs and squid. Makes northward migrations in summer in northern parts of its range. Its fresh/salted and dried flesh is utilised for food.

Both photos Peter Verhoog   **Canary Islands, Eastern Atlantic**

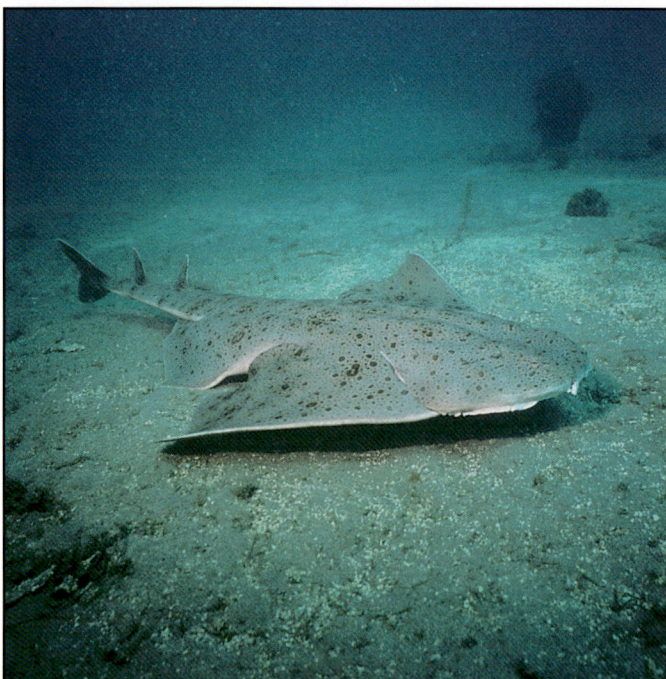

Mark Conlin   California, Eastern Pacific

Squatina californica
California angel shark

L: At birth 23, max. 150 cm. Di: Alaska to Sea of Cortez, Ecuador to southern Chile (possibly a distinct species). De: 3-100 m off California, down to 185 m in the Sea of Cortez. G: A row of small tubercles on back. No ocelli. Grey with brown flecks of varying size. Abundant near California Channel Islands, where numbers decrease during the winter months. Litter size 1-13, gestation period 10 months. Most mature females give birth in March and June. By day usually found buried in sand or mud. Nocturnally active, activity peaks at dusk and midnight. Adults are nomadic, spending days in a limited area of about 1.5 km then move to a new area several km away. Often aggregate. Prey items include bony fishes (croakers, flatfish, corbina, blacksmith) and squid. Specimens in the Sea of Cortez eat hake, croaker, peppered shark *(Galeus piperatus)*, egg cases and mantis shrimp. This ambush predator may bite if provoked. See also family account.

Rudie Kuiter   New South Wales, Australia

Squatina australis
Australian angel shark

Length: At birth 22-27, max. at least 152 cm. Distribution: Southern Australia: Western Australia (Rottnest Island) along south coast to New South Wales (at least to Sydney). Common in Bass Strait and off surf beaches on eastern coast. Depth: Intertidal to 256 m. General: Body broad. Nasal barbels and anterior nasal flaps with fringes. No lobes on side of head. Dorsal tubercles small or absent. Dull grey to greyish-brown dorsally. No ocelli on body, but with dense pattern of numerous irregular small white spots and flecks and with several medium-sized white and dark blotches on fins. Uniformly light below. Litter size up to 20, young are born in fall. Lives inshore on sand and mud bottoms, often associated with sea grass beds and rocks. Preys on benthic fishes, crustaceans and octopuses. Usually buried during the day, active at night. Very good eating, sold as angel shark or flake, mainly in Sydney and Melbourne.

Squatina japonica
Japanese angel shark

Length: Max. about 250 cm.
Distribution: Western North
Pacific: Southeastern Sea of
Japan to Yellow Sea, Japan,
Korea, northern China and the
Philippines.
Depth: Shallow waters. General: Pectoral fins broad with
rounded rear tips. Nasal barbels simple and spatulate.
Anterior nasal flaps weakly
fringed or smooth. Dermal
flaps on side of the head without lobes. A row of moderately large thorns runs along the
midline of back and tail. Upper
surface blackish-brown with
small dark and pale spots,
without ocelli. A little-known
shark of temperate waters, living on or near the bottom. Litter size about 10. Usually
inhabits sandy bottoms in shallow water, sometimes also
seen in rocky areas. Probably
preys on bony fishes and squid
as its congeners. Fished in
large numbers off China.
Utilised for human consumption and for production of shagreen (shark skin leather, the
preserved, dried skin still containing the denticles; can be
used like sandpaper) which is
used to polish and finish
woodworks. See also photo on
p.189.
　　The similar clouded angel
shark (S. nebulosa) has bilobed
dermal flaps on the side of its
head and the distance between
the spiracles is shorter than
the distance between the eyes.
Upper surface of body bluish-
brown with black and white
dots. The little-known species
is distributed from the Sea of
Japan to Taiwan and attains a
max. length of 160 cm.

All photos Kazu Masubuchi   Izu Park, Japan

This family comprises two genera with at least seven species, two of which (from Australia) have not been scientifically described yet. The genus *Pliotrema* is monospecific, the species *P. warreni* (western Indian Ocean) differs from all other sawsharks of the genus *Pristiophorus* by having six pairs of gill slits instead of five. Like in all sharks, the gills are located on the sides of the body behind the head, while the superficially similar sawfishes (family Pristidae, p.200) are true rays with gills on their underside. Sawsharks are unique in having a head with a long blade-like snout (rostrum) armed with large and small lateral 'rostral teeth'; the former are enlarged embryonic denticles, the latter are formed after birth. Contrary to the permanent rostral teeth of sawfishes (see below), those of sawsharks are shed and replaced like jaw teeth. Additionally, the sawshark rostrum bears a pair of sensory barbels absent in sawfishes. All family members are found in the Indo-Pacific with one exception from the Bahamas region *(P. schroederi)*. All are much smaller (max. length 140 cm) than sawfishes, ovoviviparous and considered harmless, but should be handled with care because their rostral teeth are very sharp. Fossil finds indicate a much wider distribution of this group in the past.

Rudie Kuiter   South Australia

Pristiophorus cirratus
Common sawshark
L: At birth about 35, max. at least 134, males mature at 97 cm. Di: Endemic in temperate and subtropical waters of Australia from Western Australia to New South Wales and Tasmania. De: 40-310 m.
G: Benthic on the continental shelf. Upper surface brownish with darker bands. Lighter areas with irregularly distributed large, dark brown spots and blotches. Rostral teeth with dark margins. Underside abruptly uniformly white. The largest and most widely distributed Australian sawshark. Occurs sympatrically with the uniformly grey southern sawshark *P. nudipinnis*, but lives deeper. Flesh good to eat, trawler bycatch.

Saul Gonor   New South Wales, Australia

Pristiophorus sp.
Eastern sawshark
Length: Females to 107 cm. Distribution: Presumably endemic to warm temperate waters of eastern Australia: Victoria (Lakes Entrance) to New South Wales (Coffs Harbour). Depth: 100-630 m. General: This greyish, little-known sawshark has not been scientifically described and named yet *(Pristiophorus* sp. A in Last & Stevens, 1994). It appears to be confined to the continental shelf and upper slope off New South Wales. Off northern Queensland there is another undescribed, similar species *(P.* sp. B, ibid.) that is smaller, yellowish-brown above, white below, and has a longer and narrower snout. Sawsharks eat small fish.

DANCE OF THE STINGRAYS

A truly amazing phenomenon takes place every year at the Poor Knights Marine Reserve off New Zealand's north east coast. Throughout the summer months hundreds of stingrays descend upon the islands and congregate in some of the many archways around the reserve. Together they perform an exquisite underwater ballet, soaring and swooping along the walls as if taking part in some ancient ritual. The reason for this summer pilgrimage has long been a mystery. Underwater photographer Glenn Edney tries to reveal some of it.

Every year in summer numerous short-tail stingrays *Dasyatis brevicaudata* congregate at the Poor Knights Marine Reserve.

Slipping below the surface I can see kelp-covered boulders on the bottom 20 metres below. Clouds of plankton feeding demoiselles swarm around me as I descend along the wall towards the cavernous opening of Tie-Dye Arch. At the entrance I stop a moment to allow my eyes to adjust to the lower light levels in the arch and to re-check my camera settings. I am hoping to photograph the annual stingray gathering at New Zealand's Poor Knights Marine Reserve.

Entering the archway I discover just how it got its name. Like an oversized hippie T-shirt the walls are a riot of colour as sponges, corals, anemones and other filter feeders take advantage of the lower light levels and the constant supply of plankton flowing through the archway. In the distance I can see large schools of fish and as I drift closer I can discern several species. Vibrant pink maomao hover in the recesses along the wall and emerge as a group to devour the rich plankton soup. A school of 30 large porae, their long pectoral fins flowing, congregates in midwater. Smaller groups of leather jackets, golden snapper and mado create a colourful contrast around the porae while the ever present sandagers wrasse dart forward to investigate this new intruder in their territory.

Remembering the stingrays I descend to the bottom of the arch and settle amongst the boulders, camera at the ready. I don't have to wait long. Almost immediately a large female ray glides past within touching distance. Suddenly she banks to the right and heads straight towards me. I have time for just one shot as she passes directly overhead, her brilliant white underside illuminated by the flash from my strobe. I can clearly see the inverted grin of her mouth with its grinding plates for crushing the shells of molluscs that make up the bulk of her diet. Behind and to each side of the mouth, two sets of five gill slits flutter as water is expelled.

Even allowing for the water's magnification I estimate her 'wing span' to be close to two metres, about a third larger than a fully grown male. The length of her tail identifies her as a short-tailed ray. The long-tailed species has a whip-like tail that can be up to twice the length of the body.

Both species have stinging barbs that can inflict potentially lethal wounds. As her tail passes just centimetres above my head I recall an episode where a diver accidentally cornered a large stingray while it was resting in one of the shallow bays in the reserve. The frightened ray lashed out as it tried to escape causing a severe wound to the diver's arm. The common sense rule of never cornering a wild animal would have prevented much anguish for both diver and stingray.

Looking out towards the entrance I can see several rays

Male stingrays glide through the water among the larger females, yet without contacting them.

silhouetted against the sun. They seem to be just hovering as if waiting for a signal before they descend once more into the relative gloom of the arch. More by good luck than design I seem to have placed myself directly under their flight path. After a few moments a pattern emerges: Most of the rays that are passing close to me are large females and I can see by their bulging abdomens that some of them are heavily pregnant. Once past my vantage point they head straight for the other end of the arch where they perform a steep vertical turn against the wall then glide effortlessly back to their starting point. The smaller males mill about close to the surface but every now and then one or two will descend behind a female and follow along her flight path. Could this be part of an elaborate stingray mating dance?

Veteran diver, writer and underwater naturalist Wade Doak has studied the marine life around the Poor Knights for thirty years. He has speculated that mating is the driving force behind the mass gatherings. Local film maker Mike Bhana shares his view. Mike has been observing and filming the stingrays at the Poor Knights for the past two summers and believes the rays are congregating in the archways to take part in a synchronised mating ritual that may be essential to ensure the viability of each new generation.

Stingrays give birth to live young after a gestation period of approximately eleven months and each female is capable of producing up to 25 offspring a year. Measuring 50 cm in length the young rays are perfect miniatures of their parents, right down to the tiny barbs on their tails. Each newborn pup is ejected in midwater with wings folded like a butterfly

This long-tailed *Dasyatis thetidis* has raised its tail with sting in a defensive posture.

Most rays propel themselves forward with graceful waving movements of their pectoral fins.

emerging from a chrysalis and just like the butterfly it takes the tiny ray a little while to stretch its wings and emulate the graceful beauty of its mother as she glides overhead.

In other parts of the reserve stingrays can be seen resting on the sand or hidden amongst the dense kelp forests. Descending towards the sandy shallows of Labrid Channel in the reserve's South Harbour I can see an enormous stingray below me, her long tail stretching out three metres behind her. Lying flat on the sand I inch my way towards the resting giant. As I get closer she slowly raises her tail and her wing tips curl up ready to provide upward thrust if she needs to escape. She is obviously becoming uncomfortable with my approach so taking the hint I stop and lie quietly on the sand, my breathing slow and steady, hoping she will realise that I am no threat to her. After a few moments she seems to accept my presence and settles her wing tips back into the sand although her tail still remains raised.

I am so close now that I could easily reach out and touch her. She watches me through eyes that were once thought to provide very poor vision, but are in fact superbly adapted for the lower light levels and lack of colour underwater. Just behind the eyes her spiracle opens and closes as she draws oxygenated water past her gills. I adjust my camera and compose the shot, all the while wondering what might happen when the flash from my strobe fires. My over-active imagination conjures up images of that three-metre-long tail whipping around to knock the strobe away and taking half of me with it. Stealing myself I depress the shutter and involuntarily shut my eyes against the certain destruction my mind has conjured up. When the expected blow fails to arrive I open my eyes to see that she hasn't moved at all. Feeling a little foolish I recompose and take several more shots, still with no reaction from the ray. Perhaps her years of visiting the marine reserve have taught her not to fear divers. Reaching out my hand I gently stroke the leading edge of her wing. Her skin has the feel of fine velvet and the way she shivers to my touch gives me the impression she is enjoying the contact. After a few moments I slowly back away and start my ascent, still buzzing from the

Oxygen-rich water is taken in through the spiracle, an opening located right behind the eye.

197

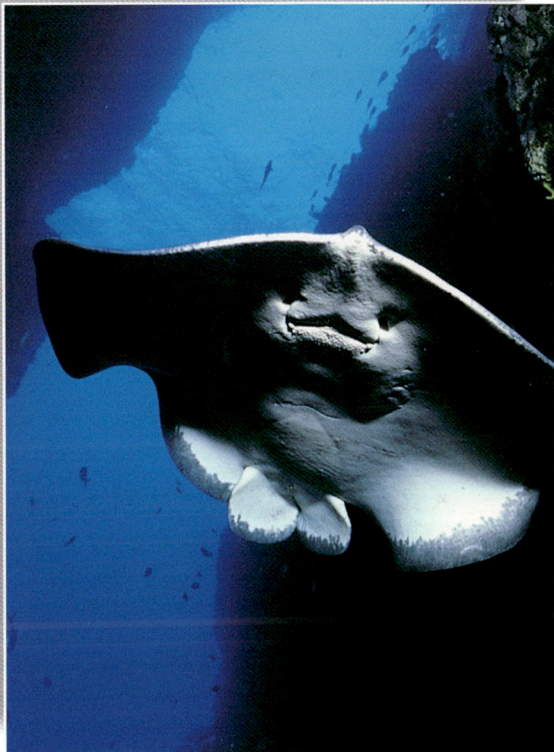

The brilliant white underside of a female reflects the flash from the strobe.

encounter and looking forward to our next destination, Northern Arch.

It's hard to imagine a more dramatic landscape than this craggy finger of sun-bleached, volcanic rock protruding from the northernmost part of the islands. The roof of the arch is only five metres above sea level but the sheer walls plunge forty metres to the boulder-strewn floor below. Easterly winds have pushed deep ocean water into the islands and visibility in the archway is forty metres. Dropping down to thirty metres I am completely awe-struck at the vista above me. Silhouetted against the surface at least forty stingrays crowd into the narrow gut. Surrounding the stingrays huge schools of maomao and demoiselles form giant moving clouds. Large snapper, trevally and kingfish cruise amongst them but my attention is focused on the stingrays.

I hope to be lucky enough to observe some mating behaviour. A few days before, Mike Bhana and his team witnessed a stingray courtship while finishing filming for their Discovery Channel documentary. They ran out of bottom time before the two lovers consummated their affair but were left in little doubt as to the eventual outcome. Once a female has given birth she is almost immediately ready to mate again. Several males will follow her until one and sometimes two are able to get a firm grip on the trailing edges of her 'wings' with their mouths. Thus attached they will hitch a ride until the female finally succumbs. Maintaining his grip on her wings the successful male contorts himself so that he is now swimming upside down and in a perfect position to insert one of the two claspers at the base of his pelvic fins into her genital opening. Several of the females cruising past me have fresh mating scars on their 'wings' confirming the theory that these mass gatherings are, at least in part, a mating ritual.

Making my way back to the boat I reflect on my experiences with the stingrays of the Poor Knights. It is always satisfying to solve some of the mysteries of the ocean. That the stingrays are here to mate now seems beyond doubt. But why these islands and in particular these two archways? Perhaps such prominent 'land marks' provide an easily recognisable meeting place. Maybe the continuous flow of water through the archways helps them conserve energy during their midwater mating dance. Whatever the reasons, I am thankful that these protected islands have given me the opportunity to observe at close hand one of the ocean's most graceful creatures. Above all, I feel incredibly privileged to have witnessed the Dance of the Stingrays.

Suddenly a large female passes directly overhead! One can clearly see the inverted grin of her mouth. Behind the mouth, five pairs of gill slits flutter as water is expelled.

This unusual family of rays comprises two genera and about six species. All are characterised by very much elongated, flattened rostra (saws) with large teeth (see below) on both lateral edges. Rostral barbels (as present in the superficially similar sawsharks of the family Pristiophoridae, p.194) are lacking. In sawfishes, the gill slits are located on the ventral side of the body as in all rays and skates. They have two dorsal fins which are about equal in shape and size, the origin of the first is located approximately over the origin of the pelvic fins. The lower lobe of the caudal fin is not well developed.

The genus *Anoxypristis* is monotypic. *A. cuspidata,* distributed from the Red Sea to Japan, New Guinea and northern Australia, is characterised by the longest and narrowest rostrum of all family members. It bears 18-22 pairs of lateral teeth which are unevenly spaced along the rostrum (more widely spaced towards the head). Furthermore, the rostral teeth do not reach the front of the head as in other sawfishes, all of which are placed in the genus *Pristis.*

Sawfishes are ovoviviparous, their mode of reproduction is yolk sac viviparity. Some species aggregate in river mouths to give birth communally there. Why are pregnant sawfish females not injured during the birth of their young? Nature has found a way by covering the rostral teeth of full-term embryos with a membraneous sheath. Additionally, the saw itself is soft and flexible and can bend far backward. Both features prevent the rostrum from injuring the mother while giving birth. The saw cartilages harden soon after birth. The dorsoventrally flattened saw teeth are modified dermal denticles, grow throughout a sawfish's life, but are not shed like the teeth in the jaws; if one is lost, it is not replaced (other than in sawsharks, which replace lost rostral teeth, see p.194). The tooth-bearing rostrum of sawfishes resembles a double-sided comb, the rostral teeth are almost uniform in length, while those of sawsharks are not, what makes their saws look jagged. Sawfishes do well in large public aquaria. Their dried heads and saws are frequently depicted on old paintings or etchings of naturalia collections and may be also found in curio shops today.

Reliable data on habits and biology of all family members are scarce for several reasons. Most important, the widely distributed group is badly in need of revision, misidentifications in the past have led to a data mix of several species. Also these generally large and heavy animals are not easily handled and investigated. Last, but not least, their habitats (e.g. on the coasts of West Africa) are not easily accessible together with the necessary equipment for studying large elasmobranchs. Most family members readily enter brackish environments or freshwater and at least one (*P. microdon*) may spend its entire life cycle in freshwater.

There are reports of sawfishes slashing humans and damaging tankmates with their saws. A large specimen was reported to inflict blows with the rostrum to any shark in the aquarium that ventured too close; guitarfish and whiptail stingrays on the other hand were ignored. Specimens have been observed to approach sharks resting on the tank bottom and smack them with their saws. On record are sharks in the wild with blow marks or scars from sawfish rostra. Also the death of a dugong is attributed to a sawfish wound and old accounts report of fights between crocodiles and sawfish in estuarine habitats. They are not known to be aggressive towards divers, but encounters are very rare anyway. There is one report on a bannerfish (*Heniochus acuminatus*) that actively removed trematode flatworm parasites from a captive sawfish.

With reported total lengths of significantly more than 700 cm, most sawfish species rank among the largest of all living elasmobranchs. Unfortunately, they have restricted habitats, are extremely vulnerable to all sorts of fishing gear (flesh utilised for human consumption) and habitat degradation and require protection.

Pristis pectinata
Smalltooth sawfish

Length: At birth 60, usually about 300, max. 760 cm. The saw makes up about a quarter of the total length.
Distribution: Tropical and sub-tropical Atlantic. E-Atlantic: Morocco to South Africa. W-Atlantic: New England to Brazil. Records from Mediter-ranean, Indian and Pacific Oceans are uncertain.
Depth: 1-15 m.
General: Pectoral fin bases broad with bluntly rounded tips. First dorsal fin originates over pelvic fins. Saw narrow, each side with 21-34 pairs of laterally protruding, small ros-tral teeth. Most common in estuaries, bays and lagoons, also in rivers and lakes. Seen in the shallows with fins breaking the water surface. Seems to be adapted to water temperatu-res of 16-30°C in the western Atlantic.

Coloration uniformly dark grey or brown on back and sides, withish below. Litter size 15-20. Gives birth in shallow bays and estuaries. Feeds on schooling bony fishes by swim-ming into a school, violently shaking head and saw in a quick succession of rapid blows from side to side and thus stunning prey fish, which are subsequently retrieved and eaten. Saw is also used to dig up crabs and mussels from the sediment.

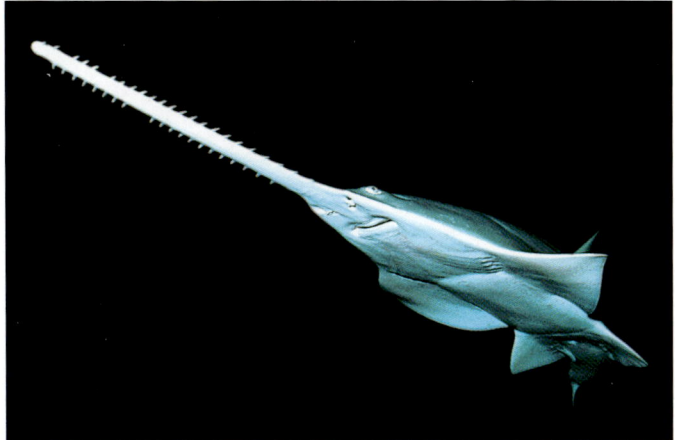

Norbert Wu   Ivory Coast, West Africa

Doug Perrine   Caribbean

Pristis clavata Dwarf sawfish
L: Max. 140+ cm. Di: Tropical Australia: Queensland (Cairns) to Western Australia (Kimber-ley). De: 1-10 m. G: Small, saw with 18-22 pairs of rostral teeth. Coastal, in estuaries, rivers, not in pure freshwater. Below: Similar **Freshwater sawfish** P. microdon, to 2+ m, rivers of E-India to N-Australia.

Rudie Kuiter   Queensland, Australia

SAWBLADES AND MILLSTONES

Imagine yourself sitting in front of your kitchen table. On the table there are all sorts of food you just bought and intend to process into a delicious meal. There are soft tomatoes you want to transform into a juicy salad, there is some meat for the goulash and finally, you just could not resist purchasing this really old Gouda cheese for desert. In order to chop all these different sorts of food to a size convenient for eating, you will certainly not use any knife, but select a special one tailored for each task. Tomatoes may best be cut with a finely serrated blade, meat with a very sharp one and the solid Gouda must be crushed or ground. Sharks or rays basically have the same problem, not in front of a kitchen table though, but each time they face a prey. How do the diverse species of elasmobranchs grab, hold and chop their prey? The author peers deep into their mouths and shares his 'dental impressions'.

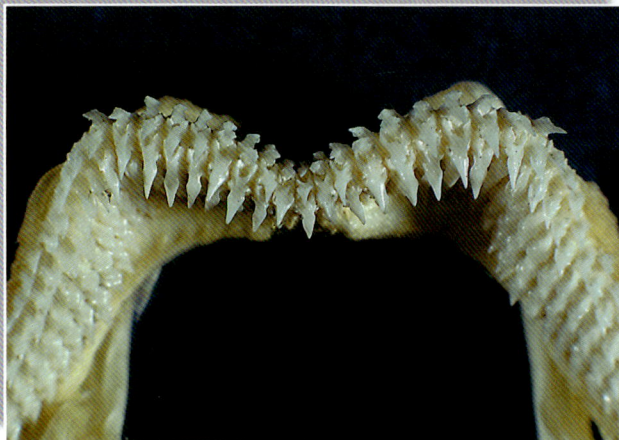

Portuguese dogfish *Centroscymnus coelolepis,* subadult female, TL (total length) 88 cm, NE-Atlantic. Occlusal view of upper jaw. Revolver dentition with 4 to 5 transverse tooth rows in 'working position'. Note distinctly arrow-shaped crowns. Largest tooth 4 mm. All jaws shown have been collected and prepared by the author and are more or less deformed after drying.

Teeth appear about 80 million years after the advent of the first (jawless) fishes in the fossil record of vertebrates, in late Silurian times about 420 million years ago. Among the first fish-like animals with this novelty in their jaws are the oldest known sharks. The two main parts of a typical shark tooth are the **crown** which is exposed in life and covered by a very hard, shiny, enamel-like layer (enameloid), and the bony **root** that is firmly anchored in connective tissue (Sharpey's fibres) surrounding the jaws. The tooth's interior is hollow and - to keep this special organ alive - filled with blood vessels and nerves which enter the hollow (pulp cavity) via one or several holes and canals in the root.

But where did jaw teeth come from? They didn't just grow like that in the mouth of early fishes, but are thought to have been derived from an already existing part of early vertebrate anatomy. The skin of most sharks and rays resembles the chain armour of a medieval knight: It is covered with zillions of minute so-called dermal denticles. Each denticle is - depending on species - from only fractions of a millimetre (most species) up to 15 millimetres (bramble sharks) long and built exactly like a jaw tooth with crown, root and pulp cavity. By movement of enlarged and specially shaped skin denticles from the margins of the mouth opening onto the jaws, 'real' teeth might have evolved during the course of evolution. The dermal denticles of a shark or ray species usually look very different from its jaw teeth and may be much more variable in shape, the latter varying with the position of the denticles on body or fins. Gener-

ally speaking, one could say that also the teeth of the most highly evolved vertebrates - mammals, including man - have originally been derived from the skin denticles of early shark-like ancestors. Think about that!

Jaw teeth and denticles are produced by the thousands throughout the lifetime of an individual ray or shark because they are regularly shed, no matter if they are worn or

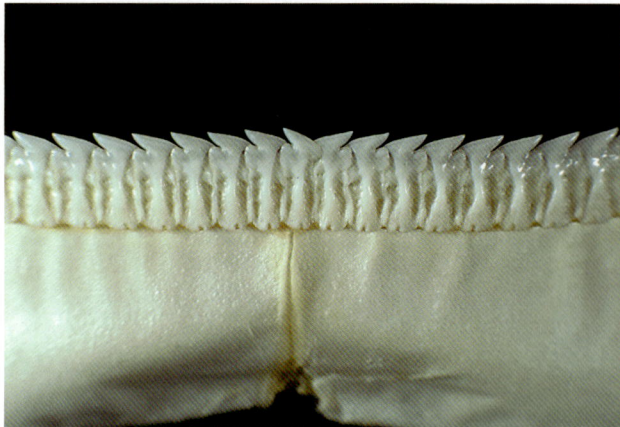

Portuguese dogfish *Centroscymnus coelolepis,* same specimen as above. Labial view of lower jaw with only one tooth row in 'working position'. Note long roots (dull bone) and short crowns (shiny white enameloid) with single, strongly oblique cusps. Largest tooth 6 mm.

not. It is this ancient model of continuously replacing hard parts by new ones that makes elasmobranchs laugh at the concept of the dental profession. The proverbial **revolver dentition** of shark jaws consists of one or more **transverse tooth rows** in 'working position' and usually several **replacement rows** lying flat behind the active rows. The replacement teeth are covered by a fold of mucous skin with the youngest, yet incomplete teeth deepest in the fold (dental lamina). A **longitudinal tooth file** hence starts with the tooth germ deep in the skin fold and ends with the frontmost active tooth which soon will fall out. All teeth are sunk with their roots into a layer of connective tissue that very slowly moves

Scaly gulper shark *Centrophorus squamosus*, female, TL 129 cm, NE-Atlantic. Labial view of lower jaw with interlocked teeth. The left half of the oldest tooth row has already been shed, while the right half is still adhering to the jaw.

- like a conveyor belt - forward towards the upper rim of the jaw as long as the animal is alive and healthy. Whenever an active tooth is lost while being used (ripped out by shearing forces) or simply falls out because its time has come, the next one to follow behind in the same file is ready to take its position. The file continuously moves on and another replacement tooth starts to erect. A dream for everyone who hates to see the dentist! Experiments with lemon sharks in captivity and conclusions drawn from indirect evidence found in free-ranging white sharks have shown tooth replacement rates to range from about one week (lemon) to 110 days in juvenile and 235 days in adult white sharks. This means, a captive lemon shark loses the front tooth of each file after approximately one week, an adult white after about 8 months. Whether these data are significantly influenced by the captive condition of the lemon sharks investigated, is not known.

Some sharks, especially the squaloids (spurdogs, gulpers, lantern sharks, birdbeak sharks and the like), shed their teeth not singly but as an entire transverse row at a time. This is possible because the teeth of one such row are overlapping and interlocking at the sides of their roots with each of their neighbours, much like the shingles on a roof. When its time comes, an entire row moves down the front margin of the jaw, while the next generation already stands upright and in working position. The oldest tooth row typically splits in the middle of the jaw and both halves are shed whenever mechanical forces tear them off.

The minerals involved in building-up the teeth, mainly calcium phosphate and calcium fluorite, are taken up with food. As they are completely lost whenever teeth are shed, the elasmobranch way to avoid encounters with doctors of a certain kind simply is a waste of resources. But the principle itself is obviously unbeatable, as it has been established hundreds of millions of years ago and remains successful ever since. In environments where the necessary minerals are scant, sharks came up with a simple, but effective recycling method. In the deep sea, where food is usually scarce, they swallow shed teeth instead of letting them fall out of their mouth! An individual may not catch all of them, but this unusual diet helps relieve the pressure of acquiring new 'bony' food. It is interesting to note that the mentioned squaloid pattern of shedding entire, still partly connected transverse rows, neatly coincides with the need of deep sea organisms to recycle minerals as effectively as possible. Yet another example was discovered only a few years ago: Some of the first unborn white sharks ever to be examined had their own teeth in their stomachs. As most lamnoid sharks (e.g. sand tigers, threshers, makos, mackerel shark, salmon shark) heavily prey on their siblings and undeveloped eggs

Centroscymnus coelolepis, same specimen as shown on the left. Lingual view of lower jaw with one tooth row in 'working position' and 7 replacement rows. Note interlocking roots and successive development of teeth from bottom to top.

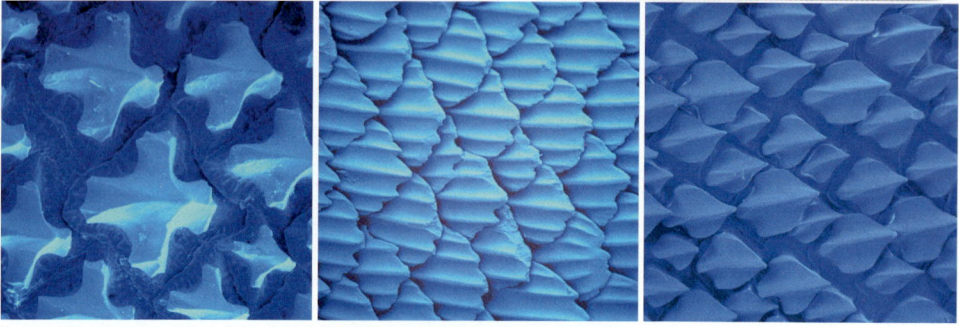

Dermal (skin) denticles of (from left to right) Port Jackson shark *Heterodontus portusjacksoni*, a requiem shark *Carcharhinus* sp. and tiger shark *Galeocerdo cuvier*. Scanning electron microscopy (SEM) photographs. Largest denticle about 0.3 mm.

in the mother's uteri, these three to four tooth sets have already been used and recycled before the user has even been born!

Skin denticles are not 'moving forward' like the jaw teeth are, they are simply replaced by successors that emerge in the skin from below. The jaw teeth of eagle and cownose rays, who mainly prey on hard-shelled molluscs, are interlocking to form tooth plates that wear in the front and are replaced from behind.

While ordinary people only own two sets of 20 and 32 teeth respectively during their life, the production of teeth - and denticles - of an individual elasmobranch may reach hundreds of thousands or even more. A set of teeth of the white shark comprises about 50 teeth at a time. Using the replacement rate cited above, a 10-year-old white shark thus would have produced about 15 x 50 = 750 teeth in its life so far.

The convenience of a never-ending supply of teeth demands an adequate uptake of food containing sufficient minerals. The main function of jaw teeth is crystal-clear, they are tailored for the purpose to grab, hold and eventually partition animal prey. But prey comes in many sizes and shapes, and so do elasmobranch teeth. One glimpse at a shark's maw can tell us a lot about the principal diet of its owner. Despite the great diversity of their crowns' shape, elasmobranch teeth can be grouped into six basic types of dentition corresponding to the food preferences of the species. The classification presented here generally follows that of Henri Capetta (Montpellier, France),

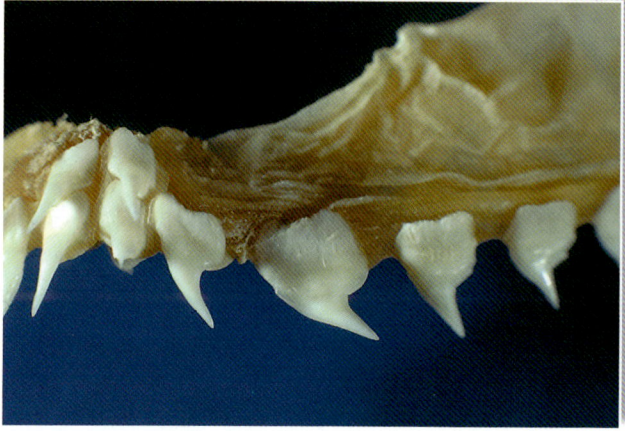

Sixgill shark *Hexanchus griseus*, juvenile female, TL 137 cm, NE-Atlantic. Labial view of upper jaw symphysis with slender, unicuspid parasymphyseals and anteriors which laterally become broader and multicuspid, an example of monognathic heterodonty. Largest tooth 9 mm.

a renowned specialist for fossil elasmobranchs of mesozoic and neogene age.

Unfortunately, for a better understanding of the elasmobranch dentition a few **technical terms** have to be defined first. Each cartilaginous jaw, upper and lower, consists of two more or less curved halves. The central part, where the

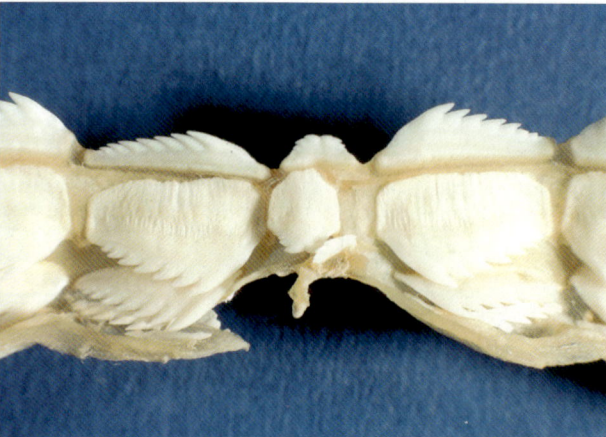

Sixgill shark *Hexanchus griseus*, same specimen as above. Lingual view of lower jaw symphysis with broad, multicuspid anteriors and small symmetrical symphyseals. Note replacement tooth rows, folded back on inner face of jaw cartilage, visible after preparation. Note different shape of teeth when compared to those of the upper jaw shown above, an example of dignathic heterodonty. Largest tooth 14 mm.

front ends of each half meet in the middle, is called **symphysis**. There, many elasmobranch dentitions have a file of symmetrical teeth which consequently are termed **symphyseals**. In some species there also are **parasymphyseals**, a few files of distinct asymmetrical teeth immediately to the right and left of the symphysis. The tooth rows in the front of the jaws are termed **anteriors**, those at the sides of the jaws are termed **laterals**. In several species (e.g. some lamnoid sharks) we find distinct **intermediates** between anterior and lateral tooth files. The often small teeth in the back of each jaw are termed **posteriors**. All the different tooth types defined above may occur in one dentition, e.g. that of the sixgill shark; such a dentition

1) Clutching type dentition. Swell shark *Cephaloscyllium ventriosum*, California. All teeth of both jaws are similar (dignathic homodonty). Largest tooth about 2 mm.

HOWARD HALL

is called **heterodont**. But there also are dentitions with only one tooth type, e.g. that of the spiny dogfish; such a dentition is called **homodont**, all teeth of a set of jaws are more or less of the same shape, their size not regarded. **Monognathic** refers to only a single, upper <u>or</u> lower jaw. **Dignathic** refers to both, upper <u>and</u> lower jaws. If a dentition has different tooth types in one, upper or lower, jaw, this condition is called monognathic heterodonty (seen in sand tiger, sixgill and sevengill sharks for example). If there is only one tooth type found in the upper jaw and a second type in the lower jaw, this dentition is dignathic heterodont, but likewise monognathic homodont (most requiem sharks, many dogfish sharks). **Monocuspid** teeth have a crown with one cusp only (white shark, makos). **Multicuspid** teeth have a crown with more than one cusp (frill shark, many catsharks). **Lateral cusplets** are small cusps to either side of the main cusp (mackerel shark, sand tigers). If the lateral cusplets are large when compared to the main central cusp, there may be a continuous gradient from monocuspid teeth with lateral cusplets to multicuspid teeth even in one jaw (nurse sharks). **Lingual** (lingua, Latin for tongue) refers to the inner, 'tongueward' face, **labial** (labia, Latin for lip) to the outer, 'lipward' face of a tooth. After such boring business, let's plunge into the kitchen tool box of elasmobranch evolution.

1) Clutching-type dentition. The tooth shape in this type generally varies little, hence it is a homodont dentition. The teeth are small, with more or less numerous lateral cusplets which facilitate holding prey. The teeth are set in several functional rows and are often folded on their lingual and labial faces to reinforce the tooth walls. This type is often found in benthic sharks such as catsharks, nurse and angel sharks; in the latter, the teeth do not have lateral cusplets. Also, the dentitions of male skates and many stingrays belong in this category (see below). Such a dentition is ideal for grabbing and holding a wide variety of smaller prey from slippery fishes to shelled crustaceans and molluscs which can be swallowed whole.

2) Tearing-type dentition. This type is characterised by teeth with narrow cusps on the anterior files, but enlarged cusps on more lateral files. On the anterior part of the jaws, there are many functional rows, e.g. up to four in mako sharks or even more in goblin sharks. Generally, the teeth have very distinct cutting edges and often one to several pairs of small lateral cusplets. This type of dentition first appeared in Triassic times and many living lamnoid sharks (makos, mackerel shark, sand tigers) are placed in this group. Such a dentition is ideal for grabbing and holding slippery prey - mainly fishes - which are swallowed whole.

2) Tearing-type dentition. Shortfin mako *Isurus oxyrinchus*, subadult male, TL 185 cm, Red Sea. Complete lower jaw with long, slender anterior and broad, triangular lateral teeth. Note replacement rows, exposed after preparation, toothless symphysis and small posterior teeth. Largest tooth 26 mm.

3a) Cutting-type dentition. Tope *Galeorhinus galeus,* adult female, TL 130 cm, Helgoland, North Sea.
Left: Part of right upper jaw with anterior and lateral teeth. Symphysis is to the right. Largest tooth 7 mm.
Right: Symphyseal region of lower jaw, note small symmetrical teeth of single symphyseal file. Largest tooth 6 mm. Note general similarity of all teeth (dignathic homodonty) and simple serrations on cutting edges.

3) Cutting-type dentition. Two sub-types (cutting and cutting-clutching) have to be recognised here. The cutting sub-type is characterised by a pronounced monognathic and dignathic homodonty, the teeth are flat and stand in one functional row forming an almost continuous sharp blade. The white shark and many requiem sharks are good examples for this type. While the teeth of the white shark are very wide and have a straight, high and thin crown, those of the tiger shark have cusps slanted much toward the rear. The spiny dogfish can also be placed in this group. The efficiency of this type is considerably increased by acquisition of serrated cutting edges; these occur in nearly all forms of this category. The serration may be simple (white shark) or double (tiger shark), the large main serrations being themselves serrated. Such a dentition is ideal for grabbing large prey like fishes and other marine vertebrates which are cut to pieces and swallowed.

The cutting-clutching sub-type is characterised by dignathic heterodonty. In one jaw, the teeth become wider and flatter towards the corner of the mouth, in the other they retain a high and rather narrow cusp. This sub-type is found in different families. In requiem sharks the teeth with wide, flat crowns are those of the upper jaw, in six- and sevengill sharks and many dogfish sharks the wide teeth are found in the lower jaw. The cutting-clutching sub-type is an improvement over the cutting sub-type. The pointed teeth of one jaw enable the predator to hold its prey, while the flat, often serrated teeth of the other jaw act as a blade to cut the prey to pieces. Also the squaloid cookiecutter sharks belong in this category; the upper jaw clutching dentition fixes the small shark to the skin of its victim and then the lower jaw teeth cut nice round plugs of meat while the predator rotates around its longitudinal axis.

4) Crushing-type dentition. This type is found mainly in species living on or near the seafloor. The number of tooth files and also the number of tooth rows is great. Together the bulging crowns do not produce an even, but rather an embossed surface. This type of dentition typically processes prey like shelled molluscs, small fish and cephalopods. It is found in skates, guitarfishes, dasyatoid stingrays and the smoothhound sharks. The dentition of the latter shows a striking similarity to the dentition of guitarfishes which are true rays. Such development of the same anatomical features in basically non-related groups is called convergence.

3b) Cutting-clutching type dentition. Bignose shark *Carcharhinus altimus,* subadult female, TL 154 cm, northern Red Sea.
Left: Part of right upper jaw with anterior and lateral teeth. Note simple serrations on cutting edges. All upper teeth are generally similar to each other (monognathic homodonty) and form a cutting blade. Symphysis is to the right. Largest tooth 15 mm.
Right: Symphyseal region of lower jaw, note tiny symphyseal tooth. Largest tooth 10 mm. Note different shape of the more slender, erect, smooth-edged teeth (clutching dentition) when compared to those of the upper jaw (dignathic heterodonty).

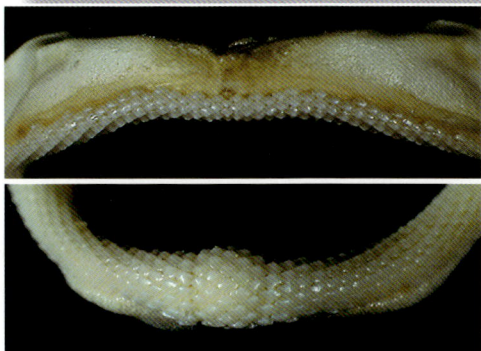

4) Crushing-type dentition. Left: Moses shark *Mustelus mosis,* adult female, TL 74 cm, Red Sea. Largest tooth 1 mm.
Right: Smooth skate *Raja batis,* female, about 50 cm disc width, North Sea. Upper and lower jaws with multiple tooth rows and files which together form an embossed surface, ideal for crushing small mollusc and crustacean prey. Largest tooth 2 mm. Note general similarity of both dentitions, a fine example of convergent development in different animal groups during evolution.

5) Grinding-type dentition. This type is found in species like eagle and cownose rays which forage on the seafloor. It is designed to handle hard invertebrate prey with heavy, resistant shells. The teeth have a high crown with a polygonal outline and form a dental plate with a nearly plane surface in each jaw. Forms appearing early in the fossil record have numerous tooth files, advanced forms less, the extant eagle ray has only one tooth file, it is the most evolved form of the grinding type dentition. The dental plates of these rays are - like millstones - able to grind down even thick mollusc shells. However, the enameloid covering such teeth is often very thin and rapidly lost by wear. This type of dentition already existed among some hybodont sharks in mesozoic (dinosaur) times.

5) Grinding-type dentition. Fragmentary fossil dental plate of *Myliobatis* sp., North Sea. Miocene. Largest tooth about 15 mm.

The teeth of elasmobranch fishes are not only ideal tools for acquiring and preparing meals, but are also used during activities related to reproduction. A special purpose of the jaw teeth of male elasmobranchs is to ascertain a firm grip on a female's pectoral fin during courtship and copulation. The male needs to fix himself to one part of

6) Clutching-grinding-type dentition. This mixed type of dentition is found exclusively in members of the horn shark family. The anterior teeth have a main cusp with lateral cusplets and are thus of the clutching type, the lateral teeth are flat and wide and of the grinding type. In young individuals, all teeth are of the clutching type. The lateral grinding teeth gradually develop during growth of the individual toward adulthood along with strong jaws and muscles capable of attacking tough prey. The teeth, especially the flat lateral ones, have a thick coat of enameloid. Such a dentition is ideal for processing a wide variety of shelled invertebrate prey.

6) Clutching-grinding-type dentition. Horn shark *Heterodontus francisci,* Sea of Cortez. Anterior teeth with main cusp and lateral cusplets; posterior grinding teeth not visible. Largest tooth about 3 mm.

HELMUT DEBELIUS

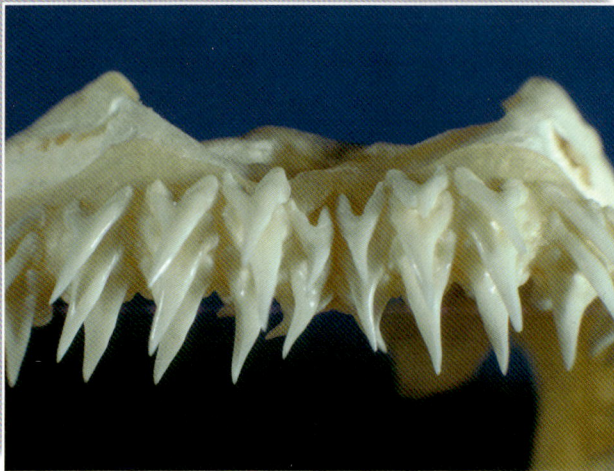

Kitefin shark *Dalatias licha,* adult female, TL 147 cm, NE-Atlantic. Labial view of upper jaw symphyseal region. Note parasymphyseal teeth. Largest tooth 7 mm.

the female in order to be able to insert one of his claspers into the female's cloaca. Because the teeth of many shark species are sharp and pointed, they naturally cause deep cut wounds on the female's hide. To compensate for this crude treatment, the skin of female elasmobranchs (e.g. blue shark) is several times thicker than that of their mating partners.

An interesting phenomenon is seen in species with flat-crowned, crushing-type teeth, namely skates, many stingrays and also smoothhound sharks: While the females have blunt tooth crowns, males have more or less sharply pointed, high crowns. This anatomical difference (dimorphism) between the sexes allows a firm grip on females during sexual activities. Hence the dentition of males is of the clutching type, while females have crushing-type dentitions. Additionally, such males and females may not share exactly the same prey preferences.

Also serving the purpose of clutching to a female is a less pronounced variant of sexual dental dimorphism found in requiem sharks, catsharks and some squaloid deep sea sharks: The teeth are of the same shape in both sexes, but the main cusps of their crowns are more erect in males than in females.

In conclusion, there is much more to say and learn about shark and ray teeth than is commonly expected. The living elasmobranch fishes are representatives of evolutionary old lineages in the tree of life of animals with a skeleton. Adaptation to the challenges of changing environments, prey availability and rival species has produced a multitude of sizes and shapes for teeth. Some types have vanished completely and are known as fossils only, others have survived unchanged for many millions of years until today. Whenever you have the possibility to see shark tooth collections other

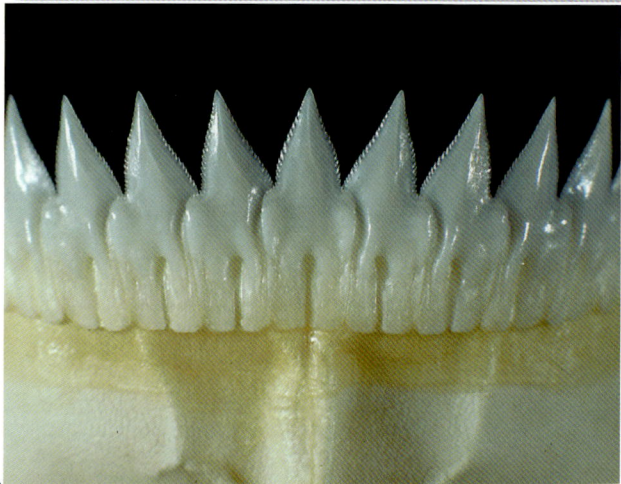

Kitefin shark *Dalatias licha,* juvenile female, TL 76 cm, NE-Atlantic. Labial view of lower jaw. Note symphyseal tooth and simple serrations on cutting edges. Largest tooth 9 mm. This dentition closely resembles those of the famed cookiecutter sharks of the genus *Isistius* which indeed are close relatives of *D. licha.*

than at the front end of a live 2-metre-long reef shark underwater or as a dust-covered, smelly curio pinned to a wall in a smoky harbour bar, go see them (in private collections or museums) and try to imagine what passed their mouths while their owner was still alive. And do not envy sharks for their habit of ignoring dentists completely, just keep brushing...

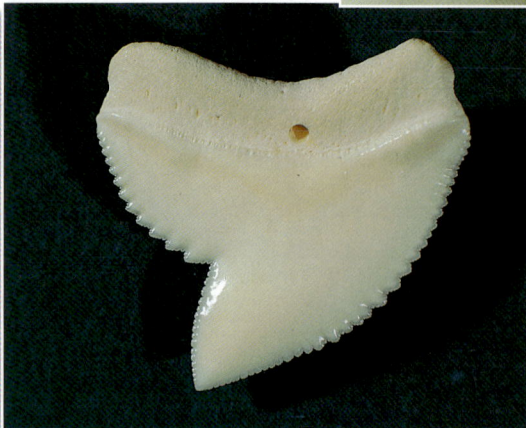

Tiger shark *Galeocerdo cuvier.* Upper jaw tooth. Note double serration on cutting edges. Largest width 26 mm. A hole has been drilled into the root of this specimen in order to sell it as a necklace decoration. Purchased in a tropical souvenir shop.

Rhina ancylostoma   **Mauritius, Indian Ocean**

This family is represented worldwide in tropical to warm temperate waters and comprises two genera with five or more species, most of which are found in the Indo-West Pacific region. The genus *Rhina* is monospecific, *Rhynchobatus* comprises at least four species. All family members grow to a relatively large adult size, some reaching well over 300 cm in total length. Their first dorsal fin originates over the pelvic fin base. The lower lobe of the caudal fin is well developed in contrast to the members of the following family, the guitarfishes, which only have a well-developed upper caudal lobe. The anterior part of the body and the large dorsal fins create a shark-like appearance, the animals look like a mix between shark and ray; individuals are often confused with sharks, especially when seen underwater for the first time. Maldive fishermen, for example, simply refer to them as sharks. Some species are regularly encountered on sandy bottoms in the vicinity of coral reefs and rocky shores in warm waters. The mode of reproduction of all family members is ovoviviparous (yolk sac viviparity).

Sharkfin guitarfishes do well in large public aquaria; one species is known to breed in captivity. A specimen of the bowmouth guitarfish *Rhina ancylostoma* has survived almost seven years in captivity. The habits and biology of this unique elasmobranch are little known. It is usually seen swimming near the ground, but also high in the water column and has been observed being cleaned by cleaner wrasses of the species *Labroides dimidiatus*. In the Australasian region, it can be a nuisance to shrimp fishermen: In the confined space of the small trawlers, the bulky, rough and spiky animal is difficult to handle and may also damage the valuable commercial catch. This peculiar guitarfish itself is caught on all kinds of gear, and is marketed fresh for human consumption regionally in the western Pacific region.

Mark Strickland   Andaman Sea

Rhina ancylostoma
Bowmouth guitarfish
L: At birth 45, max. 270 cm. Di: Tropical Indo-West Pacific: Red Sea and South Africa to Australia and southern Japan. De: 1-20 m. G: Snout broad and rounded. Pectoral fins shark-like, triangular and positioned far forward on the body. Prominent ridges above eyes and along centreline on head, bearing large protective dermal denticles. Coloration: Juveniles brown to bluish-grey with large white spots, partial eye-spots (ocelli) on pectoral fins and black bars between eyes. Adults grey with small white spots, black bars between eyes faint, without ocelli. Litter size 4. Preys on crabs, shrimp and clams. See also family account and photo on previous page.

Rhynchobatus djiddensis
Giant guitarfish

Length: At birth 43-60, max. 310 cm (230 kg).
Distribution: Tropical Indo-West Pacific: Red Sea, South Africa, Seychelles and Maldives to Japan, Australia (New South Wales), New Caledonia and Marshall Islands.
Depth: Intertidal to 50 m.
General: Snout pointed. Base of pectoral fins often with eye-spots (ocelli). Coloration tan to black with distinct white spots above, light below. Found in the surf zone, estuaries, lagoons and on sand near reefs. Feeding and resting on the reef flat during high tide, sometimes together with roundnose stingrays. Litter size up to 10. Gives birth during the summer months in South African waters. In some regions, birthing occurs in brackish estuarine environments. Prey includes crabs, lobsters, clams and small fishes. Some of the food items are dug up from the sand. Difficult to approach underwater. Has successfully been kept in public aquaria for more than 5 years.

both photos Mark Strickland   Andaman Sea

Rhynchobatus luebberti
Lübbert's guitarfish

Length: At birth about 50, max. 300, usually about 150 cm.
Distribution: Eastern Atlantic: Senegal to Congo, West Africa.
Depth: 1-20 m.
General: Similar to the previous species *R. djiddensis*, but with a more dense and regular white spotting on the brown back. Usually encountered over sandy bottoms in shallow coastal waters. Caught locally with all kinds of line gear and nets and utilised dried and salted for human consumption.

Bernard Séret   Senegal, West Africa

JUST FOR THE FINS

The statistics on total (i.e. worldwide) catches of elasmobranchs published by the FAO (Food and Agriculture Organisation of the United Nations Organisation UNO) show 'Third World' countries top-ranking, among them Indonesia, where in 1997 alone 'officially' about 100,000 metric tons of sharks and rays were caught. During a coincidental meeting, underwater photographer Joerg Adam was able to document why these animals are mercilessly slaughtered by the thousands.

Staring in bewilderment, I watch Indonesian fishermen cutting off first the dorsal and then the caudal fin of a medium-sized requiem shark.

During summer 2000 and far off the usual tourist trails, we are diving the pristine coral gardens along the northern coast of Irian Jaya, the Indonesian part of New Guinea. Here, virgin diving locations are still abundant and fortunately the global coral bleaching has not affected these reefs yet. With the impressions of this paradise still buzzing in our heads, we surface and head for an uninhabited island to reload our cameras. At about noon, our boat glides through the mangroves and we finally land on a sandy beach. Everything here seems to be beautiful and dreamlike.

A few steps away from the shoreline we notice a large lizard. It is ripping pieces of flesh from a carcass of unknown identity while completely ignoring us. Only now we start to give our surroundings a closer look. Then, suddenly, more dead animal bodies become obvious in the shallow water and on dry land: They are the remains of sharks of diverse species, all with fins missing. In some of the carcasses also the vertebral column (spine) has been removed. We are shocked. Even in this sparsely populated coastal region these elegant creatures are not safe any longer. We examine the island more thoroughly and find a primitive hut with a fire place. Cut-off fins of all sizes are spread on long wooden planks for drying. We agree on delaying our afternoon dive and hope for the return of the 'culprits'. After some time, a motorised canoe approaches and three fishermen start to unload last night's catch.

The Indonesian fishermen are not bothered at all about our presence and feel free to talk to us. They are not residents here, but are using the island as a home base for their fishing activities which often last several weeks. Their main target are sharks which they catch on longlines

with countless hooks. Every few metres along the line a hook is fastened to it. Now we become witness of what we only could have guessed until then. I take photographs of how fins and spine of a previously caught shark are removed with a few long cuts. The bulky remains, the body and head of the chondrichthyan fish, are thrown back into the sea without any further consideration. The elegant shark had lost its life for less than ten percent of its total weight. At the same time it was taken out of the food and reproduction chain without replacement. When questioned, the fishermen tell me that they sell the dried fins for a few US dollars per kilogram to middlemen in the town of Sorong. These merchants resell them, e.g. to dealers in Hong Kong, where prices of up to several hundred US dollars are paid for each kilo of dried fins, the price depending on the quality of the product.

I finally get the impression that the penniless fishermen I am talking to are merely victims of the exploitation by unscrupulous businessmen. The wholesalers in the shark fin trade take advantage of these fishermen in order to gain maximum profit while investing a minimum amount of money. Aided by our indigenous translators we soon realise that for this reason a change in the attitude of the fishermen within a short period of time - only after a few talks and without offering alternatives - is practically impossible. Deeply shocked we leave the 'scene of the crime'. Soon afterwards we describe our impressions and disappoint-

These fins are spread for sun-drying on primitive wooden frames specially constructed for this purpose.

ment to Dutchman Max Ammer who lives in this part of Indonesia. He is the local diving pioneer and knows about the diving spots in the North of Irian Jaya like nobody else. He tells us about his own year-long intense explanatory work and even partially successful efforts while talking to the local fishermen about the protection of sharks and rays. I can only hope that his efforts to protect these wonderful chondrichthyan fishes will find even more resonance in the future. Besides such enlightening work, the only effective way to stop the commercial slaughter of sharks and rays as soon as possible is the total boycotting of shark products by each and every consumer.

Dried shark fins densely packed and waiting to be transported to Hong Kong.

KURT AMSLER

The family at present comprises 4 well defined genera with more than 40 species, most of which are found in the Indo-West Pacific region. All are characterised by the first dorsal fin originating behind the pelvic fin base. The caudal fin has no lower lobe, other than in members of the family Rhynchobatidae. With about 30 species *Rhinobatos* is the most speciose genus. Many guitarfish species are indistinguishable under water, but characteristics like snout shape and length, shape of spiracle and nostril, presence or absence of spines along back and tubercles on snout, colour pattern and orientation of nostrils relative to the mouth can help identifying species. All family members are ovoviviparous (yolk sac viviparity), the largest reach a length of well over two metres. Their mating behaviour has not been observed yet, but males closely follow females during courtship. Guitarfishes are most often seen while resting on sand or among seagrass. They swim in a shark-like manner with lateral strokes of tail and caudal fin, not by undulating their pectoral fins like most rays do. Most species are typical inhabitants of marine coastal habitats, some frequent estuaries and bodies of freshwater, but they are rarely seen near oceanic islands. Several species are known to segregate by sex and make seasonal migrations. Some guitarfishes adapt well to a life in captivity, while others do not.

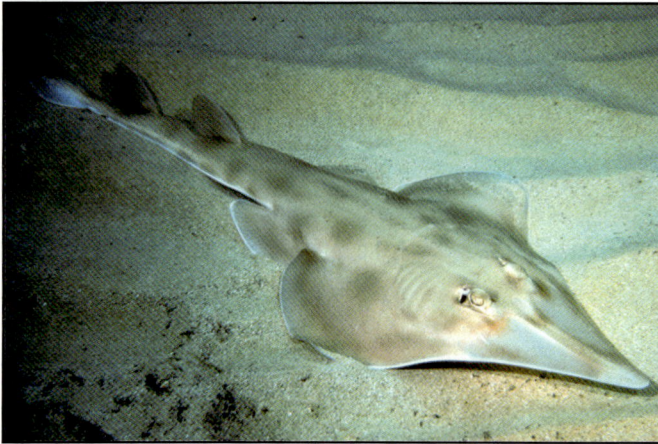

Rudie Kuiter   **New South Wales, Australia**

Aptychotrema rostrata
Eastern shovelnose ray
L: Max. 120, mature females 85, mature males 72 cm. Di: Eastern Australia: Queensland (Moreton Bay) to NSW (Jervis Bay). De: 1-50 m. G: Rarely dark around eyes. Nearshore and in estuaries. *A. bougainvillii* is probably a synonym. Below: Plain coloured specimen.

Clay Bryce   **Shark Bay, Western Australia**

Aptychotrema vincentiana
Western shovelnose ray
L: Max. to at least 79, mature males about 65 cm. Di: Western and southern Australia (Port Hedland, Western Australia, to Kent Islands, Bass Strait). De: 1-32 m. G: Yellowish brown with dark blotches, usually with black bands diagonally from eyes to snout. Juveniles plain in colour except black snout tip. Found nearshore in temperate (southern) waters, deeper in tropical (northern) waters. Slight shape differences between both populations, may be two species. Genus members prey on crustaceans, molluscs and fish.

Rhinobatos schlegelii
Schlegel's guitarfish
L: Max. at least 75 cm, mature at about 55 cm. Di: Western Pacific: Japan to Viet Nam and Philippines. D: Shallow inshore waters. G: Biology, habits and utilisation little known.
Below: **Clubnose guitarfish** *R. thouin*, to 280 cm, Red Sea to New Guinea and Japan.

Kazu Masubuchi   Izu Park, Japan

Rhinobatos halavi
Halavi guitarfish
L: At birth about 30, max. 170, males mature at 83 cm. Di: Indo-West Pacific: Red Sea to Gulf of Oman. Nominal records from the Persian Gulf, India, Myanmar, Viet Nam, China and the Philippines need to be confirmed. De: Shallow inshore waters. G: The biology and habits of this moderate-sized tropical guitarfish are little known. In the northwestern Red Sea, females are common along sandy shores from May to October, when they seek the shallows to give birth. Preys mainly on crustaceans, especially shrimps.

Friedhelm Krupp   Arabian Gulf

Rhinobatos salalah
Salalah guitarfish
Length: 55 cm. Distribution: Western Indian Ocean: Arabian Sea. Depth: Shallow inshore waters. General: A row of medium-sized blunt thorns (enlarged denticles) along mid-line of back. Coloration above tan with scattered bluish-white, dark-edged spots (pupil-sized or larger) on outer part of body disc, pectoral and pelvic fins. Described in 1995 from a single specimen found in the fish market in the town of Salalah, Oman. Placed in the subgenus *Acroteriobatus* due to shape differences when compared to other *Rhinobatos* spp.

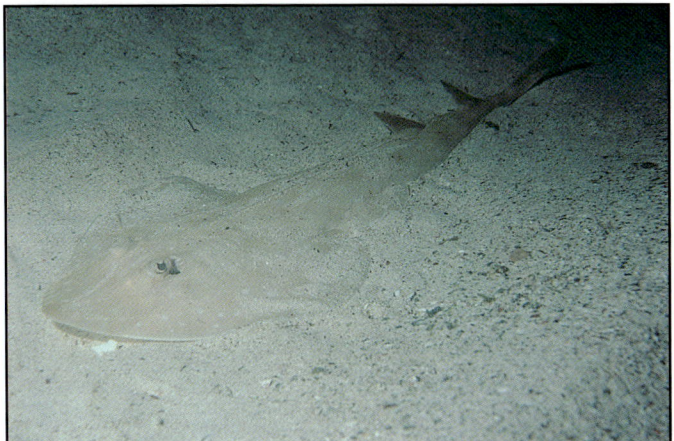

Phil Woodhead   Oman, Arabian Sea

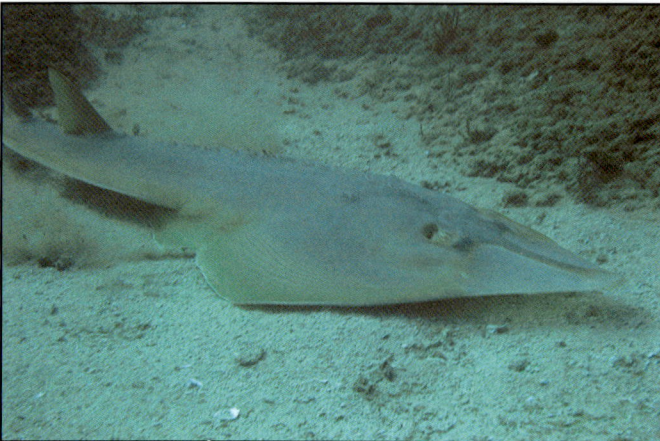

Phil Woodhead   Gulf of Oman

Rhinobatos granulatus
Sharpnose guitarfish
L: Max. 215 cm. Di: Temperate and tropical Indo-West Pacific: Arabian Gulf and Gulf of Oman to India, Sri Lanka, Andaman Islands, Myanmar, Thailand, Viet Nam. De: Intertidal to 120 m. G: Large, found inshore and offshore. Has broad band of enlarged denticles with a series of large thorns on midline of back. Transverse row of about 6 thorns at level of greatest width of pectoral fins. Coloration greyish-brown to tan above, snout translucent, edge of pelvic fins and dermal folds on either side of tail white. Biology little known.

Oliver Kirsch   Gulf of Oman

Rhinobatos punctifer
Whitespotted guitarfish
L: Max. at least 80 cm. Di: Red Sea to Gulf of Oman. De: 35-70 m. G: No large thorns, but only minute denticles on rostrum, over eyes, on shoulders and midline of back. Yellowish-brown above with scattered white spots (pupil-sized or smaller) on head, back, pelvic fins and tail, additionally with more or less prominent diffuse, large, dark grey blotches. Light below. Described and named only in 1987 when assumed to be restricted to northern Red Sea. Shown is first photo of living specimen from outside the Red Sea.

Bernard Séret   Ivory Coast, West Africa

Rhinobatos irvinei
Irvine's guitarfish

Length: Max. 100, usually 50-80 cm.
Distribution: Eastern Atlantic: Senegal to Angola. Depth: 20-50 m, one record from 110 m. General: Has a patch of enlarged dermal denticles (thorns) set in rows between and above eyes. Back brownish with rows of ocellus-like spots extending on tail. Occurs in inshore coastal waters on sandy and muddy bottoms but is not known to enter estuaries or mangrove areas. Preys on bottom-dwelling shrimps, crabs and molluscs.

Rhinobatos annulatus
Ringed guitarfish
Length: Max. 140 cm, usually to 80 cm.
Distribution: Eastern Atlantic: Namibia to South Africa.
Depth: 1 - 70 m.
General: This is a common inshore species that also enters estuaries. It feeds on small sand-dwelling invertebrates including crustaceans, mussels, polychaete-worms and small bony fishes.

While many guitarfishes enter the brackish water of estuaries, they generally do not penetrate into bodies of pure freshwater such as the upper reaches of rivers and lakes. There is, however, one possible exception: The giant guitarfish *R. typus* is reported to permanently live and breed in freshwater. It is greyish-brown to olive above with paler fin margins, reaches a max. length of at least 270 cm and is found in the warm temperate to tropical Indo-West Pacific from India to New South Wales, Australia, occurring from the intertidal down to about 100 m on the continental and insular shelves. Young individuals live inshore on sandy bottoms, around atolls and in mangrove areas, where they move inshore with the rising tide to feed on molluscs. Adults inhabit deeper offshore waters. May be the most important commercially utilised guitarfish in its range.

Simon Chater   South Africa, Eastern Atlantic

Rhinobatos rhinobatos
Common guitarfish

Length: Max. 100+ cm.
Distribution: Eastern Atlantic: Southern Bay of Biscay to Angola. Also Mediterranean.
Depth: Intertidal to 100 m.
General: Yellowish brown above, lighter below. Litter size 4-10, 1 or 2 litters per year. Guitarfishes have protrusible jaws which enable them to pick up prey items directly from the bottom. This species preys on benthic invertebrates and fishes. It is caught with all kinds of gear and marketed fresh or dried-salted, especially in West Africa.

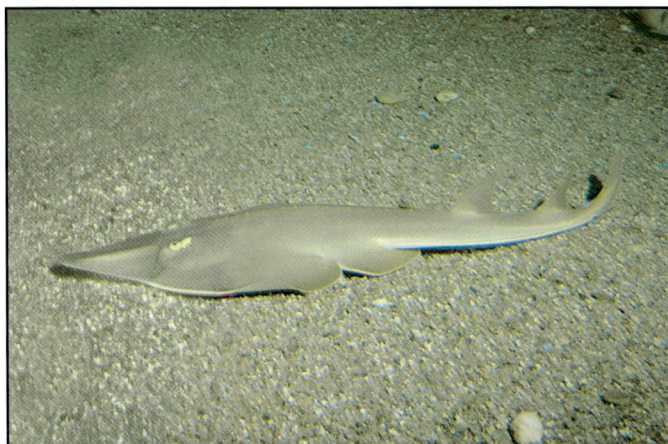

Helmut Debelius   Galicia, Spain, Eastern Atlantic

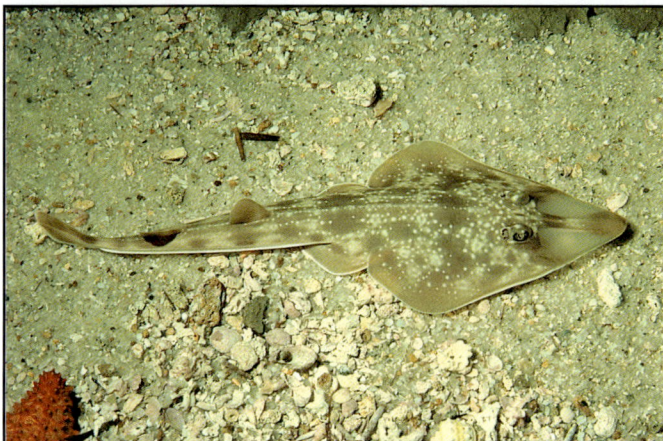

Helmut Debelius   **Curaçao, Netherlands Antilles**

Rhinobatos lentiginosus
Atlantic guitarfish

Length: At birth about 20,
max. 76 cm.
Distribution: Western Atlantic:
North Carolina, USA, to Yuca-
tán, Mexico, including Gulf of
Mexico. Depth: 1-20 m.
General: Body disc almost as
broad as long. Rostral (snout)
cartilage expanded at tip.
Snout with tubercles on tip.
Brown to grey above with
many (several hundred) small
white spots, rarely spots are
lacking; lighter below. On sand,
mud and in seagrass meadows.
Litter size 6. Prey includes
molluscs and crustaceans.

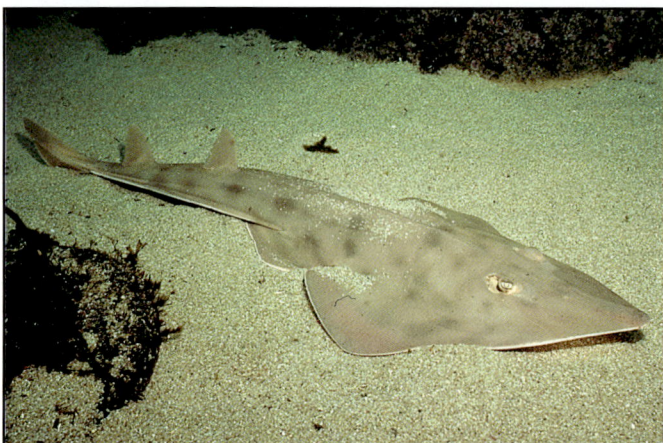

Mark Conlin   **California, Eastern Pacific**

Rhinobatos productus
Californian guitarfish

Length: Max. 170 cm.
Distribution: Eastern Pacific:
California (San Francisco) to
Gulf of California.
Depth: 1-25 m.
General: Benthic on rocky bottoms in inshore coastal waters,
also in bays and the brackish
water of estuaries. Sexes segregate during part of the year.
Sex ratio of over 560 specimens caught during a 7-year-
period in Monterey Bay, California, was 2:1 males to
females. Preys mainly on small
crustaceans, occasionally also
on small fishes.

Clay Bryce   **Panama, Eastern Pacific**

Rhinobatos glaucostigma
Speckled guitarfish

L: Max. 76 cm. Di: Eastern
central Pacific: Southern Baja
California and southern Gulf of
California to Ecuador and
northern Peru. De: Intertidal
to at least 25 m. G: Greyish-
brown above with symmetrical
series of large, darker blotches
and whitish spots arranged in
wide circles. Bottom-dwelling
species that is found in inshore
coastal waters. Does not enter
brackish estuaries or freshwater streams. Inhabits sandy and
muddy bottoms where it preys
on benthic invertebrates (crustaceans and molluscs). Biology
and habits little known.

Trygonorhina fasciata
Southern fiddler ray

L: At birth 26, max. 126 cm. Di: South to Western Australia, including Tasmania. De: 1-50 m. G: Pectoral fins broad. Snout obtusely rounded. Coloration of back brownish with network of broad white lines with dark edges. Distinguished from the very similar eastern fiddler ray by details of colour pattern on head: Eastern fiddler ray with triangular pattern behind eyes, southern fiddler ray with three parallel bars. Common in coastal waters, often seen near jetties, in estuaries, bays and seagrass meadows. Ovoviviparous.

Rudie Kuiter   South Australia

Trygonorhina sp.
Eastern fiddler ray

L: At birth 25, max. 120 cm. Di: Eastern Australia: Southern Queensland to New South Wales. De: 1-100 m. G: Distinguished from the very similar T. fasciata by pattern on head (see previous species). On sand, in seagrass. Ovoviviparous. Egg cases golden, each may contain 2-3 embryos. Does well in public aquaria.

The magpie fiddler ray T. melaleuca (southern Australia) has an overall bluish-black body coloration with irregular white margins to body disc and pelvic fins and white dorsal and caudal fins.

Rudie Kuiter   New South Wales, Australia

Zapteryx exasperata
Banded guitarfish

L: Max. about 90 cm. Di: Eastern Pacific: Southern California to Panama. De: Intertidal to at least 22 m. G: This genus is intermediate in body shape between the guitarfishes of the genera Rhinobatos and Trygonorhina: The outline of body disc and pectoral fins is not as pointed (sharply angular) as in the former genus and much less rounded (i.e. oval) than in members of the latter genus but rather broadly angular. A row of large thorns is running along the midline of the back. **Continued on next page.**

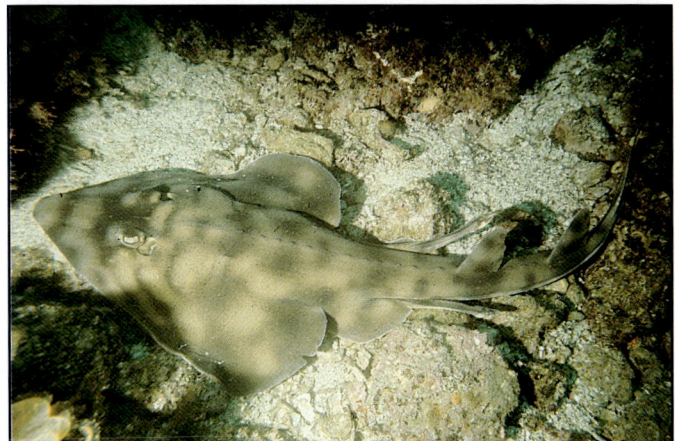

Helmut Debelius   California, Eastern Pacific

Mick McMurray / IV   Sea of Cortez, Mexico

Zapteryx exasperata
Banded guitarfish
Contd. from previous page.
It is greyish-brown above and has dark brown bands and merging series of blotches transversely crossing the back. Some specimens additionally have pupil-sized yellow spots which are outlined in black. The common species is often found in caves or under ledges on rocky bottoms. Young specimens have also been reported from tide pools. The banded guitarfish rarely buries in sandy substrate, most often it is encountered lying exposed on rocks or gravel, relying on its camouflaging pattern for protection. Contrary to another genus member (shortnose guitarfish *Z. brevirostris,* Brazil), the posterior margins of both dorsal fins are shorter than their bases and the rostral ridges on the snout are parallel to each other and do not converge towards the tip of the snout as in its more plain coloured relative.

Howard Hall   California, Eastern Pacific

Platyrhinoidis triseriata
Thornback ray
L: Max. 91 cm. Di: E-Pacific: Central California to Baja California, Mexico. De: 1-50 m. G: Body wider than long, 3 rows of large thorns on back and tail. Uniformly brown above, most fin margins whitish. Solitary or in small groups. By day, buried in sand or mud, among kelp. Parturition and mating in late summer. Preys on worms, molluscs, crustaceans.

The family comprises 3 genera with 3-5 species which look similar to guitarfishes, but have 2 rostral cartilages (only 1 in guitarfishes), a rounded snout and 1 or 3 rows of thorns on back and tail. Thornback rays (max. length 1 m) are thought to be intermediate between torpedo rays and guitarfishes. All are benthic on soft bottoms, most in deeper offshore waters. Ovoviviparous, one species presumably oviparous (laying skate-like egg cases). Prey on invertebrates that dig in the substrate (infauna).

MIND YOUR STEP!

Only few people in Europe and North America know about the occurrence of rays, especially stingrays, in warm freshwater habitats all over the world. But in South America, these animals are among the most feared fishes. They are much more feared than the piranhas, even more than the dangerous candiru - a tiny catfish that can enter the urinal ducts of bathers and only be removed surgically. Freshwater ichthyologist Frank Schäfer talks business.

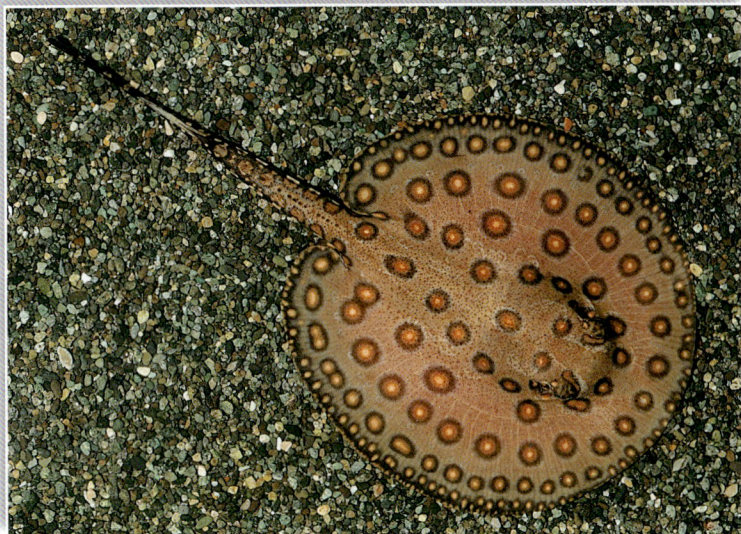

H. NAKANO

An orange colour morph of *Potamotrygon motoro,* a common stingray of South American rivers.

Freshwater stingrays can grow to a large size, a disc width of up to 2 metres has been confirmed; more common, however, are sizes around 60 cm disc width. On the back of their tail these brutes carry one or more large poisonous spines. In the shallow, often murky waters of South American rivers it is not unusual to step on a resting ray which immediately will lunge forward with its tail and sting. Most often feet and lower limbs are the target of the retaliating blow, hands and upper body parts may also be hit, especially when fishermen remove stingrays from nets. Wounds inflicted by such a barbed sting can be deep and have irregularly jagged margins, where infection easily sets in - all in all, a stingray wound is no pleasant welcome to a jungle river bank at all. And although stingrays never attack humans without being provoked, a sting of such a ray might finally be fatal in the jungle, where no medical help is available.

The river stingrays are placed in the family Potamotrygonidae, all members of which live exclusively in fresh water and are restricted to the river systems of South America. One of the peculiarities of the species that make up this family, is the immense variability of their colour patterns and markings. Each individual fish is uniquely coloured and patterned. From aquarium-bred specimens it is known that the individual pattern is retained as long as the fish lives. Obviously all this presents problems as regards the systematics of these rays. Many species were scientifically described from single specimens or just a few individuals, and in many cases the only character cited to differentiate them from one another is their coloration. The situation is made even worse by the fact that rays, being relatively large fishes, are not easy to preserve.

Tail of an estuarine stingray (*Dasyatis* sp.) with its long, barbed and poisonous sting.

N. WU

Often the preserved specimens are incomplete and important taxonomic characters are missing. Thus nobody in the whole world knows the exact number of freshwater stingray species. 36 family members have been described by scientists, 22 of which are regarded to represent valid species. About half a dozen more species are already known, but not described scientifically yet.

Some, like the most common species *Potamotrygon motoro,* or the so-called black rays (*P. henlei* and related

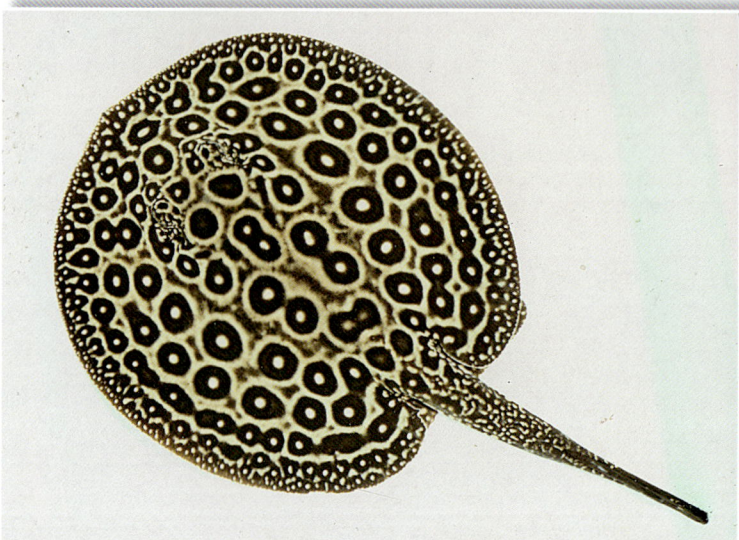

An undescribed species of the genus *Potamotrygon* from Brazil with a fantastic colour pattern.

forms), are known to be relatively aggressive and thus dangerous for humans. Others, like the widely distributed *P. orbignyi,* hardly use their sting, even when netted. However, as in all poisonous fishes, nobody can foresee the effect of a wound inflicted by a stingray, and it is better to treat such an accident always as serious. First aid in such a case means to place the wound in water as hot as the victim can bear. This procedure destroys most of the toxin (a special protein) which is contained in a sheath of mucous tissue covering the sting. People who have been stung by a *Potamotrygon* describe the sting as being extremely painful and report that the wound needs at least months to heal completely.

But freshwater stingrays do not exclusively occur in South America. In Florida, for example, the coastal marine species *Dasyatis sabina* developed populations that live and breed in freshwater lakes. And the largest of all the *Dasyatis* species, *D. brevicaudata,* is also said to enter the lower reaches of rivers quite frequently. This giant among the stingrays grows up to 4.3 metres long and about 2 metres wide. The base of its tail, which bears a single poisonous spine, can supposedly be as thick as a man's thigh, and a blow from it has sufficient power to drive the spine through the wooden planks of a boat! For that reason this stingray, which is widespread in the Indo-Pacific Ocean and ranges from Africa to Australia, is much feared by many fishermen.

Stingrays living in pure freshwater also occur in Africa, e.g. *Dasyatis garouaensis* and the mystic *D. ukpam.* The 'ukpam' (the collective term used by West African natives for this and other stingray species) of West Africa is in some ways the freshwater ray equivalent of the Yeti in the Himalayas! Like the equally gigantic species *Himantura fluviatilis* from India, it is little known to science. But in contrast to the Yeti and the Indian giant freshwater ray, there are at least a few museum specimens of the ukpam, even though at the time of the first description of the species in 1863 the only known specimens were two embryos from the Old Calabar River. Matters remained thus until 1978 when further

Potamotrygon henlei, one of the so-called 'black rays' of South America, can attain a disc width of up to 40 cm.

222

individuals were seen in the Ogooué river; additionally, a specimen collected in 1937 in the lower Zaire (now: Congo) river was identified as belonging to this species. But to the present day it remains unclear what monstrous freshwater ray was seen by a Mr. Sanderson in 1937 in Mamfe Pool, Cross River (Nigeria); that individual, which had a disc width of more than 140 cm, remains the only known evidence of the existence of a giant freshwater ray in Nigeria to date. The spine on its tail was supposedly more than 40 cm long! A spine of that size is unknown in any of the described African species.

In Asia, many species and populations of stingrays are found in bodies of freshwater. Most of them belong to the genus *Himantura*. Some of these stingrays are dwarves, but there also occur gigantic forms like *H. chaophraya* for example. This is a species from Thailand which has been scientifically described only very recently and attains a huge size - the disc width of the largest preserved specimen is almost 2 metres. But even larger individuals are said to occur in nature, and specimens weighing up to 600 kg have been reported in local daily newspapers! The type specimen was caught in freshwater more than 100 km from the sea. It is very interesting to note that in a country as well investigated ichthyologically as Thailand, and in which freshwater fisheries constitute an important element of the economy, such a

The freshwater whipray *Himantura chaophraya,* a giant stingray of Australasian rivers, was described as recently as 1990 from Thailand. It is reported to reach a disc width of almost 2 m and a weight of 600 kg!

gigantic fish could remain undiscovered until 1983. It is probable that this species is migratory, and may also be found in brackish water. It may be identical with *H. fluviatilis*. At present, the giant rays of New Guinea (Fly River system), Borneo (Mahakam basin), and Australia (Gilbert River, Queensland; South Alligator rivers, Northern Territory; Ord and Pentecost rivers, Western Australia) are assigned to *H. chaophraya*. Newly-born juveniles have a disc width of about 30 cm. A female weighing 30 kg, captured in January, gave birth to 4 young.

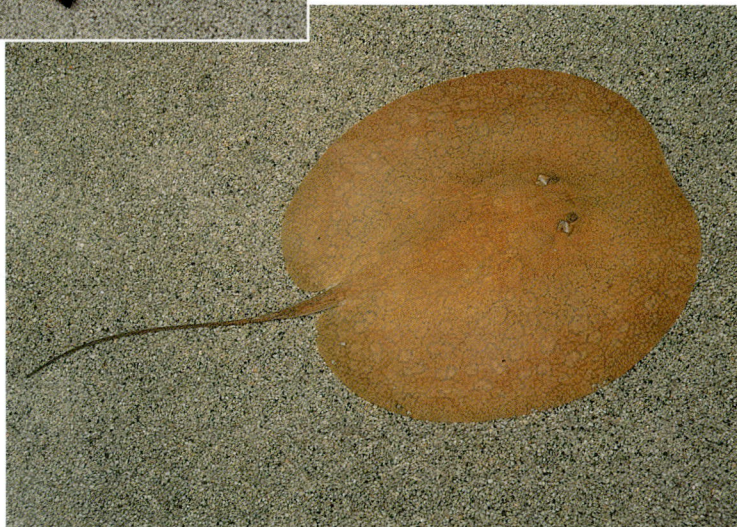

Paratrygon aiereba represents a different type of South American freshwater stingrays. It is distinct from species of *Potamotrygon* by smaller eyes and simpler colour pattern. Max. disc width 120 cm.

So now you know a little bit more about the most common elasmobranch fishes in the freshwater of the tropics and subtropics. Whenever you step into a river or lake in South America, Africa or Asia and simply happen NOT to wear the casual iron boot of a medieval knight, keep your eyes open and watch out for those excellently camouflaged flat fellows and their stinging appendices!

This ray family contains only one genus with about 17 species. All family members are characterised by a round to oval flabby body disc, two dorsal fins, a straight or notched anterior disc margin and a short tail with a well developed caudal fin. Some species are hard to identify under water, as the colour pattern, which may be variable in certain species, is the only non-anatomical aid to identification. Also helpful are position of first dorsal fin relative to pelvic fins and presence and number of papillae on the spiracle margins. This family is in need of taxonomic revision, especially the species occurring in the western Indian Ocean; there may exist several different species, each endemic to one of the island groups there.

Torpedo rays are found in all temperate and tropical seas where they inhabit inshore as well as offshore waters. Their mode of reproduction is ovoviviparous (yolk sac viviparity), but additionally the embryos are nourished by a secretion ('uterine milk') of the mother's uteri. This fluid is rich in nutrients and is ingested by the developing young inside the uteri. There is only little information on courtship and mating behaviour.

All torpedo rays possess two kidney-shaped electric organs. They are derived from muscular tissue and are embedded under the skin on each side of the head, where their outlines are visible in most specimens. These electrogenic organs are capable of producing electrical discharges in excess of 200 volts in some species. They are used to stun prey, to ward off predators and possibly also serve in communicating with conspecifics. The sensation of being electro-shocked has been described by waders and divers as if being hit by a very strong blow from a fist. Divers might become unconscious after being shocked, but there are no confirmed reports on such cases. The prey of torpedo rays mainly consists of bony fishes, which are more or less covered with the pectoral fins, stunned or killed with electric discharges and finally eaten.

Find more on the **feeding behaviour** of torpedo rays on p.227.

Herwarth Voigtmann   Maldives, Indian Ocean

Torpedo fuscomaculata
Blackspotted torpedo ray

Length: Max. 64 cm.
Distribution: Western Indian Ocean: South Africa, Mozambique, possibly also Zanzibar, Madagascar, Mauritius and Seychelles to southwestern India.
Depth: 2-200 m.
General: Similar to the scalloped torpedo ray T. panthera (p.226), but eyes much nearer to the spiracles than to the front margin of the disc.

Coloration of upper side very variable: Background colour dark brown or rusty brown to light grey, without or with dark spots (popular name), black lines or white flecks (photo left). Dark spots usually numerous, may be smaller than diameter of eye and densely distributed or about as large as eye or larger, often fused into blotches, and more sparsely distributed (these two colour forms are found in South Africa).

See also pattern on photo on facing page (Mozambique).

Torpedo sinuspersici
Variable torpedo ray

Length: Max. 130, usually about 90 cm.
Distribution: Western Indian Ocean: Red Sea and South Africa to Arabian Gulf and India. Depth: 1-200 m.
General: First dorsal fin only slightly larger than second dorsal fin. Eyes very near to spiracles, the latter with 9-10 papillae. Dorsal colour pattern variable (compare photos): Golden or cream circles or marbling on a dark reddish, brown or blackish background. Some juveniles with up to seven large, pale-edged, dark spots (ocelli) on the back. Encountered on sandy bottoms and coral reefs. Usually solitary, but may aggregate during the reproductive season. Litter size 9-22. In South Africa (small photo below), females give birth in the shallows, including brackish estuaries, in the summer. Newborn young are occasionally stranding on the beaches there during low tide. Adults are reported to survive for hours after being stranded. Preys on bony fishes.

both photos Phil Woodhead   Gulf of Oman

Torpedo panthera
Scalloped torpedo ray

Length: Max. 100 cm.
Distribution: Red Sea.
Depth: 2-55 m.
General: The correct identification of specimens of this species is often hard because the colour pattern on the back is quite variable (compare photos).
    The photographer of the photos on the facing page reported a very unusual behaviour of a specimen of *T. panthera* he found around midday in a depth of 7 m near the entrance of a cave.
Continued on next page.

Helmut Debelius   Egypt, Red Sea

226

Torpedo panthera
Scalloped torpedo ray

Contd. from previous page. After half an hour observing the specimen in the posture depicted at the right and below, he gently pushed the ray into a horizontal position. But soon afterwards the ray assumed its upside-down posture again and kept it until the observer's air supply ran out. Currently, there is no explanation for the ray's behaviour.

Feeding behaviour of torpedo rays may occur close to the bottom or up in the water column, depending on species. The small jaws of torpedo rays are greatly distensible and enable them to eat large prey items relative to their body size, for example a 120-cm-long Pacific torpedo ray *T. cali-*

fornica (p.230) has eaten a 60-cm-long silver salmon. Several hunting strategies are employed to catch prey. Lying buried just beneath the substrate's surface, they lunge up from the bottom when a fish comes within striking distance. Most torpedo rays emit an electrical discharge to immobilise or disorient prey just prior to the attack itself. They also stalk their prey by slowly drifting over the substrate or by 'creeping' along the ground. Despite their powerful electric organs torpedo rays are occasionally eaten by sharks.

The photo at the right shows a specimen of *T. panthera* eating a poisonous scorpionfish *Scorpaenopsis* sp.

all photos Klaus Hilgert　　　　　　**Egypt, Red Sea**

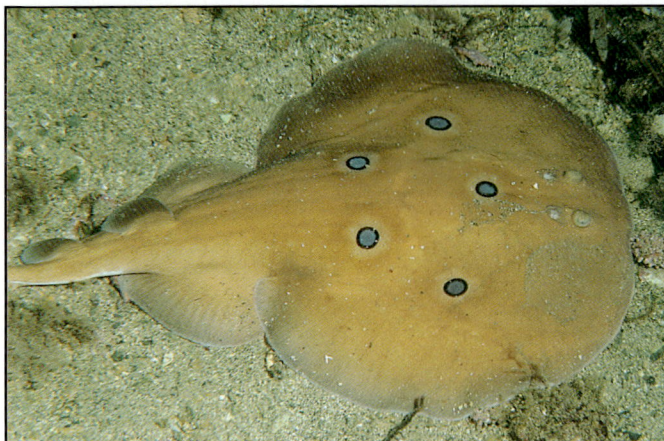

Juan Calvin   Spain, Mediterranean

Torpedo torpedo
Common torpedo ray
L: At birth 9, max. 60 cm. Di:
Eastern Atlantic: Southern Bay
of Biscay to Angola, and Mediterranean. De: 2-70 m.
G: First dorsal fin only slightly
larger than second. Spiracles
with small, knob-like papillae.
Uniformly dark to light brown
above, with 1-7 (often 5) ocelli
with blue centres. Litter size
3-21. Migrates inshore in fall to
mate, moves offshore in summer to give birth (Mediterranean). Buries in sand, shocks
bony fishes that come within
striking distance. Newborns
generate about 4 volts, four-
months-old rays 26 volts.

Bernard Séret   Ivory Coast, West Africa

Torpedo bauchotae
Bauchot's torpedo ray

Length: Max. 25 cm.
Distribution: Eastern Atlantic:
Senegal to Ivory Coast (West
Africa). Depth: 5-60 m.
General: This rare torpedo ray
has a tan background colour
above with numerous darker
flecks which are bordered by
almost black spots; interspersed are lighter blotches.
Caudal fin plain. Found in
coastal waters on sandy or
muddy bottoms. Preys primarily on small bottom-dwelling
bony fishes. Only a few specimens are on record in museum collections.

Bernard Séret   Ivory Coast, West Africa

Torpedo mackayana
Mackay's torpedo ray

Length: Max. 40 cm.
Distribution: Eastern Atlantic:
Senegal to Angola (West
Africa).
Depth: 5-150 m, usually 25-
70 m.
General: This little-known torpedo ray has a greyish-brown
background colour above with
numerous, close-set spots and
less numerous white flecks. It
is usually found in coastal
waters on sandy bottoms but
also on muddy bottoms in the
vicinity of river estuaries. This
species preys primarily on
small bottom-dwelling fishes.

Torpedo marmorata
Marbled torpedo ray

L: At birth 20, max. 60 cm, females grow larger than males. Di: Eastern Atlantic: North Sea to South Africa, and Mediterranean. De: 2-200 m. G: Spiracles with 6-8 long papillae, eyes very close to spiracles. Light brown to cream with dark brown mottling above, sometimes uniformly brown; white below. Solitary among seagrass, on rocky or soft bottoms. Litter size 5-32, gestation period 8-10 months. Preys mainly on bony fishes (conger, hake, sea bass, goatfish, mackerels, gobies, sole), but occasionally also on shrimp and squid. A 41-cm-long specimen from Scotland had devoured a 34-cm-long three-bearded rockling! When threatened by a predator, this ray will bend its body, swim off the bottom by undulating its pectoral fin margins, which results in the ray looping in open water, and deliver an electric shock. The bending of the body places an opponent that tries to grasp its tail in the focus of the high voltage field.

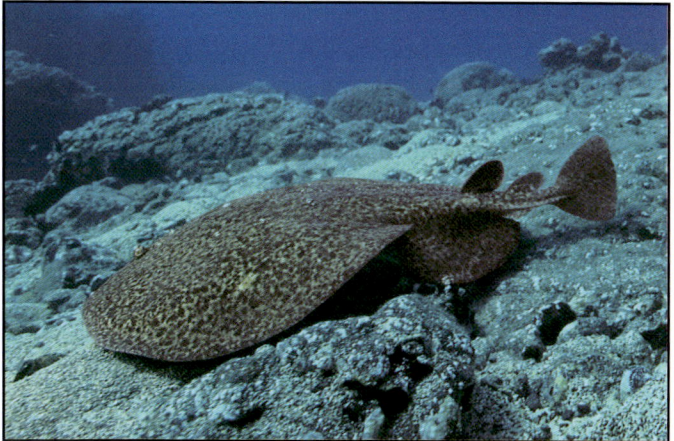

Jürgen Warnecke   **Canary Islands, Eastern Atlantic**

Helmut Debelius   **Spain, Mediterranean**

Torpedo nobiliana
Atlantic torpedo ray

L: Max. 180 cm. Di: Eastern Atlantic: Scotland to South Africa, and Mediterranean. Western Atlantic: Nova Scotia to Florida and Cuba. De: Shallow inshore waters to 460 m. G: First dorsal fin clearly larger than second. Spiracle margins smooth. Plain brown to dark violet above, also with darker markings or white spots; white below. Juveniles benthic, adults (semi-)pelagic, solitary, migratory. Can deliver a powerful shock. Observed to enclose prey fish with its pectoral fins while swimming well off the bottom. Litter size up to 60.

Ricardo Santos   **Azores, Atlantic**

Torpedo californica
Pacific torpedo ray
Length: Max. 140 cm.
Distribution: Eastern Pacific: British Columbia to Baja California. Depth: 3-200 m.
General: Margins of spiracles smooth, without papillae. Coloration grey above with scattered black spots. Preys exclusively on bony fishes (kelp bass, anchovies). Seen on sand among kelp (photos) where burying and ambushing prey during the day. Emerges to stalk and stun prey fishes at night. Usually solitary, but several specimens may bury in the same area. Nomadic, without home range. Its watery, flabby body tissue and oily liver make this ray almost neutrally buoyant. May shock divers and attack with open jaws when molested.

The small photo below shows that rays readily bite on hooks (here baited with squid).

both photos Howard Hall   California, Eastern Pacific

Hypnos monopterygium
Coffin ray
L: At birth 8-11, max. 60, mature male 24 cm. Di: Western Australia to Queensland, endemic. De: 1-240 m. G: Tan to brown above, light below. On sand, mud, reefs, seagrass. Weak swimmer, body flabby. Buried by day with only spiracles visible. Preys on fishes, molluscs, crustaceans, worms. Females give birth in summer. Arches back, opens mouth and swims in loops when molested. Family monospecific, syn. *H. subnigrum*. Teeth tricuspid (unicuspid in torpedinids).

Fred Bavendam   South Australia

Narcine entemedor
Giant numbfish

L: Max. 76 cm. Di: Eastern
Pacific: Southern Gulf of California to Panama. De: 1-20 m.
G: Benthic on soft bottoms in
shallow, protected waters.
Preys on worms, occasionally
also on sea squirts. Coloration
of adults and juveniles distinct
(see photos).

The family of numbfishes
comprises 4 genera with about
18 species. Their body discs
have a broadly rounded anteri-
or edge, are oval and longer
than wide in most species. The
two dorsal fins are of similar
size and positioned in the middle of the tail. The caudal fin is
well developed, the tail has lateral skin folds. The electric
organs of narcinids can produce up to 37 volts. However,
more harassment is usually
necessary to provoke electrical
shocks than in torpedinids.

Their mode of reproduc-
tion is ovoviviparous (yolk sac
viviparity), but additionally the
embryos are nourished by a
secretion ('uterine milk') from
the interior wall of the mother's uteri. This fluid is rich in
nutrients and ingested by the
developing young.

All family members are
harmless to divers, although
they will shock if molested.
Most of them are not suited
for a life in captivity, unless
they are provided with living
natural prey, i.e. worms.

Doug Perrine   Revilla Gigedo Islands, Mexico, Eastern Pacific

Clay Bryce   Golfo de Chiriqui, Panama, Eastern Pacific

Narcine brasiliensis
Brazilian numbfish

Length: At birth 11, max. 45,
usually 25-38 cm.
Distribution: Western Atlantic:
North Carolina to northern
Argentina, including Gulf of
Mexico. Rare at Florida coast
and Bahamas, common in
southern Caribbean.
Depth: 1-36 m.
General: On sand and rubble
bottoms, among seagrass,
common in bays. Migrates into
deeper waters in winter. Preys
mainly on burying worms
which are dislodged from the
substrate by undulating movements of the pectoral fins.

Paul Humann   Caribbean

Avi Baranes   Northern Red Sea

Narcine bentuviai
Ben-Tuvia's numbfish
L: Max. about 20 cm. Di: Red
Sea endemic. De: 80-220 m.
G: A small, uncommon numbfish of deeper waters, living on
the slope below the coral reef,
described only in 1989. Without external sexual dimorphism. Below: An undescribed
Narcine sp. from Oman.

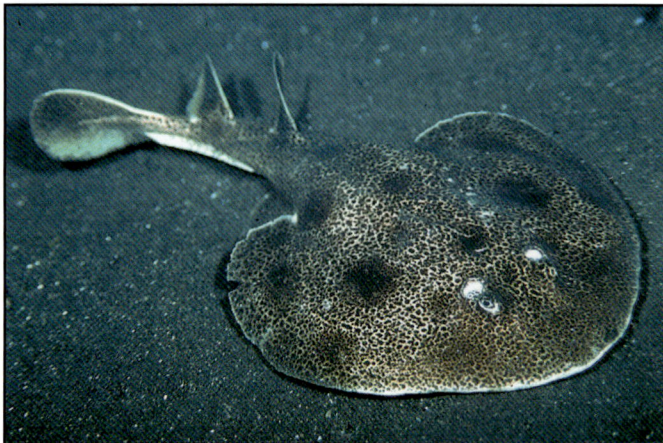

Rudie Kuiter   Bali, Indonesia

Narcine sp.
Indonesian numbfish
Length: Max. at least 30 cm.
Distribution: Northeastern
Indian Ocean: Off south coast
of Java and Bali, Indonesia.
Depth: Around 60 m.
General: This not yet scientifically described and named
numbfish species is known
from offshore waters of the
continental shelf. A female
specimen of 30.5 cm in total
length was presumably adult
and sexually mature. Biology
and habits unknown. Many
family members seem to have
very restricted distributions
and several more may be discovered in the future.

Rudie Kuiter   Tasmania, Australia

Narcine tasmaniensis
Tasmanian numbfish
Length: At birth about 9, max.
at least 47 cm. Distribution:
Australia: New South Wales
(Coffs Harbour) to South Australia (in the west at least to
Beachport, probably to the
Great Australian Bight) and all
around Tasmania. Depth: Shallow inshore waters to 640 m.
General: Brown above, margins paler; white below, sometimes with a few darker
blotches. Upper side of juveniles often with dark blotches
and a dark stripe on midline of
back. Common, but biology little known. The photo was tak-
en in a depth of 25 m.

Diplobatis ommata
Ocellated numbfish

Length: Max. 25 cm.
Distribution: Eastern Pacific:
Southern Gulf of California to
Panama.
Depth: 1-65 m.
General: Brown above, often
spotted or marbled, always
with distinctive ocellus on
back. Benthic on soft bottoms
in shallow, protected coastal
waters. Nocturnally active,
when displaying the unusual
technique of 'hopping' over the
substrate by using its pelvic
fins as stilts. Preys primarily on
crustaceans (amphipods,
shrimp), but also on worms.

Clay Bryce   Panama, Eastern Pacific

SLEEPER RAYS   NARKIDAE

Narke japonica
Japanese sleeper ray
L: At birth 10, mature males
23-37, max. at least 37 cm. Di:
W-Pacific: Japan to South China Sea. De: 10-30+ m. G: Little
known but locally common ray
of temperate to subtropical in-
and offshore waters.

The poorly studied family
of sleeper rays comprises 4
genera with about 10 species.
Number of dorsal fins is diagnostic for distinguishing genera: Species of *Heteronarke*
have two dorsal fins, those of
Narke and *Typhlonarke* have
one, those of *Temera* no dorsal
fin. Some family members have
minute (reduced) eyes, in the
'blind' species of *Typhlonarke*
the eyes are hidden by skin.
Most family members live in
temperate seas. They are usually found in deeper waters on
the continental shelf and slope,
only a few in shallow waters.
Little is known about their
habits and biology. The 'blind'
Typhlonarke aysoni feeds on
small fish and crustaceans, the
onefin sleeper ray *N. capensis*
(South Africa) on worms.

Kazu Masubuchi   Izu Park, Japan

233

KILLER MOVIES

Hardly anyone would think of domestic pigs or honey bees as deadly monsters. Nevertheless, almost 400 people die from bees' stings or are trampled to death by pigs in the USA each year. However, only 13 human beings were (probably) killed by sharks during the past year. But each of these shark-related fatalities became a feast for the media. Nothing seems to be as easy as to fan the flames of fear of sharks, says Helmut Debelius.

Bruce in action: When the diver pushes a button, the shark dummy puts on a fierce grin.

PHOTOS: JACK MCKENNEY

It is mainly the Hollywood film industry that is responsible for such an attitude. Today, every kid in the street knows which film is one of the greatest box office successes of all times: Yes, you guessed right, it is the spectacle 'Jaws'. Produced in the 1970's and directed by Steven Spielberg, the movie 'Jaws' was so profitable that the film industry did not hesitate to produce several sequels starring the 'vengeful monster'. Again, hundreds of millions of US $ came rolling in.

When a professional abalone diver was attacked and killed by a great white shark in South Australia several years ago, his colleagues did not go back to work for nearly six months. It was not so much that their buddy had died while diving for abalone. Divers permanently crippled or even killed by decompression sickness ('the bends') are much more numerous than those attacked by sharks. But getting the bends is the diver's own fault; he himself is responsible for that. If you were careless, you got bent. The comparatively frequent cases of decompression sickness or drowning had little influence on other divers. But this accident was different: It was an unprovoked attack by a great white shark. There was nothing the abalone diver could have done to prevent it. He had been totally helpless.

Such a death caused South Australia's abalone divers to back away from the water in quiet horror. There is something primordially terrifying about shark attacks. The thought of being devoured alive while completely defenceless seems somehow far worse than being crushed in an automobile or drowning. Evidence of this pervasive fear is the unparalleled success of Peter Benchley's novel 'Jaws', and the motion picture which followed. 'Jaws' had two major effects upon society. First, it made people collectively aware of an almost instinctive fear of sharks. Second, it made the great white shark the definitive monster of the seas; an almost supernatural demon which must be hunted down and destroyed before anyone could safely go near the ocean again.

The movie script authors of 'Jaws' cleverly played with horror and implied that an animal driven by instinct like the white shark could have feelings and seek revenge like humans. Therefore, the movie monster patrolled not far from bathing beaches although the real shark is hardly ever seen by people in its natural habitat. In the 1970's, good footage of white sharks was practically non-existing. Hence, the visions of horror for the cinema crowd had to be evoked differently: Oversized white shark dummies were constructed, including a replica of the distinctive head for close-ups.

Another dummy was constructed for scenes showing the entirety of the shark in the Hollywood movie 'Jaws'.

Jack McKenney, one of the underwater cameramen for the 'Jaws' movies, was very amused by the first head dummies made by the studio technicians. They looked much more like a rabbit than a shark! But 'Bruce' (the mechanical monster's nickname) finally succeeded in generating the desired horrors. Each time the diver who pushed the shark dummy through the water pressed a button, Bruce would open its jaws full of oversize teeth or roll its eyes. Even cameraman Jack got his share of horror while shooting close-up scenes. But Bruce wouldn't always cooperate like the director wanted him to. The idea was to film the full-size dummy while it was approaching a boat full of youngsters. But Bruce's electronics failed and he tumbled - jaws wide agape and slowly rotating around its long axis - towards a reef in 15 metres depth where he crashed. Well, making underwater horror movies can be costly...

The effect 'Jawsmania' had on white sharks was devastating, especially in South Australia and the United States. Sport and commercial fishermen began hunting those monsters with a vengeance. Sport fishermen paid huge sums for expeditions dedicated to the destruction of whites. Marine parks paid even greater sums for the bodies of great whites which were frozen and put on display. One such display at Sea World San Diego, USA, increased park attendance by 30%! In Australia white shark jaws could be sold for up to US$ 5,000 and individual teeth for US$ 200 each.

One witness of shark killings is Rodney Fox. For 30 years now he has been guiding diving tourists to places along the coast of southern Australia where they can film white sharks in their natural habitat from the safety of a cage. Rodney's main problem developed during the last decade. His quest for white sharks became more and more futile and he also knew why: Sport anglers used Rodney's knowledge, followed his ship and went after white sharks themselves as soon as he left the area. Today, the great white shark is protected by law in Australia after Rodney successfully fought for the rights of these magnificent animals.

Did you know that the Jaws movie monster was all fibreglass, pneumatics and electronics?

A SOFT SPOT FOR SHARKS

A young lady in a silvery suit of armour kneels on the sandy bottom just off the shore of the Caribbean island of Grand Bahama. She is surrounded by sharks. At least 20 fully grown Caribbean reef sharks circle her, and one after another suddenly darts in her direction to get at the proffered food. A group of astounded recreational divers stand by in a semi-circle with their cameras ready and observe what is going on. Professional underwater photographer Kurt Amsler is among them and gives us his impressions of the happening.

This shark demands being caressed and this is exactly what Bea, the shark tamer of Grand Bahama, does.

Every once in a while a diver ducks down respectfully as one of these animals swims by close enough for its tail fin to brush his hair. From this perspective, the spectators of this submarine performance can get a perfect view of the rows of sharp teeth in the sharks' gaping jaws. Calmly, and apparently obeying some kind of hierarchical order, the sharks take the food from the young lady's extended hand. One of them, a pregnant female, ignores the tidbit and nestles up to her stomach. The shark wants to be rubbed! It appears to be enjoying the caresses and makes no effort to swim away again.

The show has been going on for about 30 minutes now. The young lady has approached one diver after another so that every one gets the chance to observe the sharks from close quarters. Eventually the remaining mackerels in the black container will all have been fed to the sharks. As soon as the last fish is snatched from the young lady's hand, the elegant predators leisurely return to the depths of the Caribbean. Once the smell of fish and blood has dissipated they quickly lose interest in humans. As always, the young lady will be the last one to climb back on board. Even though she has been working with these animals for several years, she still enjoys every moment of the encounter. The sharks have become an integral part of her life.

The shark lady's name is Bea Bachmann. She is 25 years old and hails from the Lake Zürich region. Most of the American visitors who witness the shark feeding shake their heads in disbelief at the activities of the "girl from Switzerland," as they commonly refer to her afterwards. What is a delicate little blonde like her doing in the company of the "terror of the seas"? Is it the thrill of danger, exhibitionism, or one more way of exploiting wild animals for the sake of profit? All of these speculations have little to do with the former banking clerk's motivation, however. Before she ended up working as an au pair girl on the Caribbean island of St. Maartens, which is where she learned how to dive, her only knowledge of sharks had stemmed from watching Cousteau's films. The novice diver soon advanced to instructor. She landed her first job on the island of Grand Bahama. It was here that she learned how to handle sharks from underwater naturist Ben Rose. She quickly realised that, contrary to popular beliefs, sharks were not man eaters.

Ben's philosophy made sense to Bea: We have to redeem the reputation of these animals! Humans have to be made to realise that sharks are not the killing machines they are constantly made out to be. A shark feeding performance helps to illuminate the true nature of sharks. Every one who has ever witnessed how 20 fully grown animals remain perfectly calm in spite of the blood in the water no longer falls for such horror stories. In addition, it is very likely that the diving tourist, who usually either photographs or films the performance, will pass his experience on to others. Since shark feeding performances have caught on all over the world, appreciation for these animals is also increasing among non-divers.

But living sharks are also an asset for diving tourism. Most divers would love to come close to one of these elegant swimmers. Since sharks are extremely shy animals under normal

Bea feeds tasty mackerels to the Caribbean reef sharks.

diving conditions, this will seldom occur. And so divers from all over the world journey to diving centres like this one, where a close encounter with sharks is practically guaranteed. This even makes sense to government administrators, even if for commercial reasons only, and soon restrictive measures are passed to guarantee the survival of local shark populations.

She has never really been afraid of the sharks, Bea assures us, even when there have been so many of them swimming around her that she was being pushed around like rubber ball by their steely bodies. The fact that she still wears a chain mail suit has other reasons: Sharks have skins like sandpaper. The structure

of the suit makes it safe for sharks to shoot by her at speeds exceeding 60 kilometres per hour. Since the instructors feeding these animals come into constant contact with them, a regular wetsuit would soon be useless. Bea works with the sharks in order to promote the survival of sharks. Commerce helps ecology in this case - a sensible contradiction.

For her, watching all those sharks milling about is daily routine.

SKATES
RAJIDAE

Comprising well over 200 species, this family is by far the most speciose family of rays. In all species, head, body and the broad pectoral fins are fused to form a rhomboid disc from which the tail is clearly separated. The latter is about as long as the disc, relatively stout, has lateral skin folds, carries two small dorsal fins near its tip and serves in steering while undulating movements of the pectoral fins provide lift and forward thrust. There is no barbed sting on the tail, but dorsal and/or ventral surfaces of most species are covered with enlarged dermal denticles (thorns), arranged in longitudinal rows (especially on the midline of back and tail) or patches (alar thorns on the 'wings' = pectoral fins). The thorn patterns change during growth from juvenile to adult, are different between adult males and females and play an important part when describing new species. Additionally, the adult female is often about a third larger than the male, hence in earlier times two species have been described for each sex of one species. In some species, weak electric sensory organs (different from those of torpedinids) are present in the tail. Most skates live on the bottom of the continental shelf and slope in cold and temperate regions; in the tropics, they are found in colder deepsea waters. Like other benthic rays, skates have large spiracle openings - located directly behind the eyes and rhythmically closed by fleshy flaps - through which respiration water is pumped into the gill cavity. Reproduction is oviparous. Copulation has been observed in several species: The male grabs a hold on the female with his pointed teeth (blunt in females, see dentition on p.207) and inserts one of his claspers which are large and swollen during courtship. Alar thorns and hook-like denticles on the claspers keep the partners together during mating which can last for hours. Each fertilised egg is enclosed in a rectangular capsule with stiff, horn-like projections at each of its four corners that serve in anchoring it to the substrate. After a few months, when the yolk sac is fully absorbed, the juvenile hatches from one end of the capsule. Skates have camouflaging colour patterns, prey on all kinds of benthic life which is not heavily shelled and are themselves eaten by sharks.

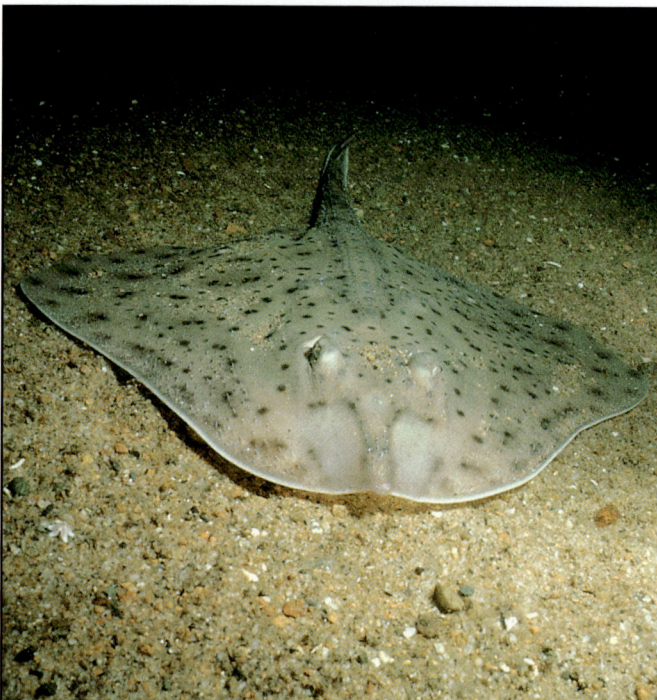

David Hall | New York, Western Atlantic

Raja ocellata
Winter skate

Length: At birth about 12, max. 100 cm.
Distribution: Western North Atlantic: Nova Scotia and Newfoundland to North Carolina.
Depth: 2-200 m, most abundant around 70 m.
General: Coloration light brown with dark blotches and 1-4 distinctive white-edged dark spots. Benthic on sandy and rocky bottoms. Migrates into coastal waters during the winter months (popular name) and retreats back into deeper, cooler waters in summer. The species reproduces throughout its entire range. Greenish to brownish egg-capsules (5.5x 8.5 cm) are deposited during the summer and autumn months. Preys mainly on crabs and shrimp, also on other invertebrates and bony fishes. A common species that is commercially exploited: Its fleshy pectoral fins ('wings') are marketed for human consumption.

Raja eglanteria
Clearnose skate

Length: Max. 85 cm, usually to 65 cm. Distribution: Western Atlantic: Nova Scotia to Florida. Probably also Iceland. Depth: Inshore waters to at least 100 m. General: The photo shows a regularly spotted young specimen. Adults feature a more irregular pattern of blotches and stripes. Copulation observed, may last for hours! Egg-capsules are deposited during the summer months. The empty four-tipped cases of skates are often washed ashore and are called 'mermaids' purses'. Preys on fishes, squid, crabs and shrimp.

Ingo Vollmer   Nova Scotia, Canada, Western Atlantic

Amblyraja radiata
Starry skate

L: Max. about 60 cm at lower latitudes (living in moderate depths) and about 90 cm at higher latitudes (living in deeper water). Di: Polar Sea and North Atlantic: Spitzbergen, Norway and Iceland to English Channel, northern North Sea and western Baltic; Nova Scotia to South Carolina. Also reported from deep water off Cape Town, South Africa. De: Inshore waters to 1,000 m, mainly at 50-100 m. G: Common, on various bottoms. Egg-cases max. 6.6x5.3 cm. Preys on all kinds of benthic animals. Caught in large numbers.

Florian Graner   Lofoten Islands, Norway

Raja clavata
Thornback skate

L: Max. about 90 cm. Di: Eastern Atlantic: Iceland and Norway to probably South Africa and into southwestern Indian Ocean, including North Sea, western Baltic, entire Mediterranean, western Black Sea. De: 1-300 m. G: With extremely variable pattern of different types of small and large thorns on back. Common in most areas. Preys on all kinds of benthic animals, preferably on crustaceans. Large females lay up to 150 egg-cases (7x9 cm) during one year! Commercially important in NW-Europe, Mediterranean.

John Neuschwander   Bantry Bay, Ireland

Raja brachyura
Blonde skate
Length: Max. about 120 cm.
Distribution: Northeastern
Atlantic: Shetlands, English
Channel and western North
Sea to Morocco, Madeira and
western Mediterranean. The
photo from the Canary Islands
extends that range further.
Depth: Inshore waters to
about 100 m. General: Benthic
on sandy bottoms. The popular name relates to the uniformly ochre upper surface
with numerous small dark
spots which reach the extreme
margins of the disc (different in
the otherwise similar *R. mon-
tagui*, see bottom of page).

Fernando Espino — Canary Islands, Eastern Atlantic

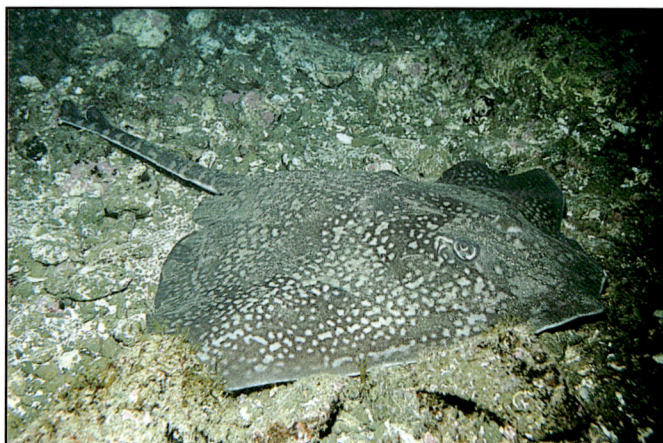

Raja maderensis
Madeira skate

Length: Max. about 80 cm.
Distribution: Eastern Atlantic:
Known only from Madeira and
the Azores. Further records
from the Canary Islands and
the African coast have been
misidentifications, mainly of
R. clavata (previous page).
Depth: Inshore waters to
150 m. General: A moderately
common skate in its range, living on muddy and sandy bottoms on the insular shelves,
but with only a few verified
records. Oviparous, no details
known. Preys on all kinds of
bottom-dwelling animals.

Peter Wirtz — Azores, Atlantic

Raja montagui
Spotted skate
Length: Max. about 80 cm.
Distribution: Northeastern
Atlantic: Shetlands, southern
North Sea and extreme western Baltic Sea to Morocco and
western Mediterranean.
Depth: Inshore waters to
100 m. General: Upper surface
with numerous large dark
spots which do not reach the
extreme disc margins (different
in *R. brachyura*, top of page),
sometimes with two faint ocelli. Egg-cases (max. 4.6x7.7 cm)
are laid in summer. Preys mainly on bottom-dwelling crustaceans. Moderately common,
regularly marketed in its range.

Helmut Debelius — Galicia, Spain, Eastern Atlantic

Raja miraletus
Twineye skate
Length: Max. about 60 cm.
Distribution: Northeastern
Atlantic: Portugal, entire Mediterranean, Morocco, Madeira
and Canary Islands to Namibia.
South Africa from False Bay to
Durban (SW-Indian Ocean).
Depth: 17-300 m, most abundant in 50-150 m.
General: Easily recognised by
two large distinctive ocelli, one
each on the upper surface of
the inner part of the pectoral
fins. Preys on small bony fish,
shrimp, crabs, mysids etc. but
has no specific preferences. A
common skate that is utilised
throughout most of its range.

Florian Graner   **Croatia, Mediterranean**

Raja undulata
Undulate skate

Length: Max. about 100 cm.
Distribution: Eastern Atlantic:
Southern Ireland and southwestern England to Mauritania,
including Canary Islands and
western Mediterranean (mainly
along African coast). Depth:
Inshore waters to 200 m.
General: This attractively patterned skate lives benthic on
muddy and sandy bottoms on
the continental shelf. Most
egg-cases (5x9 cm) are laid
from March to September.
Preys on diverse bottom-
dwelling animals. Displayed in
public aquaria (see p.108).

Helmut Debelius   **Galicia, Spain, Eastern Atlantic**

Dipturus batis
Common skate
L: Max. 250 cm. Di: NE-Atlantic: Iceland, N-Norway, North
Sea and western Baltic to
Madeira, Morocco and most of
Mediterranean. De: 10-600 m,
most abundant <200 m. G:
Common, marketed in northern Europe. Large egg-cases
(to 15x25 cm) are laid on the
ground in the spring and summer months. Preys on all kinds
of benthic animals, large individuals seem to prefer fishes.
About 30 Dipturus spp. live
worldwide (except in the tropics) on shelves and slopes.
They range among the largest
of all skates (length 200+ cm).

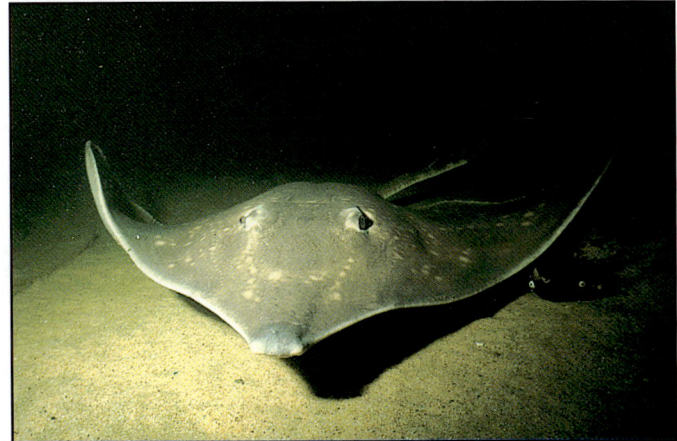

Lawson Wood   **Moray Firth, Scotland, North Sea**

Dipturus australis
Sydney skate

Length: Max. about 50, mature males 43-48 cm.
Distribution: Eastern Australia: Queensland (Moreton Bay) to New South Wales (Jervis Bay). Depth: 50-180 m.
General: The Sydney skate is the most common skate in the continental shelf waters off central eastern Australia. When compared to similar family members, its tail is conspicuously broad and flattened over most of its length. The photo shows an adult female. Size at maturity for females is about the same as for males.

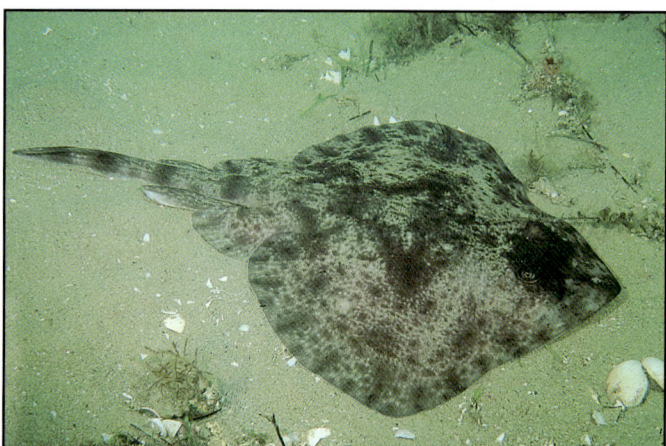

Saul Gonor — Jervis Bay, Australia

Dipturus lemprieri
Lempriere's skate

L: Max. at least 52, mature males 39 cm. Di: S-Australia: South Australia (Beachport) to NSW (Jervis Bay), Tasmania. De: 1-170 m, usually <40 m. G: Most common skate in S-Australian inshore waters. Blackish patch under snout tip. Male left, female below.

Rudie Kuiter — both photos Portsea, Australia

Dipturus whitleyi
Melbourne skate

Length: Max. at least 170 cm. Distribution: Southern Australia: Western Australia (Albany) to New South Wales (Wollongong) and all around Tasmania. Depth: 5-170 m.
General: While the Melbourne skate is most abundant in shallow inshore waters, it has also been recorded from the deep continental shelf but not from the slope. The largest skate of Australian waters may reach over 50 kg in weight. Large specimens caught in nets should be handled with care as their tails are clad with thorns.

Rudie Kuiter — Bicheno, Australia

CREATURES OF THE METEOR SEAMOUNT

As diving technologies develop and new breathing gas combinations are being established, divers go deeper and the border of exploration is pushed further, allowing dives to go down to 130 m and beyond. With dive expeditions to new habitats come new encounters with rarely seen marine organisms, including sharks and rays. Marine biologist Boris Frentzel-Beyme had the opportunity to glimpse into a world less well-known than our own solar system from aboard a modern research vessel and reveals some of the mysteries of deep sea elasmobranchs.

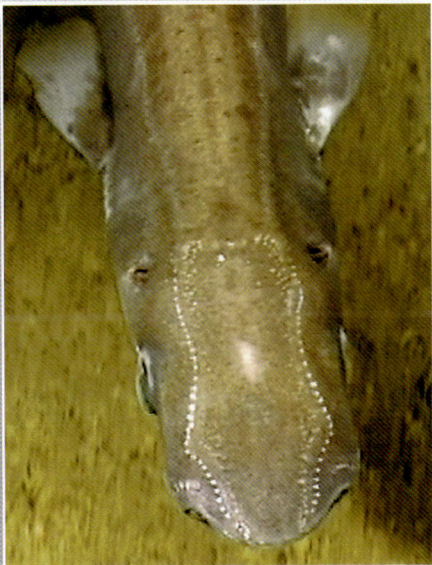

Head of the luminous shark *Etmopterus spinax* showing rows of pressure sensory pores (extended lateral line system).

In certain places, when diving at night to about 60 m depth and deeper, some of the inhabitants of deeper waters can be witnessed rising up to their shallowest vertical distribution every night, sharing a bit of their range with modern hi-tech divers. Technical diving may expand the range of humans in the oceans, but it still merely allows to scratch the 'surface' of the deep sea. The seafloor can be reached only within the range of the continental or insular shelves surrounding the coast, where the landmass gradually slopes from the shoreline toward the shelf's edge (or shelf break), which is mostly situated in a depth of around 200 m. Most marine life we are familiar with lives in this so-called neritic or littoral zone on and above the shelves, where daily and seasonal cycles of light, temperature and nutrient availability determine the settings for all forms of life. Extending from the shoreline to the shelf's edge, the shelf zone can be only a few kilometres wide, but also hundreds of kilometres in other regions of the world. From the edge of the continental (or insular) shelf, the seafloor slopes down more or less steeply to the deep sea plains, which level off at a mean depth of around 3,700 m. Hence, this part of the ocean floor is called continental slope. The deep sea plains constitute the overwhelming part of the world's oceans, roughly about 2/3 of the earth's surface is water above the deep sea. Covered mostly by fine sediment, the deep sea plains resemble a desert landscape - vast, virtually empty, monotonous and seemingly hostile to life.

In some places, the deep sea plains are interrupted by more or less extensive seamounts and Guyots that rise some hundreds of metres or even kilometres like mountains towards the surface. Some of these stand alone, others form groups or big aggregations of many individual seamounts. Some are pointed and small, others are like huge plateaus with oddly-shaped summits. In the middle of the Atlantic Ocean, stretching from Iceland in the North almost to the Antarctic in the South, is a huge mountain range called the Mid Atlantic Ridge. Similar structures can be found in the Pacific as well as in the Indian Ocean. There, new ocean floor is constantly building up by material rising from inside the earth and spreading out in opposite directions. In other regions, the ocean floor in turn is pushed against solid continental plates and forced to slip underneath them, resulting in deep faults and chasms where the oceans reach their greatest depths, 9,000 m and even more. The deepest spot of the world's oceans is located in the Mariana Trench in the Western Pacific and is 11,034 m deep!

Although the view of the deep sea as a habitat seems to show a diverse landscape that comes in different shapes with many niches, all life down there has to arrange with some simple and uncomfortable facts: The pressure is bone-crushing, the temperature is low, there is no light and there is little food to find. Additionally, these settings are permanent, with little or no seasonal changes compared to those of shelf waters. Life in the abyss is primarily dependent on the organic matter (debris) sinking down through the water column from the upper water layers. Planktonic algae, tiny crustaceans, faecal droplets, dead fish or whales - one way or the other, their final destination is always the ocean floor. And there, many hungry mouths await their arrival: Filtering sessile animals, burrowing worms and crustaceans, bony fishes, cephalopods and - of course - sharks, rays and skates.

One seamount that has just recently received some attention by scientists is the Great Meteor Seamount located in the Central Eastern Atlantic, some 600 nm (nautical miles) south of the Azores and 800 nm west of the Canary Islands. Its oval-shaped plateau summit spreads out about 30 by 40 km in a depth of around 300 m, rising steeply from the surrounding depths of 4,000 m in the Canary Basin. With these characteristics it forms an extraordinary, shelf-like habitat within a truly oceanic surrounding, posing an isolated oasis for the lifeforms trapped on it, unable to travel the great distances over the deep water to the next continental shelf. This is not to the least part what makes remote seamounts so interesting to science.

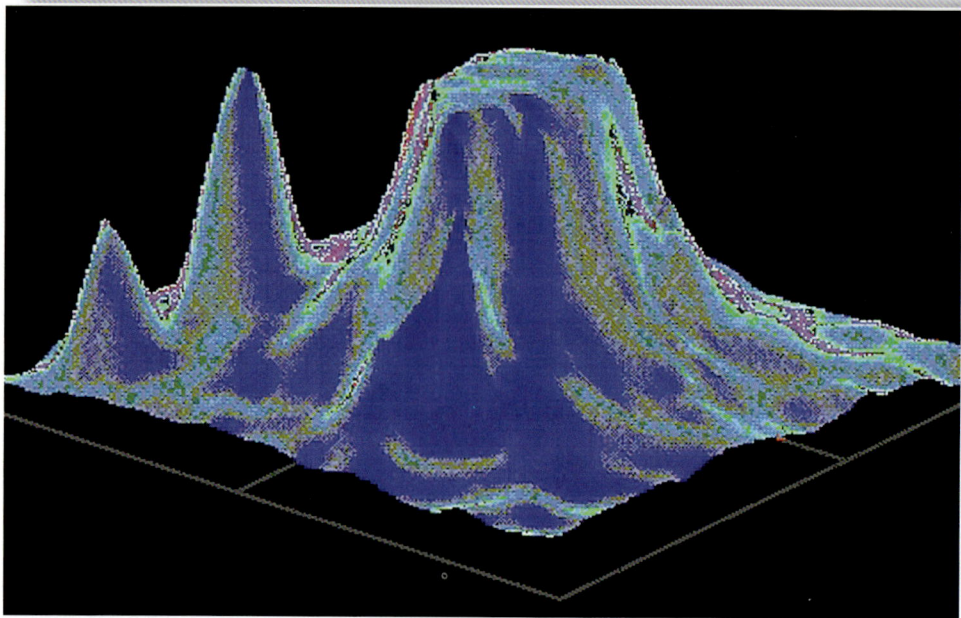

Computer image of the Great Meteor Seamount in the Central Eastern Atlantic. Magnification of profile about 20x.

On the plateau of the seamount, among such very abundant bony fishes like small snipefishes *(Macroramphosus* spp.) and boarfishes *(Capros aper)*, horse mackerel *(Trachurus picturatus)*, conger eels *(Conger conger)*, sailfin dory *(Zoenopsis conchifer)* and others, some species of elasmobranchs can be found. Besides single longnose skates *(Raja oxyrhinchus)* and Atlantic torpedo rays *(Torpedo nobiliana)* present in bottom trawl catches conducted there, high numbers of sharpnose sevengill sharks *(Heptranchias perlo)* revealed this shark to be the dominant predator on the seamount plateau. This moderately-sized, slender shark (maximum length about 140 cm) is known to inhabit the seafloor on the outer shelves and upper slopes of most of the world's oceans. It is mainly found in depths of 200-800 m, but occasionally also in inshore coastal waters.

Primarily a bottom-dwelling species of deep waters, the sharpnose sevengill shark is an agile and active predator with powerful jaws and teeth that enable it to grasp and swallow small prey as well as dismember larger prey organisms. On the Great Meteor Seamount, it was found to feed on small fishes and octopus as well as some larger bony fishes, rays and also other sharpnose sevengill sharks. Most of the sharks found in that region were juvenile or subadult, which lead to speculations whether the adult animals are really absent or simply not caught by the fishing gear used. It is also possible that the seamount serves as a nursery ground for this shark species, with adults migrating to and from the seamount.

Apart from those on the plateau in about 300 m depth, further catches (using vertical longlines) were made in about 800-1,100 m depth along the slope of the seamount. The shark species found there included Portuguese dogfish *(Centroscymnus coelolepis)*, birdbeak dogfish *(Deania calcea)*, arrowhead dogfish *(D. profundorum)* as well as rough longnose dogfish *(D. hystricosa)*, all in the size range of about 100 cm. These shark species are in many ways different from the sharpnose sevengill sharks from the plateau and bear some common characteristics and can be considered typical for deepwater sharks.

When comparing their body shape to that of the commonly-known, agile, stream-

The sharpnose sevengill shark *Heptranchias perlo* is the top predator on the seamount plateau.

This (male) arrowhead dogfish *Deania profundorum* is not luminous like the (smaller) species of the genus *Etmopterus* but also inhabits the deep sea environment described in the text where large eyes are a distinctive feature of many predators.

lined 'classic' sharks like reef sharks, blue sharks or the sharpnose sevengill shark, they seem to be sluggish and the consistency of their body is rather soft and watery compared to the firm and muscular flesh of their relatives. Due to the lack of light in the deep, their colour mostly ranges from uniformly dark brown to dark grey with no distinct patterns or markings on their bodies. Their dermal denticles are relatively large and often spiky and hence their skin is rough to the touch, the individual denticles having bizzare shapes that certainly do not enhance the waterflow over the skin as in fast-swimming species. Most deep water sharks are small to moderate in size, less than or around 100 cm. "This reflects the low availability of food in this environment" states Brad Wetherbee, US shark biologist and expert on deepwater sharks. "Furthermore, these sharks can actually increase their size despite such low food levels." he adds - "By substituting some proteins in their tissue through incorporation of water, they can reach bigger sizes without investing the energy needed to use organic components. And the larger you are, the fewer predators are there that can eat you. Hence also the soft consistency of their tissue."

Also their fins tell a story: They are soft and small, lacking the characteristic rigidity and size as seen in other sharks. All this makes sense when looking at their habits and environment. In a cold and completely dark world where food is scarce, everything moves more slowly. Most of the time, all animals are rather in a stand-by or energy saving mode than swimming about, actively searching for prey. Upon encountering either living prey or a large carcass, e.g. of a dead dolphin sinking onto the bottom, quick movements and active feeding are possible to ascertain a maximum food uptake - it probably could be the last opportunity to do so for a prolonged period of time. Additionally, all energy accumulated should be saved and spent on a minimal basis. In this context, another feature of deepwater sharks plays an important role: Their huge liver that contains large amounts of oil. In some other sharks, the liver provides some positive buoyancy and thus reduces their tendency to sink (sharks lack the swimbladder of bony fishes). For deepwater sharks, their enormous oily liver provides both - literally neutral buoyancy, freeing them from expending energy while swimming in order not to sink to the bottom - and it serves as an energy storage, allowing survival in extended times without food. Consequently, deepwater sharks can just float along, moving little and if so only slowly in their environment, then reacting spontaneously to feeding opportunities and can actively swim and hunt whenever necessary. Most deepwater sharks have tiny teeth and feed mostly on small prey like fish, crustaceans and molluscs. Nevertheless, they are capable of tearing pieces out of bigger prey and carrion, as cetacean remnants in the stomachs of Portuguese dogfish sharks document.

Two more shark species inhabit the waters surrounding the Great Meteor Seamount. With a maximum length of only about 25 cm possibly the smallest shark of the world, the spined pygmy shark (*Squaliolus laticaudus*) was found living in the open water above the plateau. This oceanic dwarf species shares the upper 200 metres of the ocean near the edge of the plateau with small schooling fishes but avoids coming too close to the surface. The underside of its body is covered with photophores, tiny organs embedded in the skin that emit light and help to camouflage the shark against the dim light from above when seen by predators from below. Several other deepwater shark species are also known to possess photophores which emit light, a phenomenon called bioluminescence. Sharks of the genus *Etmopterus* are even called lantern sharks for their ability to glow in the dark by means of photophores scattered on their bellies.

Although it would seem that potential prey availability is higher on the plateau due to the more direct coupling to the animals of the upper water layers, the elasmobranch fauna of the Great Meteor Seamount seems more diverse along its slopes. With the various catching and sampling techniques applied, it is still possible if not likely that some shark species present on and around the seamount have not yet been discovered. And the species depicted here represent just a small portion of the great variety of deepwater sharks that have been seen so far. In adaptation to their habitat and lifestyle, they display a variety of shapes, sizes and habits, reflecting their success in occupying the niches of predators in their environment. However, we still know only little about this dark world and its mysterious inhabitants, and few occasions allow a better insight in the dynamics and processes of the deep sea. When probing the deep sea with up-to-date but yet clumsy scientific equipment, still many organisms will stay out of its reach and remain undiscovered. This in turn means that much more is still to be learned and discovered about the largest of all biomes of the earth.

This speciose family of stingrays comprises about six genera with more than sixty species, its systematics are constantly revised. Many family members can be seen while diving or snorkelling in tropical waters. But they are also often overlooked because many species bury in the substrate, with only their eyes protruding, to conceal themselves from predators. Some have a distinct morphology and colour pattern, others are very similar to each other and difficult to distinguish in the wild. See distinctive features of genera listed below.

These stingrays are better known than their relatives because of the many interactions that occur between them and humans. Popularisation of dive sites, where large stingrays envelope divers in their pectoral fins and feed from their hands, gave rise to an overall positive image of stingrays in recent years.

On the other hand, most of these rays can also be potentially hazardous to humans. Like many species of the order Myliobatiformes (in which this family is placed), most family members possess barbed tail stings. These are inserted dorsally on the tail, continuously replaced like jaw teeth, covered by a sheath of venomous tissue and used in defence. An individual may have from one to several stings, depending on species and size. When trodden on in shallow water or seriously molested, a stingray thrusts its tail over the body and into the attacker, no matter if it is a human foot or a hammerhead shark! Besides the wound itself, also venomous tissue associated with the sting and secondary infections affect the unlucky victim. There are several confirmed cases of human fatalities, resulting from loss of blood, physical damage and/or gangrene - few of the victims were diving, however. First aid: Remove sting and tissue remains from the wound, clean with water, immerse wound/limb in water as hot as the victim can bear (50+°C) to destroy venom and relieve pain. Jagged wound margins and infections should be treated by a medical doctor. See also MIND YOUR STEP! on pp.221-223.

While most family members are strictly marine, some are known to migrate up rivers, for example the feathertail stingray Pastinachus sephen; still others live exclusively in freshwater, e.g. the Laos stingray Dasyatis laosensis.

All family members are ovoviviparous, several are known to reproduce in captivity. Two different copulation postures (belly to belly and male mounting female) have been observed. Prey includes a wide variety of invertebrates (often dug up from the substrate) and bony fishes. Several family members have been encountered at cleaning stations, where cleaner fishes remove parasites, infected skin and mucus from their bodies. Suitability for a life in aquaria varies from species to species.

The following characteristics are important in the recognition of whiptail stingrays in their natural habitat: Disc shape, snout shape, coloration, presence or absence of dorsal or ventral skin folds on tail, presence of dorsal keel, tail length in relation to body size (tails of live specimens may be damaged and thus shorter than usual!), presence of tubercles (thorn-like spines) on body disc and/or tail, position and number of tail stings.

The genera of this family are characterised as follows:
Dasyatis (with 35-40 species): Body disc rhomboid, ventral finfold not as high as tail above it; Himantura (with about 20 species, some of them still undescribed, see H. uarnak): Body disc rhomboid to oval, tail long, slender, without finfolds; Pastinachus (with 1 species): Body disc rhomboid, ventral finfold much higher than tail above it; Taeniura (with 4 species): Body disc round or oval, ventral finfold reaches tip of tail; Urogymnus (with 2 or 3 species): Body disc round or oval, with large thorns (tubercles) on back, without tail sting.

Himantura jenkinsii   **Maldives, Indian Ocean**

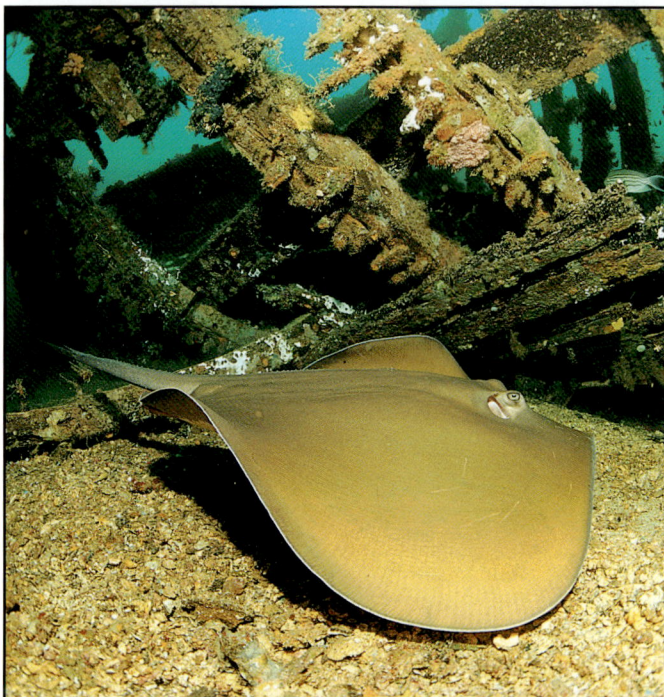

Werner Thiele

Philippines, Western Pacific

Himantura jenkinsii
Jenkins' whipray

Width: Max. 105, length to about 200 cm. Distribution: Indo-West Pacific: Southern Africa to Australia, New Guinea and Philippines. Depth: Shallow waters to about 55 m. General: A medium-large yellowish-brown stingray. Underside and disc margin distinctly white, hind part of tail greyish. Tail whip-like, without skin folds, but with one to three stings. A row of large spear-shaped thorns and a narrow band of close-set, heart-shaped denticles extend from nape onto tail. Although the distribution of this whipray species is poorly defined it can be quite common in ideal habitats which appear to be sandy or silty bottoms associated with coastal lagoons. Seen singly, sometimes in groups. A similar ray with additional dark spots along its posterior disc margin occurs in the Arafura Sea and may be conspecific or *H. draco* described from South Africa.

Himantura granulata
Mangrove whipray

W: Max. 100+, total length up to 400 cm. Di: Indo-Pacific: Maldives to Micronesia. De: 1-20+ m. G: Body disc oval. Usually two stings on tail. Found in the shallow water of mangrove areas, estuaries, and sheltered coastal bays of corals reefs. The species can best be recognised by its partly white tail (including the spine), which looks like having been dipped in white paint. Base of tail and body disc dark slate grey with scattered white spots. Preys on a great variety of shelled benthic invertebrates, which are covered by the bluntly rounded snout and sucked from the substrate to be crushed between the teeth.

John Neuschwander

Maldives, Indian Ocean

Himantura undulata
Leopard whipray

Width: Max. 150 cm.
Distribution: Indo-West Pacific:
Red Sea and southern Africa
to eastern Australia, New
Guinea, Philippines and Micronesia. Also eastern Mediterranean (Lessepsian migrant).
Depth: 3 - 175 m.
General: The (unbroken) whip-
like tail of this species has
more than twice the length of
the body disc. The snout is
pointed, the pectoral tips are
bluntly rounded. The distinct
reticulate pattern is slowly disappearing with age. This stingray inhabits sandy bottoms
near reefs; usually only parts of
the animal such as eyes and tail
are visible when it is hiding in
sandy substrate. The species
preys on crustaceans, molluscs,
including squid, and jellyfish.

The photo at the right
shows a specimen which has
the same type of colour pattern as the holotype of this
species described by the
Swedish scientist Peter
Forsskål in 1775.

The small photos below
show colour variants. See also
following page.

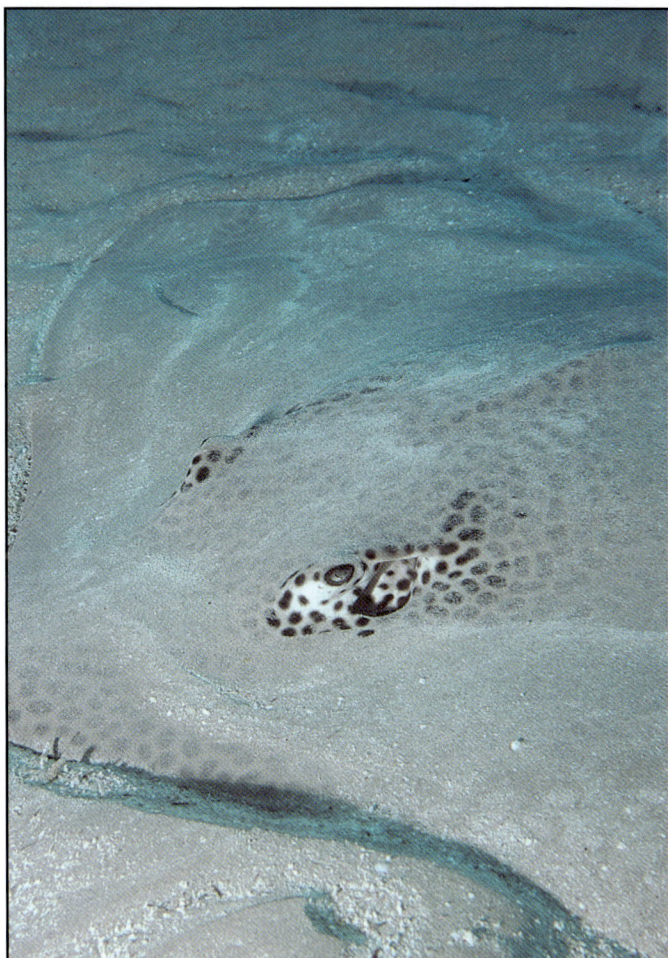

Helmut Debelius   Saudi Arabia, Red Sea

Helmut Debelius   Sinai, Red Sea

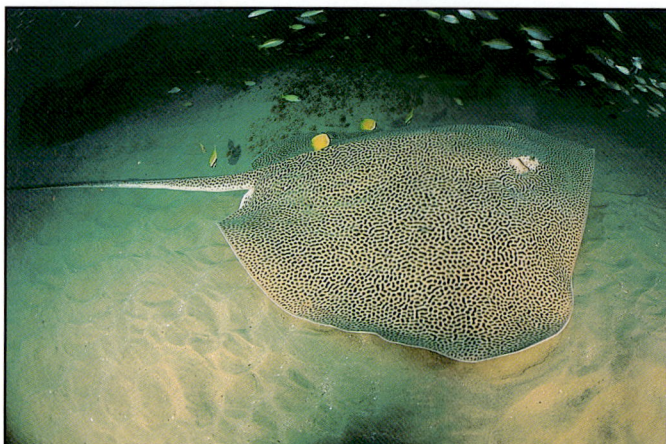

Herwarth Voigtmann   **Mozambique, East Africa**

Himantura uarnak
Honeycomb whipray

The photo shows either a colour variant of *H. uarnak* or a new, undescribed species. This species complex of similarly patterned Indo-Pacific stingrays is currently under investigation by Dr. Peter Last (CSIRO Division of Fisheries, Hobart, Australia). The true identity and relationships of the various species involved are being analysed by DNA sequencing (genome analysis). Interestingly, juveniles of these stingray species are more distinct from each other than adults (concerning disc shape, degree of spotting, denticle structure).

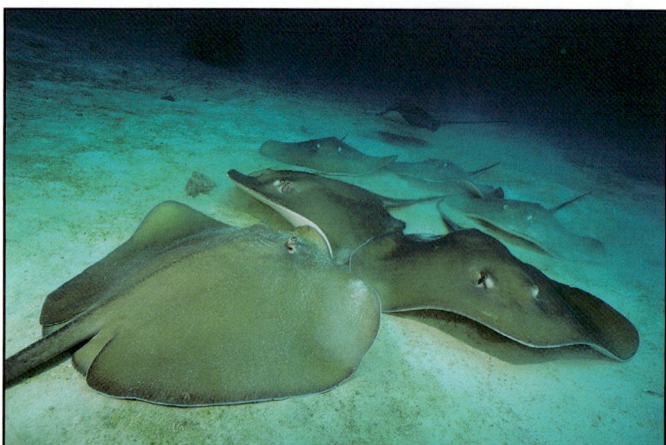

Surin Island, Thailand

Himantura fai
Pink whipray

Width: At birth about 55, max. 150+, total length up to 500+ cm.
Distribution: Not well known but possibly widespread in Indo-West Pacific from South Africa to Society Islands.
Depth: Intertidal to at least 20 m.
General: A large stingray species with a very long tail: If the latter is unbroken, total length may exceed 5 metres! Undamaged tail almost three times in disc length. Tail without finfolds and usually with one sting. Body disc rhomboid. Snout bluntly pointed. Back smooth, without tubercles. The coloration of the upper surface is uniformly tan to brownish-pink mottled with darker blotches; the underside is white. The biology and habits of the pink whipray are little known although it is not uncommon in its range. The species is mainly encountered in offshore habitats, usually around atolls in deep water. Large groups are known to occur at times on atolls of the Great Barrier Reef and throughout the Caroline Islands, Micronesia. Schools aggregate in shallow water to feed at night. The species preys on crustaceans and may have a preference for shrimps as it is often caught in the nets of shrimp trawlers.

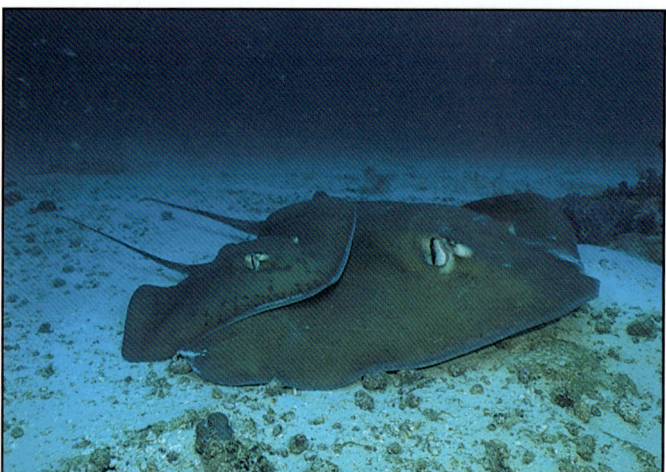

both photos Mark Strickland   **Myanmar**

Dasyatis americana
Southern stingray

Width: At birth 18, max. 150, total length to about 350 cm.
Distribution: Western Atlantic: New Jersey to Brazil.
Depth: Intertidal to 25 m.
General: Body disc rhomboid. Row of small tubercles on centreline of back. Low dorsal keel on tail. Ventral finfold long. Coloration dorsally grey to brown, without markings. Found on sand, in seagrass meadows, lagoons and reefs. Lives solitarily, in pairs or aggregations. Often buried in sand during the day, emerges at night to actively hunt, often in seagrass meadows. Has been observed near cleaning stations, where it is cleaned by diverse fish species. Preys on bony fishes (scorpionfishes, toadfishes, surgeonfishes), crabs, worms, clams, shrimp, mantis shrimp and tunicates. Clams are crushed between the teeth and the shell fragments are spat out.

A specific site at Grand Cayman Island in the Caribbean has become known for its resident population of southern stingrays and their interactions with divers. World-famous 'Stingray City' is located in a lagoon behind the barrier reef near the entrance to the 'North Sound' of Grand Cayman (photos). Since decades fishermen have anchored and sorted their catch there. The rays soon found out and started to aggregate in this place. In the late 1980s, the first divers arrived and were greeted by numerous, totally inaggressive stingrays who would slowly circle the divers and allow to hand-feed and caress them. Many popular magazine articles and films covering these encounters have appeared since. As the news quickly spread, many people came to meet the gentle rays. Consequently, some rules had to be set up: No gloves or diving knifes are allowed and the food for the rays is sardines or squid provided exclusively by the dive masters who also teach the correct feeding technique. Despite the rays' mass and strong spines, no serious accidents have happened so far!

Georgette Douwma   Bonaire, Netherlands Antilles

Helmut Debelius

Kurt Amsler   both photos Grand Cayman, Caribbean

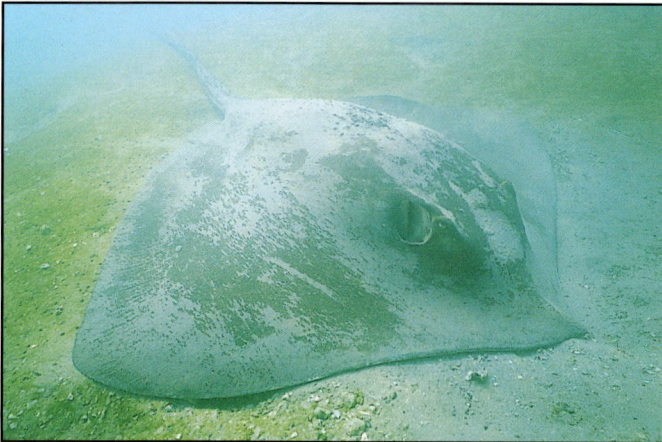

Paul Humann   Florida, Western Atlantic

John Neuschwander   Canary Islands, Eastern Atlantic

Jürgen Warnecke   Canary Islands, Eastern Atlantic

Dasyatis centroura
Roughtail stingray

Width: At birth about 35, max. 210, usually 100-130 cm. Distribution: Eastern Atlantic: Southern Bay of Biscay to Angola. Also Madeira, Canary Islands, Cape Verde Islands and entire Mediterranean where most common off Algeria, Tunisia and Sicily; absent from Black Sea. Western Atlantic: George's Bank to Florida and eastern Gulf of Mexico; Uruguay to southern Brazil. Depth: Shallow waters to about 200 m.
General: Disc rhomboid with straight front margins. Tail about twice as long as disc, with long and deep ventral fin-fold that starts at level of spine. Back with row of large thorns along midline, some thorns scattered on centre of disc. Top and sides of tail also with large thorns (popular name). Back olive-brown, underside whitish. This sting-ray lives on sandy and muddy bottoms. In West Africa, it is found not only in marine habi-tats but also in estuaries and the lower reaches of rivers. Preys on bony and cartilagi-nous fishes, bottom-living crus-taceans and cephalopods. Ovo-viviparous, reproduces annual-ly, gestation period is about 4 months, litter size 2-4, parturi-tion occurs in the autumn and early winter months.

Dasyatis pastinaca
Common stingray
W: Max. 60 cm. Di: Eastern Atlantic: North Sea and Bay of Biscay to the Azores, Madeira, Canary Islands, Cape Verde Islands and Mauritania. Also in Mediterranean and Black Seas. De: 1-200 m. G: Back smooth, without tubercles. Short dorsal and short, high ventral tail fin-fold, the latter originating at level of sting. Tail less than twice disc length. Back grey to brown, white with brown to grey margin below. On sand and mud, sometimes also in estuaries. Litter size 4-7, gesta-tion period 4 months. Preys on crustaceans, molluscs, fishes.

Dasyatis marmorata
Blue stingray

Width: At birth 15, max. 75, total length to about 150 cm. Distribution: Eastern Atlantic: Senegal, West Africa, to Cape of Good Hope, South Africa. Western Indian Ocean: Natal to the Cape. Depth: Intertidal to 110 m. General: Body disc rhomboid. Back smooth, without thorns. Tail less than twice disc length, with short dorsal and long ventral finfold, usually with one spine. Distinctive coloration of bright blue reticulations on a golden-brown background; tail darker, underside white. Often found close inshore in the surf zone, near sandy beaches and in shallow bays. Migrates from inshore waters to deeper off-shore areas in winter. Gestation period is about 12 months, litter size 1-4. Preys on swimming crabs, shrimp, mantis shrimp, mysids, amphipods, diverse worms and small bony fishes. The blue stingray is popular with surf anglers in South Africa but generally released after capture.

Simon Chater   Namibia, West Africa

Dasyatis violacea
Pelagic stingray

W: Max. 80, total length up to 160 cm. Di: Circumglobal in tropical and subtropical waters. De: Intertidal to 64 m. G: Disc angular, front margin broadly rounded. Back smooth. With long ventral finfold on tail. One very long sting on tail. Eyes do not protrude. Dark purple or black on both sides. Rare, agile swimmer of the open ocean; sometimes also close inshore. Preys on jellyfish, crustaceans, oceanic squid and small pelagic bony fishes. Uses pectoral fins to manipulate prey into mouth. Unique to family, also placed in separate genus *Pteroplatytrygon*.

Christoph Gerigk   Cape Verde Islands, Eastern Atlantic

Dasyatis akajei
Red stingray
Width: Max. 66, max. total
length at least 138 cm.
Distribution: Northwestern
Pacific: South Japan to Thailand. Depth: Shallow waters to
at least 100 m. General: This
stingray species is found in
tropical waters and lives on
the bottom of the continental
shelf. The underside of the
body disc and the pelvic fins is
whitish with a broad yellowish-
brown margin. Tail length is
less than twice the disc width.
Preys mostly on small fishes
and crustaceans and is caught
commercially in some quantities in the Gulf of Thailand.

Kazu Masubuchi

Dasyatis matsubarai
Matsubara's stingray

Width: Max. about 200 cm.
Distribution: West Pacific:
Japan to Thailand.
Depth: Inshore waters to at
least 60 m.
General: A large stingray with
a tail shorter than the body
disc. Found living on the bottom in shallow coastal waters.
Described by the Japanese
ichthyologist Miyosi in 1939
and closely related to the following species *D. brevicaudata*.
Records of the latter from
Thailand may refer to this
species.

Kazu Masubuchi   both photos Izu Peninsula, Japan

Dasyatis brevicaudata
Shorttail stingray
Width: At birth 15, max.
200 cm. Distribution: East and
South Africa, Australia, New
Zealand. Depth: 1-476 m.
General: Back smooth. Short
dorsal and long ventral finfold.
Tail shorter than disc, usually
with two stings. On sand and
rocks, also in bays. Often
aggregate in large numbers,
move into shallow waters with
rising tide. Aggregation and
mating behaviour in a New
Zealand underwater cave system have only recently been
documented on film. See also
DANCE OF THE STINGRAYS
on pp.195-199.

David Fleetham / IV   South Australia

Dasyatis kuhlii
Bluespotted stingray

W: Max. 50 cm. Di: Indo-West Pacific: Red Sea, South Africa to Samoa, Japan, Australia. De: 1-50 m. G: No thorns on skin. Short dorsal, long ventral tail finfold. Usually one sting on tail. Back reddish-brown to olive drab with iridescent blue (sometimes white) and smaller black spots; underside white. Posterior part of tail with distinctive black and white rings. Little known, but apparently common, often close inshore. Singly on sand, moves into shallower water at high tide. Possibly migratory. Eats crabs, shrimp. Difficult to approach.

Helmut Debelius   Sri Lanka, Indian Ocean

Dasyatis thetidis
Thorntail stingray

W: At birth <35, max. 180, total length up to about 400 cm. Di: Indo-West Pacific: Mozambique, South Africa, southern Australia (Shark Bay, Western Australia, to Coffs Harbour, northern NSW) and all around New Zealand. De: Inshore waters to 380 m. G: Uniformly dark brown, green or black above, underside white. Tail beyond sting (1-2) black, uniformly covered in thorns (name). Ventral finfold long, low, black. May attain 210+ kg. Often confused with the similarly distributed *D. brevicaudata* (previous page).

Rudie Kuiter   New South Wales, Australia

Pastinachus sephen
Feathertail stingray

W: Max. 180, length to 300 cm. Di: Indo-West Pacific: South Africa and Red Sea to Micronesia and Melanesia. De: 1-60 m. G: Close inshore, also in freshwater streams. Uniformly brown with darker long ventral finfold. Eats molluscs, crabs. Inquisitive, investigates divers' activities.

Klaus Hilgert   Egypt, Red Sea

255

A HUNDRED ANXIOUS MALES

We were well aware of the various inconveniences associated with filming in the IMAX format. What we could not predict was the weather we would be subjected to during our five expeditions to Cocos Island in the East Pacific. The great El Niño of 1998 created spectacularly unfavourable conditions for filming large migratory marine life at Cocos. Most of the large species simply left the Island for cooler climes. When the El Niño finally dissipated, the warm clear waters were almost immediately followed by abnormally cold and murky water. Its evil sister, the La Niña phenomenon, had replaced the El Niño. For months we had been wishing for cooler water. Did underwater filmer Howard Hall succeed in catching huge schools of sharks and rays on film in the end?

Divers filming with the first-ever IMAX 3-D underwater camera in the eastern Pacific Ocean.

MARK CONLIN

The good news was that the great predators returned to Cocos in time for us to capture them on film. Hammerhead sharks once again filled the waters over the Alcyone seamount. Mantas glided gracefully past Dirty Rock. However, we were still waiting for the marbled stingrays (*Taeniura meyeni*) to gather and mate. This event had been in my original script, but as the months wore on, I decided our chances of seeing this phenomenon, let alone filming it, were becoming negligible. So I deleted it from the script. Then, during the last week of our final expedition to Cocos, marbled stingrays began gathering in the pass between Cocos and Manualita (a large islet on the northern side of Cocos). Courtship had begun. In one day we would capture one of the most spectacular sequences for our film.

The day began before breakfast when the scouting divers returned from a dawn dive to report marbled rays gathering in the pass at the southern tip of Manualita. One of them was beside himself with excitement. "It's happening!" he cried from the skiff as I looked on from the *Undersea Hunter's* deck while warming my hands on a hot mug of tea. "It's happening now and we gotta go right now!" Twenty minutes later, the camera was ready and the crew loaded into the skiffs, most without breakfast. Drifting with the strong current, we passed over a large boulder to discover a cloud of huge marbled stingrays milling about on the other side. Over a hundred of the about one-metre-wide rays were hovering in the current in the centre of the pass. As I descended lower, I noticed a pile of rays in a cave at the base of the boulder, which lay adjacent to the current pass. I knew that all of the rays would be sexually very excited males. All except one, that is. Buried in the cave would be one extremely large, somewhat disgusted female.

I keyed the microphone on my communication system and called the camera support boat. "Surface copy, send the camera," I squawked sounding like a cross between Daffy Duck and Porky Pig. "Surface

A large number of male stingrays follow a huge female. Male elasmobranchs can easily be identified by their two penis-like claspers.

copies," was returned from the surface. "Lance is on the way with the camera." A few moments later, I saw him kicking like a madman toward the bottom, pushing the bulky camera ahead of him. The camera boat had dropped him a hundred yards upstream of our location. The goal was to get the camera down to us before the current swept him passed the spot.

I nestled down behind the six-metre-high boulder where I and the squadron of rays were protected from the current. Then Lance swam the camera down to where I had wedged myself in place between two rocks. Holding the massive IMAX camera steady was a major problem. The slightest surge or current would simply sweep the camera away dragging its hapless operator in tow. Looking at the 5-inch video monitor

This photo shows only a few of all the males milling around in a narrow pass and desperately searching for a mating partner.

The huge female emerged from the cave, paused at the boulder and then moved into the pass.

that served as the IMAX camera's viewfinder, I saw a spectacular sight. Marbled rays filled the frame gliding into view from all directions. The 2,000-watt movie light system illuminated their white bellies as they passed overhead contrasting beautifully with the dark blue water. With literally dozens of marbled rays passing within about a metre of the camera, some actually brushing against the dome port, the image captured by the viewfinder was breathtaking. I set my focus, adjusted my aperture, and turned on the camera.

Normally, it takes us about an hour and a half to shoot one roll of IMAX film. But at times we seemed to actually race through film. Thirty minutes after dropping over that boulder to find the courting marbled rays I was out of film. I called the surface on my Buddy Phone. "Surface copy. Retrieve the camera and give me a fast reload," I said. The female marbled ray was still lying where we had found her in the cave at a depth of 25 metres. Suddenly, there was movement in the cave. Sand billowed out of the cave opening as a dozen male marbled rays lifted off the bottom and took flight. Then, looking down, I saw an enormous marbled ray emerge from the cave. It was the female. At about two metres in diameter she dwarfed the males that hovered about her. She paused at the base of the boulder then moved into the pass. We swam as hard as we could. The three of us huffed and puffed through our rebreather hoses like marathon runners. After twenty days of diving, we were all in terrific physical shape. With all the effort we were expending, you'd think we would be moving through the water like Navy Seals. Not so. We crept over the bottom at a snail's pace. You can only swim two speeds with the IMAX system: Very slowly and extremely, pathetically slowly. We were moving at top speed, which is to say we were moving very slowly. The only good news was that the female marbled ray was also moving very slowly. Still, we couldn't catch her. Our only hope was that she might double back. After a half-hour of hard swimming, she did just that. I turned the camera away from her and triggered the run switch. Then I pointed the camera down as she passed beneath me with a hundred anxious males trailing behind.

My film crew enjoyed brunch at 2:30 that afternoon after the female and her admirers finally outdistanced us, or exhausted us, or actually both. Surprisingly, she returned to the same cave during the night, which allowed us to repeat the exercise the following day. In the end we had exposed 10 rolls of film on marbled ray courtship. In the months ahead, these 30 minutes would be edited down to create a sequence that would last about 2 minutes on the giant eight-story-high IMAX screen. This was the last such sequence we would capture for 'Island of the Sharks' and I was finally confident that we had a great film in the can.

Taeniura meyeni
Blotched stingray

Width: Max. 180 cm.
Distribution: Indian Ocean.
Depth: 5-100+ m.
General: A large, common stingray, reaching 300 cm in total length, if the tail is unbroken. Its coloration is highly variable, from pale grey to almost black with many irregular dark blotches. The tail usually carries one sting, which is set well away from the base of the tail.

The species is encountered along the base of drop-offs or on sand flats in or near coral reefs. In order to find prey, it often excavates large holes by blowing water from the mouth, previously taken in through the spiracles. In such a way, the ray dislodges molluscs and crustaceans from the sand, which are taken up and eaten (see characteristic feeding posture in top photo). Small photo below: By undulating their pectoral fin margins, rays are able to swim in any direction: Forward when undulations run from front to back and vice versa.

Klaus Hilgert   Egypt, Red Sea

Helmut Debelius   Western Australia

Taeniura grabata
Round stingray

Width: Max. about 100 cm.
Distribution: Eastern Atlantic: Madeira, Canary Islands and Cape Verde Islands to Senegal and Angola. Western Indian Ocean including Red Sea. Also in southern Mediterranean: Gulf of Gabes, Tunisia, to Egypt. Depth: 30-100 m.
General: Disc almost circular, tail less than disc length. Grey, brown or olive with darker blotches above, yellowish-white below. Since this widespread species is only rarely found in the Mediterranean, it most probably is an immigrant from the Red Sea.

Helmut Debelius   Canary Islands, Eastern Atlantic

259

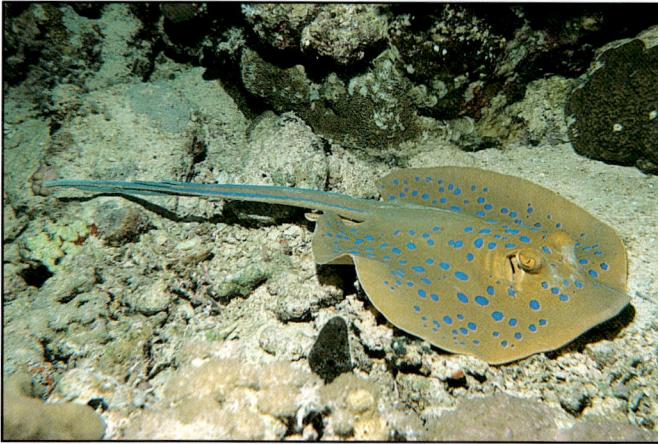

Helmut Debelius   Sudan, Red Sea

Taeniura lymma
Bluespotted ribbontail ray

Width: Max. at least 30 cm. Distribution: Tropical Indo-Pacific: Red Sea and South Africa to at least Solomon Islands. Not recorded from Japan. Depth: 2-20+ m. General: The common species reaches a total length of at least 70 cm. Old reports of a total length of 240 cm are most probably incorrect. This stingray can easily be identified by many large, brilliantly blue spots on its yellowish- to red-dish-brown body disc (under-side white). The tail sting (often two) is set well back on the tail which has a small, fringe-like caudal fin and is longer than the smooth, oval body disc. The species is encountered in reefs, resting on (rarely buried in) sand under corals and in crevices by day, swimming about and hunting for prey at night. Preys on worms, shrimps and hermit crabs. *Dasyatis kuhlii* (p.255) is similar, but has fewer, dark-ringed blue spots and its tail tip is ringed in white and black. Small photo: Sinai, Red Sea.

Helmut Debelius   Alor Island, Indonesia

Urogymnus asperrimus
Porcupine ray

W: Max. 100+ cm. Di: Red Sea and S-Africa to Fiji, also tropical West Africa. De: 1-30 m. G: Disc oval with numerous sharp thorns but tail without spine. On coral rubble, shallow sand flats and in seagrass beds. Preys on fishes (wrasses sleeping in the sand) and crabs. Rare.

Helmut Debelius   Maldives, Indian Ocean

SEX IN SHARKS AND RAYS

Little is known about the reproductive biology and behaviour of elasmobranchs - in spite of their popularity. For example, only few matings have been observed in the wild. During his studies of elasmobranchs, the author has glimpsed into some details of reproduction of various deepwater sharks and now wants you to take a peep at shark and ray sex.

Many sharks and rays are known to segregate by sex during most of the year (many dogfish sharks, large rays). But when the time comes, hundreds or even thousands of individuals of one sex gather and migrate into mating areas where they meet their partners. For example, impressive schools of hammerhead sharks that only about two decades ago - before the onset of massive shark fishing - have appeared regularly at the Galapagos Islands and in the Sea of Cortez might have been such (female)

A male zebra shark has grabbed a female's tail while trying to bring her into position for copulation.

mating aggregations. But this is speculation, nobody followed these schools all the way to really find out.

When the genders meet in the often very restricted mating area (see DANCE OF THE STINGRAYS, p.195-199), it is the male that actively pursues a prospective partner. He follows her closely, possibly percepting pheromones which signal her readiness to mate. Yet, nothing is known about such 'attractive perfumes' in elasmobranchs but when considering the vast volumes of water in which the more solitary species (white sharks, megamouth) live, it is almost certain that some kind of chemical communication must exist.

The next step towards mating is - who wouldn't know? - touching the female. Divers may have noticed the scars or fresh bite marks on body and fins of grey reef and other sharks. It has been speculated that these bites not only serve in grabbing a hold of the partner but also elicit the necessary stimuli to prepare her for copulation. As wounds heal fast in elasmobranchs, they are usually no problem. To prevent excessive damage, female blue sharks even have a protective hide that is over twice as thick than that of males.

Finally, the male grabs a firm hold of the female and swims parallel to her. Now, he has to insert (usually only) one of his two penis-like claspers into her reproductive orifice (cloaca). The claspers, located at the inner pelvic fin margins, are the best external feature to separate sexes and are swollen during the mating season. Their tips are clad with spiky modified dermal denticles which aid locking to the female. Along each clasper runs a groove that guides the flow of seminal fluid.

Mating position varies with the species and is belly-to-belly in blacktip reef sharks and many stingrays. In bat rays, however, the male swims upright under the female; a male bat ray has to rotate his clasper upwards to insert it. The slender males of catsharks literally coil around their partners to copulate. Copulation itself may last from only a few minutes to several hours. Meanwhile, seminal fluid mixed with seawater from a siphon sac, an organ found exclusively in sharks, is squeezed into the female. Note: All organs of elasmobranchs associated with reproduction are usually paired, except in a few species e.g. the basking shark.

The partners separate and the spermatozoa wander from the cloaca into the oviducts. There they are stored in the nidamental glands where fertilisation of eggs arriving from the ovary takes place and the egg case is formed around the fertilised egg. Female blue sharks for example are known to store active sperm for an extended period of time in these glands, a practice also known from many other animal groups including cephalopods and mammals.

Elasmobranchs have three different main strategies of reproduction. Oviparity (or simply egg-laying) comprises enclosing the fertilised yolk-rich egg in a tough 'horny' case and expelling it out into the open. Among others, catsharks, horn sharks, skates and chimaeras are typical egg-layers. The embryo inside relies

ROLF SCHMIDT

A female grey reef shark rests after an exhausting mating session. The impressive gashes on her back tell the gruesome tale of shark sex but will heal soon.

solely on its yolk reserve and swims in a sort of miniature aquarium which is connected to the outside via slits in the egg case in order to get fresh, oxygenated water for breathing. The embryo grows using up its yolk reserve to become a miniature replica of its parents and hatches after two months to over a year depending on species and ambient water temperature. See also A MATTER OF PATIENCE, p.100-101.

Ovoviviparity (also called aplacental viviparity, see below) is found in many rays, dogfish sharks, some nurse sharks and the whale shark (with over 300 young per litter, the largest of all fishes has only recently been recorded as the most fertile elasmobranch). Like in the former category, young and yolk are enclosed in a case but are growing up in the safety of their mother's 'womb'. After hatching still inside the uterus, the young are born at a relatively small size - roughly a tenth of the mother's length (yolk sac viviparity of dogfish sharks, in which several young are 'stacked' inside a thin, candle-shaped egg case).

A subvariant of this strategy is staying inside the uterus after hatching and continue feeding on other eggs (oophagy of mackerel, thresher and probably also great white sharks) or on siblings (adelpho- or embryophagy of sand tigers). This special form of nourishment is exclusively found in lamniform sharks and results in few but large young. Hence, sand tigers have only two young per litter (one 'intrauterine cannibal' in each uterus, survival of the fittest!) which at birth are already about a third of the mother's length.

Placental viviparity is found only in the most advanced of all living elasmobranchs e.g. requiem, hammerhead and some smoothhound sharks. After the yolk is used up, the empty yolk sac firmly connects to the uterus wall. Thus, the direct exchange of nutrients from mother to embryo is guaranteed. The advantages are obvious: A fair number of young can grow to a relatively large size in the safety of the mother's uteri without wasting nutrients. While most requiem sharks have litter sizes of around a dozen young, the tiger shark has up to more than 80 young - but it also is the only non-placental family member.

This most interesting chapter of elasmobranch biology is far from being complete. How little we know becomes obvious when considering the following facts: Until about fifteen years ago, when a pregnant female white shark was landed in Japan, no one had ever seen an embryo of that species - today, the most pessimistic view is that the worldwide population has already been decimated too much to guarantee the survival of the species. The huge whale shark was long thought to be oviparous after an aborted egg case containing a fully coloured embryo had been picked up in the Gulf of Mexico in the 1950's - today, the species is seriously in danger of being exterminated by Asian fishermen selling fins and meat. Paradoxically, unborn young of the heavily exploited basking shark are still unknown; the gentle giant of temperate waters was long hunted for its liver oil and meat (Ireland, Norway) until populations collapsed and made fishing unprofitable. So how can we dare to drive animal species that have been so successful in the fight for survival over the past 400 million years to the verge of extinction within a few decades of so-called 'modern science' without even knowing a bit more than the basic facts about their alien ways of sexual reproduction?

GERLINDE APPL

The two uteri of a pregnant female Arabian houndshark. Each uterus contains two young sharks of 13 cm in length, note their dark eyes. Visible at the right are oviducts, nidamental glands and ovary.

ROUND STINGRAYS OR STINGAREES   UROLOPHIDAE

This family of stingrays comprises 3 genera *(Trygonoptera, Urolophus, Urotrygon)* with about 40 species. They are similar to the whiptail stingrays but generally smaller in size with a tail which is about as long the body disc and has a distinctive caudal fin. The floor of their mouth usually has papillae which are important for species identification. They have numerous small teeth and no enlarged denticles or thorns on their back. Most live on the bottom in coastal shelf waters, especially those of Australia, where several undescribed species occur. Some enter the brackish water of estuaries, the family is probably most closely related to the South American freshwater stingrays. Their tail has one or more stings and they may sting waders in shallow water, but no fatalities have been reported yet. During courtship, females may also ward off males with their sting. All family members are ovoviviparous, 'uterine milk' is secreted from the uterus walls of the mother to nourish the young after their yolk reserve is used up (so-called uterine viviparity, see also SEX IN SHARKS AND RAYS, previous pages). Litter size is 2-4, the gestation period about 3 months. They prey on crustaceans, molluscs, worms, small bony fish and smaller family members and fall themselves prey to tiger sharks and large bony fishes. Some are hardy and reproduce in captivity.

Urolophus concentricus
Bull's-eye round stingray

Length: Max. at least 60, width to about 30 cm.
Distribution: Eastern Pacific: Sea of Cortez to Panama, including Galapagos Islands.
Depth: Intertidal to 20 m and possibly deeper.
General: This stingray has a distinctive colour pattern of concentric dark grey bands on a lighter grey background on its back. The band next to the rim of the round body disc is the narrowest one and often consists of a series of spots, while the innermost bands are broader and fused by bars to form a cruciform (cross-like) pattern. This species is found on sandy and rocky bottoms and lives sympatrically with the round stingray *U. halleri*. In rocky environments it even seems to be encountered more often than its better-known relative. The bull's-eye stingray is also frequently seen in shallow coastal bays. It preys on small crustaceans (mostly crabs) and worms. The small photo below shows an adult specimen.

Helmut Debelius   Sea of Cortez, Mexico

Mark Norman   Victoria, Australia

Urolophus gigas
Spotted stingaree
L: Max. about 70 cm. Di:
Southern Australia. De: 2-
50 m. G: Seen by divers often
partly buried in sand near seagrass. Two colour forms (eastern, western), may be distinct
species. Below: **Sepia stingaree** U. aurantiacus, to 40 cm,
Japan to East China Sea.

Urolophus circularis
Circular stingaree
L: At least 60 cm. Di: SW-Aus-
tralia: Rottnest Island to Esper-
ance. De: 5-120 m. G: Attractively patterned, uncommon,
but sometimes seen by divers
on offshore reefs, rocky bottoms and among kelp. Below:
Banded stingaree U. crucia-
tus, 50 cm, SE-Australia.

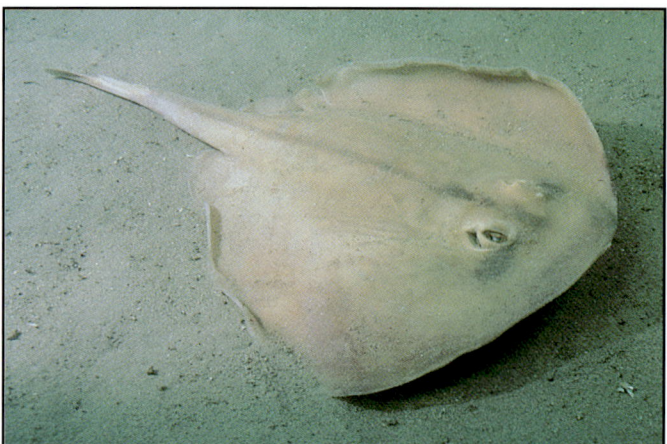

Clay Bryce   Abrolhos Islands, Western Australia

Urolophus viridis
Greenback stingaree

Length: Max. at least 44,
mature males 27 cm.
Distribution: Southeastern
Australia: Southern Queensland (Stradbroke Island) to
Victoria (Portland), including
Tasmania. Depth: 20-200 m.
General: Uniformly light green
with paler edges above. Benthic, common in deep offshore
waters, lives deeper than other family members in its range.
A similar or identical species
occurs in deep water off
southwestern Australia. Biology little known but occasionally utilised.

Saul Gonor   New South Wales, Australia

Urolophus halleri
Haller's round stingray

L: At birth 13, max. 56, width to 31 cm. Di: California to Panama. De: 1-90 m. G: On the bottom in coastal waters, bays, lagoons, estuaries. Below: **Cortez round stingray** U. maculatus, to 42 cm, Baja California and Sea of Cortez, is similar but has dark spots on both sides.

Mark Conlin   California, Eastern Pacific

Trygonoptera ovalis
Striped stingaree

L: Max. 61 cm. Di: Western Australia: Eucla to Abrolhos Islands. De: 4-43 m. G: Disc oval, small dorsal fin on tail. On sand or seagrass in coastal waters. Below: **Common stingaree** T. testacea, to 47 cm, eastern Australia, 1-60 m, similar but disc more rhomboid.

Rudie Kuiter   Western Australia

Plesiobatis daviesi
Deepwater stingray

L: At birth <50, females adult at 200+, males at 172, max. at least 270 cm. Di: Indo-West Pacific: Southern Africa to Japan, Australia, Hawaii. De: 275-680, but one record from 44 m (Mozambique). G: Disc nearly circular, eyes small, no dorsal fin, caudal fin about half of tail length, sting long. Greyish-brown to black, sometimes with spots. On soft bottoms, locally common in upper slope waters, but poorly known. Photo: 450 m (submersible).

Christoph Gerigk   South China Sea

HOOKED ON TEETH

Keen on sharks? Diving is not your business and you don't like staring at captured animals in zoos? Have read all the books, seen all the documentaries? Well, there is yet another way to get into sharks without regrets. And if you are the collector's type then this is definitely the one for you. And even better - you don't have to travel too far to get near them. The author revels in the fascinating hobby of collecting fossil elasmobranch teeth.

Typical fossil shark teeth (simple crown: *Synodontaspis cuspidata*, 'cockscomb': *Notorhynchus primigenius*) from Middle Oligocene (28 mya) sands of the Mayence Basin, Germany. Largest 3 cm long.

Sharks and their kin (rays and chimaeras) are old animals. They roam the seas - and the occasional freshwater river or lake - since hundreds of millions of years. One of their characteristic features is the continuous replacement of jaw teeth and other hard parts like dermal denticles, thorns, spines and stings. It is this replacement strategy which makes the durable teeth so numerous in all kinds of (marine) sediments laid down since Silurian times more than 400 million years ago until today.

Hence, to find a fossil shark tooth is nothing special when considering their numbers. Shark teeth are almost ubiquitous (meaning present everywhere) in fossil-rich layers of sand, clay or rock. They literally belong to what is called mass fossils, just as certain molluscs like ammonites and belemnites. It is their chemical composition that makes them ideal for fossilisation. The bone-like interior of the crown (the part exposed in life) and the root (the part anchored in the jaw's 'gum') of such a tooth is mainly composed of durable calcium phosphate. The shiny, enamel-like substance covering the crown is even harder. It contains fluorite and we all know - don't we? - that many toothpastes contain fluoride in order to harden our own teeth against bacterial attacks. If the chemistry of the soil in which the teeth are embedded after falling from their owner's mouth is not too aggressive, they will ideally be preserved without any changes except of their colour. Such sediments are prime targets of shark tooth hunters. Unfortunately, many teeth are damaged in one way or the other after they have been buried by mud on the ocean floor. Often, organic acids are to blame, especially when the fossiliferous layer is close to the top soil where plants grow. Tiny

Various shark teeth from the Tertiary, collected on a North Sea beach in the Netherlands. Largest at centre (3 cm) is extinct mako *Isurus hastalis*.

266

plant roots exude acids that are strong enough to dissolve any part of a tooth. They even carve into the enamel and often crumble the entire fossil to dust. But in cases, only the bony root and interior of the crown are dissolved and some typical fossil shark teeth (e.g. from certain deepsea bottoms) are merely hollow crowns.

Another very important factor of how such a masterpiece of natural design survives the tides of time is mechanical damage. Imagine a shark swimming about in its - let's say Miocene - habitat, about 18 millions years

The author 'at work' on a North Sea beach, using a 60x60 cm sieve with 1-cm-mesh wire.

before we pick up one of its fossilised teeth. The shark - of course we choose the most famous of all extinct species, the megatooth shark *Carcharocles megalodon* - has just found its favourite prey, a whale carcass drifting in the warm waters of the Miocene ocean, and takes the first bite with its strong, tooth-clad jaws. While the jaws close, the teeth smoothly cut through the blubber and underlying muscle but suddenly come to a grinding halt: They have just hit massive bone, in this case one of the whale's huge ribs. Despite its hardness, one of the teeth has broken. It has hit the obstacle at an angle at which the mechanical forces had been too high. This is a form of damage that may occur anytime during the life of the animal. But after the tooth has been shed and finally fossilised, it is hard to tell whether this damage has happened before or after shedding (only very rarely broken teeth are found embedded in fossil whale bone or the like). Much more frequently, the damage present in fossil teeth occur after they fossilised. The seafloor sediments containing them may fall dry during the course of time. Then, erosion by wind and weather sets in and uncovers the teeth. They are abraded by the wind, carrying myriads of sand grains which act like sandpaper in super-slow motion. But there is plenty of time and the once sharp cutting edges, fine serrations and other details are constantly worn away. Additionally, the exposed sediments may be flooded again after the sea level rises or they are washed out by rivers or the surf at the coastline. All this results in long phases of sediment transport and with the sediment also the fossils are carried away. Transport means wear: Yet another way to grind a tooth down to dust. On the other hand, such transport mechanisms during extended periods of time may also concentrate fossils. Such concentrations are especially interesting for the collector as there are many fossils per cubic metre of sediment which means you have to dig less to get more teeth. Which leads us to the practical side of collecting fossil shark teeth...

What you need first is the information on where to find them. Go to a natural history museum, read a magazine on fossil collecting, search the world wide web, or - best of all - contact somebody you know who knows somebody who collects fossils. Sooner or later you will have some clues on where to find the objects of your secret desire. In all cases, it has to be a site located in the appropriate geolog-

An anterior tooth of *Otodus obliquus*, an early ancestor of the megatooth shark, still embedded in its hard phosphate matrix. Paleocene of Morocco, height 6 cm.

Two fine lower jaw symphyseals of the extinct sevengill shark *Notorhynchus primigenius*, a rarer find in the Oligocene of Germany. From two different sites, note colours. Left tooth is 1.5 cm wide.

ical layer. You won't find a single fossil in crystalline or basaltic (lava) rock, so it has to be sediments, preferably marine ones because the vast majority of sharks lived and still lives in salty ocean water. Most popular in shark tooth hunter circles are the geologically younger sediments ranging on the time scale from the Upper Cretaceous when dinosaurs were still present through the so-called Tertiary, a period covering approximately the last 60 million years of earth's history. The reason for this preference is the abundance of shark teeth in certain layers of this period in restricted areas on the planet. In order to reach them, you don't simply start digging in your garden. You have to look for sites where others have already done the digging. These are, for example, sand pits where sand is excavated for construction purposes, banks of river beds cutting through the strata in question and shores where the fossils are washed out from the seafloor and subsequently onto the beach. Indeed there are sites where you can collect fossil shark teeth while diving, e.g. in Florida and southeast Australia. Some places have become famous for their fossils and access to them is controlled by the land owners. There also are - sometimes huge - pits where phosphate (mostly composed of the remnants of fossil bones and teeth, e.g. in Florida, Morocco and Israel) or kaoline (in southern France) is quarried but in all those cases a permit is definitely required to enter and do the collecting. Fossil shark teeth can also be found in samples obtained while probing the deep ocean floor but to be honest - who has the means to pull a deepsea trawl along the seafloor only to collect teeth? (In the warm Pacific waters around New Caledonia there is a region where some fishermen decided to do exactly that. In order to earn a living, they now sell fossil shark teeth after fishing wasn't worth the effort anymore.)

The next step after locating an appropriate site is learning how to find the 'gems'. The easiest way is walking along the shore and simply picking up the finds. For this, you only need a keen eye and adequate clothing, depending on whether you do your collecting in sunny Florida or on the shore of the sometimes not-so-sunny North Sea. An improvement of this technique known as beach-combing is bringing along a shovel and a sieve. Now, you don't have to rely on what was washed up after the last high tide (and has just been picked up by some detestable competitor right in front of your eyes) but you dig your own 'claim' and enjoy the finds left in your sieve after you have removed the sand by washing it out. Sieving can of course also be applied in any inland sand pit where you simply shake away the dry sand that obscures the more interesting stuff. Use scrapers or fork-like garden tools to loosen up sand that is too sticky to be sifted right away. If it is too wet after a rain, let it dry in the sun before passing it through the sieve. The mesh size of the latter has to be appropriate for the kind of sediment you work up and naturally also determines the minimum size of your finds. Start using mom's kitchen sieves and then - after some debate and eventual punishment - try building them yourself from solid wooden boards, mesh wire and nails available in any hardware store.

In such a way, a past-time pleasure is achieved - often for the entire family - that not only yields attractive collectable items but also takes place out in the open in fresh air. After returning home tired but happy, you can start to clean and sort the precious finds. Many sites yield solid teeth that are simply washed and dried. Teeth from such sites are often worn, however, because they have been moved around a lot during the ages. Other places yield perfect specimens which haven't been moved at all or only little after being embedded and fossilised. Such teeth and especially their roots, however, may easily break when excavating and sieving the material. Be careful when handling those finds and apply some glue if necessary.

When everything is cleaned, preserved, sorted by shape or size you can do several things: Simply put your collection in a drawer and enjoy looking at it from time to time - or start to find out which kind of shark once

Teeth from 3 different rows of *Galeocerdo latidens*, a small tiger shark. Eocene (about 45 mya), Netherlands, largest is 1.8 cm wide.

had your collectables in its mouth. This is where the homework on long, dark winter evenings really starts. Again, human tutors are best, but you will also find a lot of (mostly local) literature in natural history libraries depicting what you have picked up during the summer months. Many fossil shark teeth come from species still extant today, so books on living sharks with photos or drawings of their teeth will also be helpful, especially for beginners. Moreover, you can compare your finds with those in museums or private collections. Soon you will get acquainted with the types you collected and know which are rare or common.

When you have become a crack on fossil shark teeth there are some more aspects to look at. There is a small but established shark tooth trade, also via the internet. It is up to you if you want a 'complete' collection and decide to buy some teeth that have escaped your collecting efforts for one reason or another. Prices depend on quality and numbers, just like in the gem trade, but may be exaggerated in cases. There are no definite rules for pricing 'old stones'. Another, more communicative way of filling gaps is swapping finds with other enthusiasts, of course.

Let this be a short introduction on how to get into sharks and rays without coming close to one.

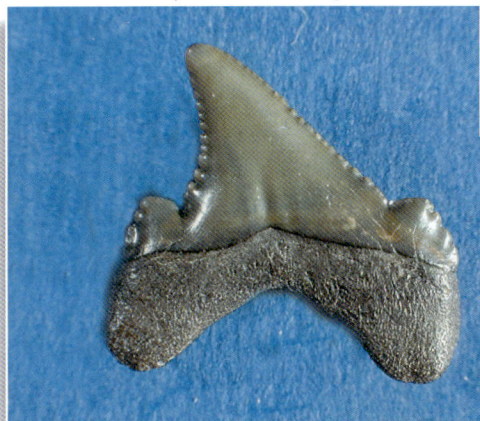

Above: Upper anterior and lower lateral tooth of the great white shark *Carcharodon carcharias*. Pliocene (5 mya) of Namibia, length 8 cm. Left: Lower lateral tooth of *Carcharocles auriculatus*, an ancestor of *C. megalodon*. Eocene (about 45 mya), Netherlands, width 2.5 cm.

In any case, you will not be alone. Hundreds and thousands of people all over the world enjoy collecting fossil shark teeth, from small children competing on who found the biggest one to serious amateur scientists with collections much more elaborate than those found in most natural museums or scientific institutes. Enjoy the glimpse into the earth's past, take home with you parts of the feeding machinery of the most effective predators of all times, turn it into a gift by attaching it to a necklace for your loved one or simply be fond of the enamel's sheen and sharp cutting edges that have miraculously been preserved through many millions of years. May the tooth be with you!

Complete fossils of cartilaginous fishes are rare because their skeleton is not as durable as bones are. This stingray (*Heliobatis radians*) belongs to an enthusiastic collector and comes from the Eocene Green River Formation of Wyoming.

WERNER GRÜTER

Rogelio Herrera   **Costa del Sol, Spain, Mediterranean**

Juan Carlos Calvin   **Madeira, Eastern Atlantic**

Paul Humann   **New England, Western Atlantic**

Gymnura altavela
Spiny butterfly ray
W: Max. 210, usually about
100 cm. Di: Eastern Atlantic:
Portugal to Angola, including
Mediterranean and Black Sea.
Western Atlantic: Massachusetts to Argentina. De: 2-60 m.
General: With short tail that
bears two barbed spines. Preys
on fishes, crustaceans and molluscs on sandy and muddy bottoms. The unborn embryos
are already equipped with tail
spines. Reported to grunt by
expulsion of air while lying on
deck after capture.

This family comprises two
genera (*Aetoplatea*, with small
dorsal fin on tail; *Gymnura*,
without dorsal fin) with about
15 species. Body disc clearly
rhomboid, up to twice as wide
as long. Tail with none to several stings and considerably
shorter than body disc. Head
not elevated above the flat
body disc. Worldwide in warm
temperate to tropical waters.
Ovoviviparous. When the yolk
reserve of the young is almost
used up, 'uterine milk' is secreted from the uterus walls of
the mother for further
nourishment (uterine viviparity). Additionally, long filaments
(villi) grow from the uterus
wall and into the embryo's
spiracles to secrete directly
into mouth and throat. Litter
size 1-7. Despite their abundance in warm inshore waters,
only little is known of the biology of butterfly stingrays.

Gymnura micrura
Smooth butterfly ray

Width: Max. 120, usually about
60-90 cm.
Distribution: Western Atlantic:
New England to Brazil.
Depth: 1-20+ m.
General: In contrast to *G. alta-
vela,* without tail spines (popu-
lar name).

The mouth of butterfly
stingrays has many small teeth
(50 rows in young and up to
over 100 in adults). Young individuals prey on small bottom-
dwelling invertebrates (crustaceans, worms, molluscs) but
larger ones also take small fish
including schooling species.

Gymnura australis
Australian butterfly ray
W: Max. at least 73, males mature at about 35-40 cm. Di: New Guinea and tropical and warm temperate waters of Australia (Dampier, Western Australia, to Broken Bay, New South Wales). Possibly also Indonesia. De: Shallow inshore waters to at least 50 m. G: Body disc greenish, greyish or yellowish above, densely covered with fine black spots over mosaic pattern. Two indistinct blotches near rear end of pectoral fins. Tail short, thin, ringed in black and white, without spine. Regularly caught by shrimp trawlers and marketed.

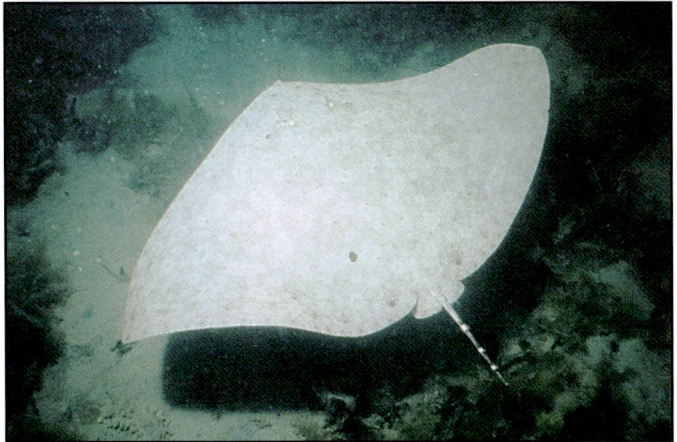

Nigel Marsh / IV   Great Barrier Reef, Australia

Gymnura poecilura
Arabian butterfly ray
W: Max. 250 cm. Di: Northern Indian Ocean and Western Pacific: Red Sea and Arabian Gulf to Philippines. De: 1-50 m. G: On sandy and muddy bottoms in tropical and subtropical shallow coastal waters, also around islands, in river mouths and estuaries. In the eastern part of its range probably sympatric with *G. japonica* and *G. australis* (see above).

Butterfly rays are heavily fished for because they are easily caught, have a smooth skin and large 'wings'. They are considered to be overexploited and endangered in the tropics.

Phil Woodhead   Oman, Arabian Sea

SIXGILL STINGRAYS   HEXATRYGONIDAE

Hexatrygon longirostra
Japanese sixgill stingray
W: Max. 100 cm. Di: S-Japan to South China Sea. De: 15-1,000+ m. G: Unique in having six pairs of gills. Snout long, pointed, with highly flexible tip probably used to probe for prey. Skin soft, smooth like in many deepsea elasmobranchs. Eyes small, flesh flaccid. Otherwise resembling round stingrays (with leaf-shaped caudal fin and 1-2 tail spines). First family member described in 1980 from South Africa. Also in Australia, Hawaii. Photo is the first shallow-water record!

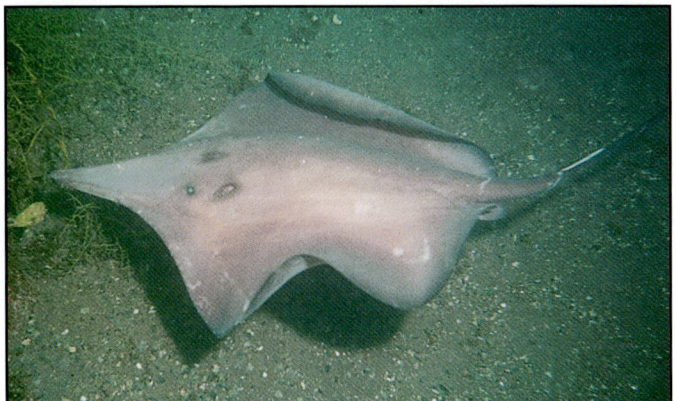

Shinichi Hara   Izu Peninsula, Japan

FORMATION FLIGHT

The rumour had been around for so long that it had already become a fact. Schools of eagle rays were observed infrequently at a certain diving location in the Maldives. But most of the time, even experienced underwater photographers have to be very patient and simply wait until the long-sought photo models finally show up in front of the camera. Chief diving instructor Danny Huisman knows exactly what that means.

Spotted eagle rays (Aetobatus narinari) in majestic 'formation flight'.

For more than seven years now I had been working in the Maldives, most of the time as a chief diving instructor in the island of Helengeli. Hence I was acquainted with the diving location generally known under the name of 'Fingerpoint'. I went diving in this spot more than once and tried for years to photograph the large schools of eagle rays which can be encountered there - without success.

One day, it seemed to become a very good one at Fingerpoint. Already early in the morning I felt that this would be the day I had been waiting for so long. Together with several guests I left our boat for the first dive during which we saw not less than 47 of the elegant eagle rays, also many grey reef sharks and whitetip sharks, not to mention all the others fishes.

After that dive I decided to return to the water immediately, this time with my uw-camera. And as soon as I had escorted our guests back to the boat, I couldn't wait any longer. One could tell from looking at the surface that the local current was yet increasing in strength and thus became an additional handicap when considering the immense drag of the camera. But there was no way out, I simply had to go for it now. My partner and I jumped and started drifting towards the reef located in a depth of 32 meters.

Normally, this particular reef at Fingerpoint is approached from the main reef, but we didn't want to waste any time and took the shortcut through the open water in order to keep our decompression time at a minimum. After arriving at our destination, we patiently started waiting for the rays. After about 20 minutes, they finally appeared. Unfortunately, they didn't approach close enough to make really excellent photos. The breathing air in our tanks started to dwindle and we had already accumulated 25 minutes of decompression time what forced us to commence the ascent. Exactly at that moment, the school arrived. There were about 30 rays at least and all were gliding towards us directly from behind. Unlike ourselves, they didn't seem to bother about the strong current at all. Now or never, I thought. I let go of my holdfast and was carried away by the current at once. Now I was approaching the school of eagle rays at an amazing speed. My only hope was that these magnificent creatures wouldn't simply disappear before I had photographed them. But, alas, they even kept their for-

mation - what bliss for a photographer! Just a few more corrections on the camera and flash settings, find them with the seeker, press the shutter release once, twice - and gone they were. The eagle rays had passed me by as if I weren't even there. I shouted with joy and hastily started to ascent back to the surface. While fumbling with my camera at our decompression stop, I looked up for a moment - and nearly lost my mouthpiece! Right in front of me, at perfect wide angle photographing distance, there it was, a huge blue marlin. I shouted again, but my partner couldn't hear me. The marlin instantly disappeared and left me behind without a chance to photograph it...

Nevertheless, this surely had been one of my top-ten dives, for nobody is witness to a perfect formation flight of eagle rays in tropical waters every day.

John Neuschwander   Canary Islands, Eastern Atlantic

Pteromylaeus bovinus
Duckbill ray
W: Max. 152 cm. Di: Eastern
Atlantic: Portugal to South
Africa including Mediterranean.
Western Indian Ocean: Zanzibar to the Cape. De: 1-100 m.
G: Disc brown with several
grey bands (may be absent).
   The eagle ray family comprises four genera with 22 spp.
Head protruding and rounded,
not indented as in cownose
rays. In contrast to the pelagic
filter feeders of the family
Mobulidae, eagle rays feed on
molluscs, worms, crustaceans
and cephalopods which are
dug out of the substrate with
the duckbill-shaped snout.

Christoph Gerigk   Côte d'Azur, France, Mediterranean

Myliobatis aquila
Common eagle ray

W: Max. 150 cm, usually less.
Di: Eastern Atlantic: British
Isles to South Africa (into
western Indian Ocean to
Natal), and Mediterranean.
De: Shallows to at least 95 m.
G: Common, lives in coastal
waters mainly on soft bottoms,
also enters lagoons and estuaries. Regularly found on fish
markets in the Adriatic and
Tunisia where the pectoral fins
are marketed fresh. Fishermen
are aware of the tail spine
which can inflict a serious
wound when picking the ray
from the net.

Rudie Kuiter   South Australia

Myliobatis australis
Southern eagle ray
W: At birth about 32, max. at
least 120, total length 190 cm.
Di: S-Australia, Tasmania, New
Zealand. De: Shallows to 85 m.
G: Common over shallow sand
flats. Adults migratory. Preys on
crabs and shelled molluscs. The
New Zealand eagle ray M. tenui-
caudatus may be the same sp.

Myliobatis californicus
Californian eagle ray, bat ray

W: At birth 29, max. 180 cm. Di:
Eastern Pacific: Oregon to Sea of
Cortez. De: Surface to 46 m. G:
Uniformly olive to brown, whitish
below. Groups mix with whitespotted eagle rays (p.278). Has unusual mating position: Male swims
underneath female with his back
to her belly, then rotates one clasper up to insert it into her cloaca.
Preys on diverse infaunal invertebrates and small fishes. Is a major
prey item of the broadnose sevengill shark *Notorynchus cepedianus*.

UW-filmer Howard Hall: The
idea was to film bat rays at the
southern tip of San Clemente
Island, California. For four days
we'd watched about 100 rays lying
in the sand - doing nothing. Then,
some start to circle high off the
ground. They're passing now from
north to south and with a little
luck I can intercept them. This
must have been only the edge of
the school! There are hundreds
here! Hundreds! I'm almost under
the school now. The image continues to develop and I pull the trigger. A ray enters the frame from
the right followed by four others,
followed by dozens more. They
are nearly 1.5 metres across. Ten
seconds into the shot there are
more than a hundred in the frame,
perhaps twice that. Then the
school turns and begins to swim
directly into the sun. They fly
against the setting sun like flocks
of soaring birds passing beneath a
sun-filled break in dark thunder
clouds! Wonderful! Magic!

Norbert Wu

Howard Hall　　both photos California, Eastern Pacific

Myliobatis tobijei
Japanese eagle ray, kite ray

Width: Max. at least 56, total
length at least 150 cm.
Distribution: Western Pacific:
Japan, Korea, Okinawa Trough,
China and South China Sea.
Depth: Intertidal to 220 m.
General: Common in inshore
and offshore waters on the
continental shelf and upper-
most slope.

For comparison, see also
the photo of a similar but
obviously yet undescribed
species on the following page.

Akihiko Mishiku　　Shizuoka, Japan

275

Myliobatis sp.

The photo shows a new, undescribed species from Japan which is not identical to *M. tobijei* (previous page) because the colour pattern includes spots on the back as well as on the underside of the disc.

The copulation of eagle rays has been observed: First, the male gets a tight hold of the female by biting into her pectoral fin. Then he slips under her, belly to belly, to insert one of his claspers. This takes place either while swimming in open water or on the bottom. The copulation itself may last from 30 to 90 seconds. A female may copulate with several males in succession.

All eagle rays feed on benthic invertebrates which are dug from the substrate. Bony fishes such as goatfishes and wrasses are attracted by this activity and follow the eagle rays as commensals, profiting from the small prey animals dug out by the rays. The typical eagle ray dentition forms massive, nearly flat plates in either jaw which are used to crush shelled prey. While *Myliobatis* spp. have several interlocking tooth rows, the members of the more advanced genus *Aetobatis* have only a single row of broad teeth *(Aetomylaeus* is intermediate). Their fossil teeth are not uncommon in marine Tertiary sediments (see p.206).

Kazu Masubuchi　　　　　**Izu Peninsula, Japan**

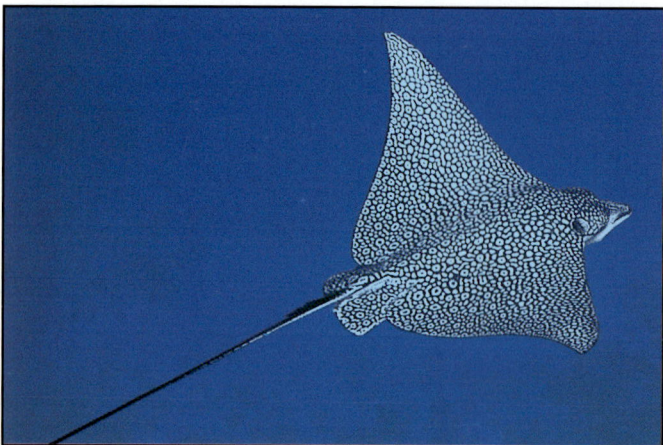

Aetomylaeus sp.

According to the ray specialist Dr. Peter Last the photo shows a new, undescribed species which is similar to *A. vespertilio* (see following page).

Small photo below: **Banded eagle ray** *A. nichofii*, width 64, length 100 cm, Indo-West Pacific, major commercial sp.

Johann Hinterkircher　　　　**Maldives, Indian Ocean**

Aetomylaeus vespertilio
Ornate eagle ray

Width: Max. 160, total length to at least 385 cm.
Distribution: Indo-West Pacific: Scattered records from southern Mozambique, Maldives, Malaysia (Pinang), Thailand (Gulf of Thailand), Indonesia (Java, Borneo), northern Australia (Arafura Sea), South China Sea (off China), Taiwan.
Depth: Shallow waters to at least 110 m.
General: Pattern distinctive. Large, spectacular, but little known. Found in muddy bays and on corals reefs from close inshore to deep water.

The photographer reports: One day in March 1998, we were diving the house reef at Mirihi (Ari Atoll, Maldives). I intended to photograph dancing shrimps in a crevice in 33 m depth. In order to save breathing air and keep decompression times short, I left the jetty in a depth of only 5 m and navigated to overhead the spot by compass. Just a few metres off the jetty I suddenly noticed a huge ray swimming directly below. Because of its size I thought of a manta ray at first. But when I noticed the thin tail, which was almost 3 m long, I knew that this one was special. The ray was too far away for my 60-mm macro lens, so I switched off the flash and took just one shot. Now the ray located a giant clam of about 40 cm in length and, covering it with the body, grabbed it and shook it violently. I approached up to a distance of one metre and portrayed the feeding ray that was not anxious at all and left behind a broken, empty shell.

all photos Erwin Steininger   **Mirihi, Maldives, Indian Ocean**

Peter Kragh   Cocos Island, East Pacific

Aetobatus narinari
White-spotted eagle ray
W: At birth 17-36, max. 250, but rarely over 200 cm. Total length with unbroken tail almost 500 cm. Females mature at 214 cm disc width. Di: Circumglobal in tropical and subtropical waters. De: 1-40 m. G: Beautifully patterned large eagle ray with a very long tail. The photos from all over the world show that this species varies in dorsal colour pattern. The most widespread form (centre) has small white spots on a black background (variations on small photos below). The form on the bottom photo has larger white spots with a black centre and the white spots merging into lines and circles. According to Dr. Peter Last, *A. narinari* is a species complex rather than a single species which currently is under investigation. As long as no new species names have been applied, this name should be used. Preys mostly on shelled molluscs but also on other benthic invertebrates, see top photo. See also FORMATION FLIGHT pp. 272-273.

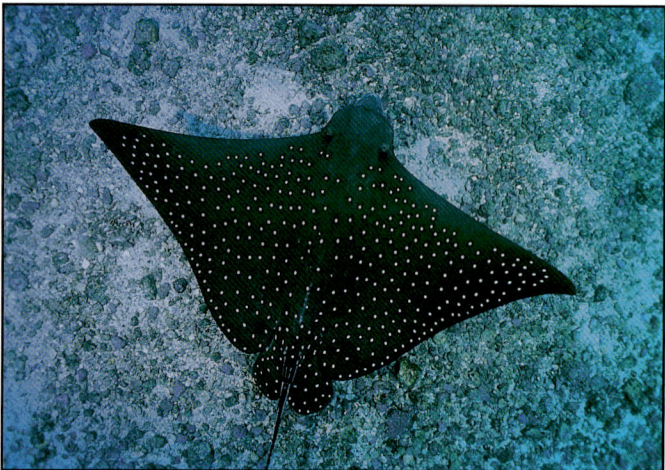
Peter Kragh   Cocos Island, East Pacific

Hawaii, North Pacific

Cayman Isl., Caribbean

Helmut Debelius   Tuvalu, South Pacific

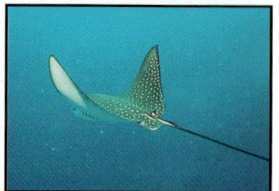
Seychelles, Indian Ocean

Rhinoptera jayakari
Indian cownose ray
W: Max. 150 cm. Di: Northern Indian Ocean including Red Sea. De: 1-40 m. G: Singly or in large schools. Enters protected brackish waters to give birth. Below: **Javanese cownose ray** *R. javanica,* 150 cm, South Africa to Philippines. Little known, in coastal bays, mangroves. In large schools.

John Hoover   Oman, Arabian Sea

Rhinoptera steindachneri
Pacific cownose ray
W: Max. 70+ cm. Di: Eastern Pacific: Sea of Cortez to Panama and Galapagos Islands. De: 1-30 m. G: In large migratory schools. Frequently leaps clear of the water. The only family member in the eastern Pacific. Below: Juveniles in a shallow nursery area at Galapagos.

Saul Gonor   Galapagos, Eastern Pacific

Rhinoptera bonasus
Atlantic cownose ray
W: At birth 36, max. 213 cm. Di: Eastern Atlantic: Mauritania, Senegal, Guinea. Western Atlantic: New England to Brazil. De: 1-60 m. G: Prey are clams, large snails, lobsters and crabs which are crushed between the dental plates. Prey obtained by stirring up sediment with the pectoral fins. Females give birth in shallow mangrove areas.

    The family comprises one genus with about 10 spp. Front margin of the protruding head of cownose rays is (like a cow's nose) indented in the middle, not rounded as in the similar eagle rays (previous pages).

Helmut Debelius   Curaçao, Netherlands Antilles

FAST FOOD FOR MANTAS

As far as Helmut Debelius is concerned, the best way to explore new diving sites is by signing on aboard a liveaboard. It ensures the shortest distance between you and the water, be it with a quick ride on the dinghy to some remote reef or on a laid-back dive right underneath the ship's hull. Helmut received an invitation to spend some time on a liveaboard. And what an offer this turned out to be: A 3,000-km-long exploratory voyage around the Cook Islands (situated between Polynesia and Tonga). An unusual destination with extraordinary possibilities.

Rarotonga is far away. At least for me, the only European in the group of selected divers. The members of our expedition, including our American host, Steve, met on the way over from Los Angeles and Hawaii. Upon our arrival on the main island of the Cook Islands we were beckoned over to the UNDERSEA HUNTER by our Australian colleague. Our sponsor had chartered this internationally renown ship for two months in order to share this South-Pacific adventure with friends. We made up the first group, which was scheduled to dive in the waters around the Cook Islands for three weeks. The two other expected groups, which were due to arrive some time after us, would then proceed to French Polynesia. Quite a distance for the UNDERSEA HUNTER to cover, but it is perfectly fitted for the task. Steve had even ordered a large shark cage as a precautionary measure, as the waters we were to dive in are said to be shark infested (even the banknotes of the Cook Islands feature grinning sharks on them!). We all felt pampered from the very beginning. And what an additional honour awaited us: Pa Ariki, the queen of the Cook Islands, who resides in Rarotonga, came down to the harbour to pay us divers a visit before our departure.

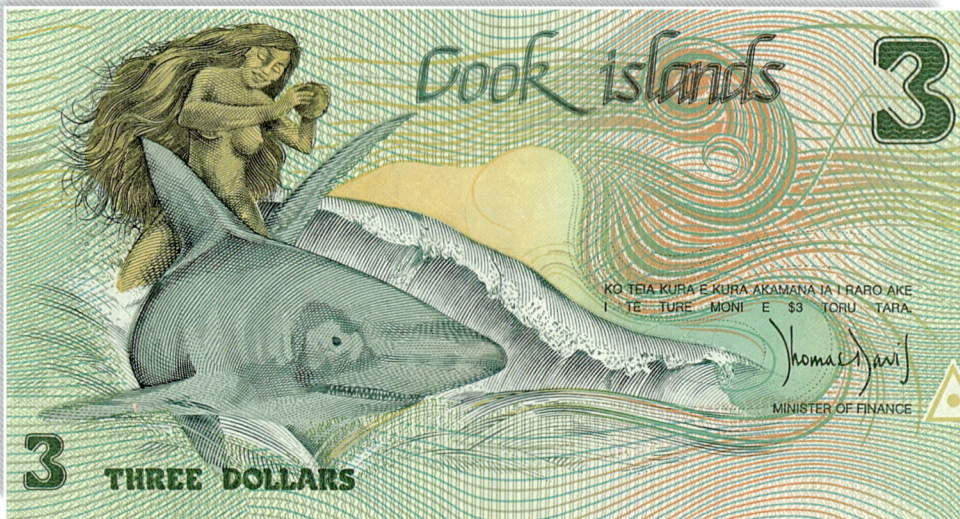

As can be seen on this money bill, the sharks encountered around the Cook Islands have a friendly attitude towards people.

Just before the Cook Islands expedition, I had been diving around the Seychelles, out in the Indian Ocean. It turned out to be an extremely depressing experience because the condition I found the reefs in during November of 1998, following the global rise in water temperatures, left me in a daze. Practically all of the hard corals had died, and their skeletons were either broken off or covered with green-black algae. It was a particularly painful experience for me, as only a few years earlier I had seen these very same reefs in their pristine state. My first dives in the southern atolls of the Cook Islands turned out to be a great relief, however: healthy coral formations wherever I looked. There are fewer filigree species than in the West Pacific, but a lot of stable coral blocks such as I had come across previously in Hawaii and its neighbouring isles. Interestingly, I found an abundance of "verdure" in the ten-metre zone, where there was an abundance of *Halimeda* and *Caulerpa* algae growing between the coral blocks even though the water temperature ranged around 24°C. Up until then I had only encountered such luscious growth of *Caulerpa* algae in temperate waters.

Queen Pa Ariki provided us with a written recommendation in which she called upon her subjects to treat the divers aboard the UNDERSEA HUNTER as guests. On our way north we anchored at a number of inhabited atolls where the people gave us a hearty welcome. The deck was frequently crowded with curious children, and some of the grownups also climbed aboard on occasion to have look at our "headquarters" or just to have a chat with other people for a change in their daily routine. The only problem we ever had occurred on Palmerston. We had just surfaced from a wonderful dive along a drop-off directly underneath the ship, when the captain frantically called for Steve. A three-man delegation awaited us on deck already and presented us with a hastily scribbled note demanding a stiff payment for diving off of Palmerston. When the other islanders heard about it, they cursed the three men. But we didn't allow this incident to depress us very much - we simply pulled up anchor and waved a friendly goodbye.

The farther north we got into the realm of the Cook Islands atolls, the better the diving got. The dis-

These rare coral fish species can only be found in the southern Pacific. Shown are, from left to right, the dwarf angelfish *Centropyge multicolor,* the lemon hawkfish *Paracirrhites xanthus,* an unusual colour morph of the butterflyfish *Chaetodon ulietensis* and the longnose sailfin tang *Zebrasoma rostratum.*

tances between the atolls, however, increased dramatically and a full night's sailing no longer sufficed to get us to our prospective destination. It took us nearly two whole days to reach Suwarrow. This atoll, which received its name from Russian explorers, has been uninhabited ever since Tom Neale, a New Zealander hermit who lived here all by himself for most of his life until he died. After the rough crossing, we were quite happy to reach the calm waters of the lagoon. Even though the channels were wide and deep enough for our ship to pass through according to the nautical charts, we sort of "stumbled" our way into the lagoon. Most of our diving took place within or around the channels, where we expected and found abundant fish life. Despite the strong currents, we felt quite safe because the crews handling the dinghies knew exactly what they were doing. My impression after a week of diving in the Cook Islands: The biodiversity is not as pronounced as it is in the West Pacific, but the quantity of fish is astonishing. Other than thick shoals of surgeon and rabbit fish, we frequently encountered huge schools of mackerels (rainbow runners, pompanos) and large groups of dogtooth tunas at the seaside end of the passes. We didn't come across any sharks until we got to Suwarrow (gray reef sharks, silver- and black-tip sharks). Our shark cage turned out to be entirely superfluous on the long run. I was much more interested in photographing the many endemic species of fish in the region, that is to say, the smaller coral fishes that only exist in this part of the Pacific Ocean.

Thousands of whitespotted surgeonfishes are magically attracted by the current channel of the Penrhyn Atoll.

It is the morning of my birthday. Our Danish dive guide Peter sets out in the dinghy to check out the current as usual and takes a few of us early risers along to go snorkelling. I swim out to the edge of the channel and, as the sun begins to rise, take my first pictures of a large group of whitespotted surgeonfish *Acanthurus guttatus* swimming at a depth of no more than 50 cm. This species is fairly common, but I have not encountered that many in the Indian Ocean, and if so then usually only solitary specimens. Good news from the dive guide: the ocean is absolutely calm and there are practically no currents, so we can actually dive the outer reef today. Fantastic visibility! The best so far on this trip and better than most places I have seen in the Indian Ocean - at least 100 metres. We dive along the outer reef drop-off in direction of the channel, where we will be picked up again. Several of the overgrown caverns around 30 metres below are teeming with species of soldierfishes, groupers and morays I have never seen before. The distance down to a pair of tilefish seems small in the clear water, but I quickly turn back after having taken a few pictures when I realise that I have passed the 60-metre range. As I ascend, four black-tip sharks approach us for a group photo. And then the climax of the day suddenly makes its appearance in

Above: The view on the 100-metre-wide channel is obscured by fishes.

Right: A single manta is disappearing into a cloud of fish eggs, shortly after spawning has commenced around dusk.

my viewfinder: a pair of pygmy angel fish of the species *Centropyge multicolor*, which I have never before encountered! The others wonder why I am breathing so hard... well, let them. Finally, we end up drifting with the surge at about 5 metres, admiring the elegantly patrolling schools of Achilles tangs with their blood-red

heart patterns set off by the brown bottom. Three manta rays, one of which is entirely black, work their way against the current into the lagoon and glide past me - a perfect way to end this venerable day. My birthday menu? Germanic portions of sushi and sashimi prepared from the Dorados and Skipjack tunas that were caught along the way, of course.

Our next destination was the inhabited atoll of Penrhyn Atoll where, among other things, the famous South Pacific black pearls are cultivated. The people there were extremely friendly and showed us their village. They were obviously quite pleased when several of the UNDERSEA divers bought their unusual jewellery. I never quite realised just how much happiness such black pearls could generated a few to a lady with a similarly coloured soul! The water inside the lagoon was too murky, so we decided to dive the outer reefs surrounding the Penrhyn Atoll. Things just kept on getting better, at least as far as endemic species were concerned. The lemon hawkfish *P. xanthus* practically blinded us with its garish colours, while the somewhat less flamboyant, totally black sailfin tang *Zebrasoma rostratum* only called our attention because of its long snout.

The sun is setting as we return to the UNDERSEA HUNTER from our late

More and more mantas arrive at the site. They remind one of aeroplanes lining up for the landing.

afternoon dive. Since the current is moving inward into the lagoon, I decide to snorkel towards the outer edge of the channel for a while and then let the current carry me back to the ship. I jump off the dinghy where the water is about 3 metres deep and can hardly believe my eyes. Rows and rows of the whitespotted surgeonfish *A. guttatus* are swimming to the Penrhyn pass, coming in from all directions. Not tens, not hundreds, but thousands of them. As I observe them in the slanted rays of the sun, it slowly dawns on me what is probably going on. In keeping with my supposition, which is based on what I read in some book, the surgeonfish swimming along the bottom suddenly begin to shoot up to the surface, where they eject a milky substance. Obviously a mass spawning session. As I try to get a little closer, the current catches me and I practically fly back into the lagoon, cursing the fact that I had not bothered to take my camera along. Back on board, I spread the exciting news of the rare natural spectacle. In the meantime it has gotten dark, but we are optimistic that the coming day will bring a repetition of what I saw because it is new moon - peak spawning time.

The following evening every diver's camera is loaded and ready. The only question that remains is just how a photographer can manage to stay put in a 5- to 6-knot current? One colleague has brought along a

For mantas, the mass spawning of surgeonfishes is a true 'fast food feast'.

steel cable to anchor himself on the reef. All the others try to hang on in the small depressions and crevices along the edge of the Penrhyn channel in the five-metre depth zone. What we get to see is incredible. More than 10,000 whitespotted surgeonfishes have made their appearance, and now I am even able to recognise the transformation of colours in the spawning fish. There are so many fish swimming about that it is practically impossible to make out the reef. The spawning spectacle begins anew. When the fish shoot up to the surface to discharge eggs and sperm simultaneously, it almost sounds like shots under water. I don't know where to look first. My legs are straining against the current, and here and there a diver is suddenly torn away from the protective reef. The surgeonfish simply ignore us. With so much spawning going on, the surroundings are soon enveloped in something like a London fog. There are so many fish bodies swirling all around us that it even becomes difficult to focus the camera.

I can hardly believe my own eyes as a dark shadow suddenly looms up beside me. As I turn around, I see a whole group of mantas coming down the channel on their way to the open sea. They remind me of aeroplanes lining up for the landing. Obviously the strong current doesn't phase them in the least. With wide-open mouths they sail down the 100-metre-wide channel, slurping up the spawn. A veritable feast for these soaring rays! The sun has almost set and the film is all used up. In the twilight I can still make out schools of tunas tearing into the agglomeration of moonstruck surgeonfish on the other side of the channel. It is difficult to imagine what we are experiencing! A colleague of mine sent me a photograph later, which shows me hunched over in the Penrhyn pass with bulging eyes and nearly-open mouth. A true attestation of fascination! There is not much talking going on board the UNDERSEA HUNTER that evening. Each one of us is still too preoccupied with digesting what we have just seen. Some of us remain in that contemplative state up until our departure from Tahiti. I am simply grateful for this unique experience, especially as I can't recall any previous dive that is in any way comparable to it.

The narrator with bulging eyes in the middle of it all at the Penrhyn pass.

This small family of highly specialised rays comprises two genera with about ten species living in all tropical to temperate seas. Devil rays are often common close inshore but do not enter the brackish water of estuaries. They are among the largest of elasmobranch fishes and prey - like the two largest sharks - on plankton organisms and small fishes which are filtered from the water. The flexible, horn-like cephalic (head) fins (popular name!) aid in directing prey into the large mouth where the minute food items are hold back by filter plates on the internal gill openings. Hence, their numerous small teeth are non-functional, in contrast to the grinding plate-dentitions present in the related eagle and cownose rays. Devil rays are agile and strong swimmers which are often seen in small groups near the surface, frequently jumping clear of the water. A few species bear small tail spines but are completely inoffensive. Especially the huge manta rays are very popular with divers/underwater photographers and, like in the case of the whale shark, their ecotouristic value exceeds by far their value as food fish for humans. But still many individuals drown in gill nets and are often discarded if not utilised. Because of their unhandy size, biology and habits of all family members are poorly known.

Leonardo Mastragostino   Greece, Mediterranean

Mobula mobular
Atlantic mobula
W: Max. 520 cm. Di: Eastern Atlantic: Ireland to Senegal and entire Mediterranean. Possibly also western Atlantic. Often confused with *M. japanica* which is widely distributed in tropical and warm temperate waters including West Africa and possibly North Atlantic. De: Surface to 20 m. G: Largest genus member. The total length exceeds the width when the long tail is not broken. Adults with 150-160 tooth rows, thorns on back and underside. One to several spines on top of tail immediately behind the small dorsal fin.

Jack Randall   Eritrea, Red Sea

Mobula thurstoni
Smoothtail mobula

Width: At birth 65-85, max. at least 180, usually to 150 cm. Distribution: Probably circum-tropical. Known from Senegal, South Africa, Red Sea, Bay of Bengal, Gulf of Thailand, (probably) Indonesia and Pacific Mexico to Panama. Depth: Surface to 20 m. General: No tail spine, front of pectoral fin margin with pronounced sinuous double curvature, underside posteriorly with two spots. In coastal waters, not oceanic. Uncommon, little known. Fleshy pectorals marketed in Thailand.

Mobula tarapacana
Chilean mobula

Width: At birth 105+, max. about 370 cm.
Distribution: Probably circum-tropical. Known from scattered localities in the western Atlantic (Venezuela), eastern Atlantic (Ivory Coast), Atlantic and Indian Ocean coasts of South Africa, northwestern Red Sea, western Pacific (Japan, Taiwan and probably tropical Australia) and eastern Pacific (southern Sea of Cortez to Ecuador and Cocos Island, and Chile).
Depth: Surface to 20 m.
General: A large, spineless devil ray, widely ranging in warm inshore and offshore continental waters. Like in other genus members, biology and habits of this species are only poorly known. It preys on planktonic crustaceans. Single individuals sometimes strand on beaches in the temperate areas of its range.
   Important differences between *Mobula* and *Manta*: *Mobula*: Mouth inferior, teeth in both jaws, with or without tail spine, ventral margin of cephalic fins never overlaps dorsal margin, when rolled up. *Manta*: Head relatively broader, mouth terminal, teeth only in lower jaw, no spine, margins overlap. In general, mobulas are much shyer than manta rays and not as easily approached by divers.

Dietmar Seifert

Avi Klapfer   **both photos Cocos Island, Eastern Pacific**

Mobula eregoodootenkee
Pygmy mobula

Width: Max. about 100 cm.
Distribution: Widespread in tropical Indo-Pacific: South Africa and Red Sea to Micronesia. Depth: Surface to 20 m.
General: Smallest genus member. Front margin of pectoral fins almost straight. Locally common but little known.

Ed Robinson   **Palau, Western Pacific**

Helmut Corneli   Mirihi, Maldives, Indian Ocean

Revillagigedo Islands, Mexico, Eastern Pacific

both photos Douglas Seifert   Komodo Island, Indonesia

Manta birostris
Manta ray

Width: At birth about 125, max. to 670, questionably reported to 900+, but commonly to 400 cm.

Distribution: Circumglobal in warm temperate and tropical waters of all oceans.

Depth: Surface to 30 m.

General: Weighing up to at least 1,500 kg, the magnificent manta ray is one of the largest of all living elasmobranchs and the largest of all rays, rivalled only by sawfishes. It is a strong pelagic swimmer, possibly able to cross the open ocean and present at many oceanic islands, but also reported to rest on the bottom. However, mantas are usually observed in the vicinity of reefs, often in small groups when in pursuit of plankton or small to medium-sized schooling fishes. The two flexible, broad, fleshy lobes (cephalic fins) on the head are unrolled and held at a downward angle to create a funnel guiding prey into the enormous terminal mouth. Feeding often occurs near or at the surface where plankton has accumulated. Appears at seasonal plankton blooms (e.g. autumn coral spawning at Ningaloo Reef, west Australia). During feeding they repeatedly somersault in the water and also break the surface. Jumping clear of the water, however, seems to be associated rather with social behaviour or ridding of parasites. As shown here, people use the placid nature of these giants to hitch a ride through open water.

The unusual centre photo shows the eyes of a manta, a remora and an angelfish. Remoras are members of the bony fish family Echeneidae (also known as shark suckers) which have their first dorsal fin modified into a sucking disc and are often seen clinging to mantas. While the remoras do not harm their host directly (apart from sore skin where attached), the manta certainly has to spend more energy in swimming. Divers have removed remoras from free-swimming mantas by force (killing them with a knife) and reported on their impression that the giant rays welcomed this service as they kept close to their helpers.

Manta birostris   continued

Manta rays have up to 300 rows of small teeth only on the lower jaw but no stinging spine and are generally harmless. Their black and white body patterns show individual variation and may - like in many dolphins - help identify individuals. The gentle giants often allow close approach but should better not be touched. Mantas appear at cleaning stations where cleaner wrasses remove small crustacean parasites from skin and gill cavities. There they can be photographed or watched at ease.

Jack McKenney   Sea of Cortez, Mexico

While only very large sharks are natural enemies of manta rays (evidenced by bite marks), man has long since harpooned them for their tasty meat, sandpaper skin and oil-rich liver. Today, many of these playful giants die a slow and often useless death in drift or set nets. In the 1990s, targeted fisheries have begun to decimate resident populations off the Pacific coast of Mexico and at the Philippines. With a probably long gestation period and only one young per litter these slowly growing rays urgently need to be protected!

Helmut Debelius   Mirihi, Maldives, Indian Ocean

EYE TO EYE WITH CHIMAERAS

An icy wind blows down from the glacier. Temperatures have dropped to minus and a thin layer of ice has developed on the surface. A snow blizzard is blowing over the mountains. Some of the deep trenches of the fjord landscape act like a wind canal. Even so - this setting can be the key to some of the most exciting diving and may be worth the risk of jumping into the water at night to explore the depth of a Norwegian fjord. Marine biologist and underwater photographer Florian Graner will take you with him on a very special dive.

A massive spine is present at the front edge of the chimaera's *(Chimaera monstrosa)* dorsal fin which can be raised together with the fin in defence. Normally, chimaeras propel themselves forward by 'waving' their large pectoral fins up and down.

Sognefjord is a fjord of the superlatives: It is the longest fjord without ice and the deepest fjord in the world. It stretches over 200 km inland and reaches its maximum depth at 1,308 m, maintaining a depth of more than 1,000 m for over 150 km. This is truly the 'Grand Canyon' of Europe, located at the foot of Scandinavia's tallest mountains and carved out by Europe's largest glacier, the Jostedals-breen. The surrounding mountains, Scandinavia's highest, the Jontun-heimen, tower up to 1,800 m above the fjord. The vertical profile continues below the surface of the sea.

It is now late in November. At this time of the year, the water tempera-

290

Despite living in total darkness, the chimaera has an amazingly complex colour pattern. Additionally, its skin has a silvery sheen.

ture is 1-4°C at the surface, but stays at 7-8°C below a permanent thermocline at about 30 m depth during winter and summer. It is therefore rather pleasant down in the water compared to the outside world. You don't have to venture far from shore, just a few metres, to find the vertical drop-off typical for a fjord. Its walls are indeed vertical, often forming overhangs, and may drop several hundred meters. Don't try to find the bottom!

Even in the shallows, the fjord is teaming with life. Reef-like structures of calcareous red algae are home to countless worms, sea stars, sea urchins and all sorts of animal larvae. On the way down, I pass walls covered with large *Modiolus* mussels and enormous colonies of the tunicate *Ciona intestinalis*. These tube-like sea squirts measure up to 30 cm and cover large areas which makes the fjord walls look a bit eerie. All over the place, massive sea urchins and sea cucumbers feed on the substrate. A variety of brightly coloured sea stars prey on

the mussels. Stone crabs (*Lithodes maja*) walk among the colonies of sea squirts and dinner plate-sized pink anemones (*Bolocera tueidae*). In fact, these large crabs often seek shelter underneath the stinging tentacles of the majestic anemones. This appears to be a kind of symbiosis (commensalism) in which the crab is protected by the anemones while the anemones may feed on the detritus left by feeding crabs. Stone crabs look like smaller cousins of the largest crab on earth - the Kamtchatka crab. The giant female stone crab can span up to one metre across its legs and carries the smaller male with her during copulation. The crabs eat mussels, various other shelled molluscs, but also feed on detritus. They seem to be rather non-selective

The corals growing in the deep water of Norwegian fjords are so surprisingly colourful and diverse that one is instantly reminded of the variety otherwise found only in shallow tropical reefs.

Thanks to their large eyes, chimaeras are able to see in the dark. Hence, some popular names translate as 'catfish'.

feeders. Often the crabs are missing legs, and in many places one can see the scattered remains of crabs that have been eaten by predators. They may be a preferred prey item for some of the deep water sharks which live in the deeper waters of the fjord.

In some of the side branches of the vast fjord, at a depth of around 50 m, there is a rare sight to be seen on a ledge covered with thick mud. Among a forest of sea pens *(Funiculina quadrangularis* and *Pennatula phosphorea)*, I discovered a few colonies of the star coral *(Kophobelemnon stelliferum)*. This soft coral is known to occur down to a depth of 3,500 m and truly qualifies as a deepwater coral. Its polyps are among the largest of any coral species and originate directly at the main stem. Also many specimens of the mud-burrowing Norwegian lobster *(Nephrops norvegicus)* live there. At night, they walk in the open over the substrate while they usually hide in their burrows during the day.

Suddenly, just as I drop below 50 m, two large eyes reflect the light of my uw-lamp, casting it back into the dark water around me, creating a ghostly green beam of light. That is all you see at first - alien eyes! As I 'sail' further down, switching my re-breather to the trimix tank, a chimaera - also called ratfish - slowly glides past me. While the ratfish solemnly moves through the water, its large pectorals gently wave up and down. Then, irritated by the light and my presence, it raises its tall first dorsal fin that is armed with a long poisonous spine and finally propels itself forward with swift sinuous movements of its long tapering tail. Now, the pectoral fins are spread like wings and the tail generates a rapid yet short burst of speed. The chimaera's body has a maze-like colour pattern of different reddish-brown hues with an additional silvery sheen brilliantly reflecting the light.

Chimaeras are distant relatives of sharks and rays. Like these, they too have a cartilaginous skeleton. They look somewhat like a mixture between shark and ray and are also appropriately called ghost sharks. Like many rays and certain sharks, they lay eggs enclosed in protective capsules. Only once have I seen a female who dragged two egg capsules through the water which were still attached to her abdomen. The capsules were long, slender and spindle-shaped.

Thanks to my noiseless diving equipment, I can approach the chimaera closely to find myself eye to eye with this living fossil. Its lateral line organ extends and branches over the entire surface of the head. Patches of sensory pores (the ampullae of Lorenzini) which are able to detect electric fields cover the nose of this strange creature. Its large eyes reflect greenish light and are excellently adapted to see even in the residual light at great depths. Each moves independently from the other in its socket. Soon, the chimaera escapes the torture of my bright uw-lights and 'jets' vertically down the wall of the fjord which means 'good bye' for now. I am in a depth of 65 m and it is time for my long ascent. The decompression stops are a good opportunity to quietly reflect on the previous encounter - it seems as if I had just met a marine dinosaur, an alien with very strange looks and behaviour. It is hard to imagine how the regular life of this animal must be like...

As soon as I reach the surface, the peaceful quietness is interrupted by a roaring snow storm. I plow my way through the deep powdery snow on the shore. My thoughts are still down in the depth of the fjord and the grim weather doesn't bother me at all.

The order Chimaeriformes comprises three extant families and several extinct groups (their fossils are known from the Devonian to the Jurassic) of bizarre fishes with cartilaginous skeletons and sometimes unique dentitions ("revolver dentition" with teeth arranged in spirals). The numerous names given to the living chimaeras demonstrate that they are strange-looking fishes. Those names relate to anatomical peculiarities or their general appearance: Ratfish (pointed snout, long tail), chimaera (a hybrid creature in ancient Greek mythology), ghost shark (pale coloration), elephantfish (form of snout). The about 43 or more living species are all marine, oviparous, range in total length from 40 to 200 cm at maturity and live in depths of a few to at least 3,000 m. They are placed in three families: Shortnose chimaeras (Chimaeridae) have a short snout, a long tapering symmetrical caudal fin and a long 2nd dorsal fin *(Chimaera* and *Hydrolagus* spp. inhabit shallow coastal to deepsea waters). All of the anatomically similar longnose chimaeras (Rhinochimaeridae) have a long, pointed snout *(Harriotta, Neoharriotta* and *Rhinochimaera,* worldwide, exclusively in the in deep sea); *R. atlantica* for example, is found in the northern Atlantic. For the 3rd family (elephantfishes, Callorhinchidae) see below.

ELEPHANTFISHES   CALLORHINCHIDAE

Callorhinchus milii
Australian elephantfish
L: At birth 15, max. 120, mature males 65 cm. Di: Australia (Western Australia to New South Wales including Tasmania) and all of New Zealand. De: Shallow to at least 200 m. G: Unmistakable, silvery with dark markings. Enters estuaries and bays in spring to breed. Egg cases elongate, golden-brown, with two broad flanges; laid in pairs (small photo below) on sand or mud, young hatch after up to 8 months and initially vanish into deep waters (not seen by divers in the shallows like many young sharks and rays). The photographer has successfully bred this sp.
    Elephantfishes or ploughnose chimaeras have asymmetrical tails like sharks, a short second dorsal fin and a plough-shaped protrusion at the tip of the snout (photos). The species of the single genus *Callorhinchus* inhabit coastal to deepsea waters of the southern hemisphere. Biology of most (deepsea) family members still little known. Only recently shallow water species are kept successfully in public aquaria. Good photos like shown here are still rare.

all photos Rudie Kuiter   **Victoria, Australia**

John Neuschwander   South Africa

Callorhinchus capensis
South African elephantfish, St. Joseph

L: Max. 120 cm. Di: Eastern Atlantic: Namibia and South Africa (to Natal, Indian Ocean). De: 1-200 m.
G: A bizarre, unmistakable fish with a hoe-like projection of the snout. Similar in looks and habits to its Australian cousin (see previous page). Common in inshore waters, often caught and marketed as "silver trumpeter". Flesh excellent, especially when marinated, with no scales and bones to worry about. Preys on sea urchins, shelled molluscs, crustaceans and small fish (dragonets).

Florian Graner   Norway

Chimaera monstrosa
Ghost shark, rabbitfish
L: Max. about 100 cm. Di: Eastern Atlantic: Norway and Iceland to Morocco and Mediterranean. De: 40-1,000, usually to 500 m. G: Benthic, sluggish, in small groups. Makes summer inshore migrations, deeper in south. Egg with narrow flanges. See also the sensational EYE TO EYE WITH CHIMAERAS on pp.290-292.
    All chimaeras have a strong, erectile, poisonous spine at the front margin of the first dorsal fin that fits into a groove when folded back.

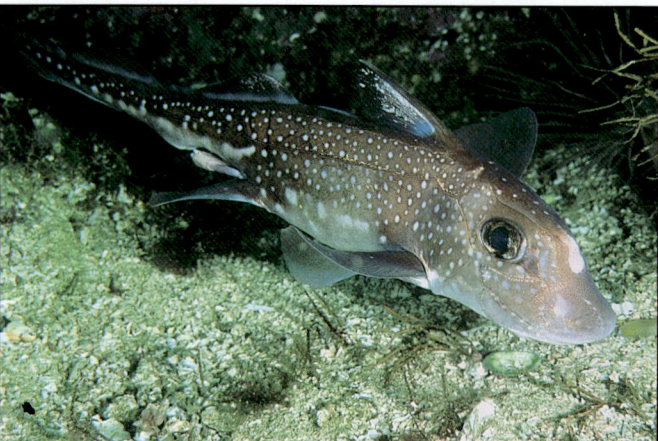

James D. Watt/IV   British Columbia, Canada

Hydrolagus colliei
Spotted ratfish
L: Max. 90-100 cm. Di: Eastern Pacific: Alaska to Baja California. De: 90-900 m. G: Brownish with light spots, sides silvery, belly white, fins blackish. Common, often in large groups near the bottom.
    The shelled invertebrate prey of chimaeras is grabbed with paired incisor-like frontal teeth (ratfish!) and crushed between strong dental plates. Males have claspers like sharks and rays but also 3 additional organs (tenacula) in front of the paired pelvic fins and on top of the head to hold onto the female during copulation.

FLYING WHITES

Great white sharks today are already rare animals and seriously endangered by people despite being protected by law in at least a few countries. Only very few privileged people have so far had the opportunity to watch this most popular - and unfortunately likewise most misjudged - of all predatory fishes, while hunting in an (even for this superpredator) most unusual manner: When making an attack run on seals, white sharks are able to literally shoot out of the sea with such an incredible impetus that they come completely free of the water together with their prey between their teeth. Underwater photographer Ralf Kiefner was not willing to let this spectacle of breaching whites pass him by and reports from Robben Island off South Africa's south coast.

Hungry white sharks are frequently curious enough to put their head out of the water to inspect a boat or the like.

For more than two hours now I am sitting on the bottom of a rolling boat, trying to make myself as comfortable as possible, permanently looking through the viewfinder, with my finger on the trigger, ready to shoot immediately and

at any time. After a while, the camera together with its lens seems to get heavier and heavier, and my arms and shoulders start aching. Questions like "Is this really worth all the effort?" come to my mind. "Will there be a white shark that is going to breach, and if so, will I then be able to capture it on film?"

Great whites are the only sharks known to breach and 'jump' completely out of the water while hunting. This spectacular hunting technique is only very rarely observed and even

Adult great white sharks simply love fat marine mammals for dinner. Around Robben (Seal) Island off South Africa's south coast the Cape fur seal (*Arctocephalus pusillus*) is one.

295

still unknown to many so-called 'shark specialists' and scientists alike. When great whites try to catch highly manoeuvrable prey such as seals, these animals desperately try to escape by changing the direction of their flight while jumping out of the water. Sea lions, harbour seals and similar pinnipeds are the most favourite prey of the great white shark. When 'Mr. Teeth' is searching for prey, 'he' patrols close to the edge of the reef. Upon finding suitable prey, the shark tries to stay unnoticed and stalks it by swimming close to the seafloor. The grey to black back, especially that of the

A great white shark as seen in its regular environment.

darker coloured younger white sharks, can hardly be distinguished from the ground when viewed from above. If the level of the seafloor lies too deep, it follows its prey at a distance, just like the proverbial 'shadow in the deep'. The slightest inattentiveness on behalf of the seal causes the fatal attack by the shark. With just a few dynamic strokes of its stiff caudal fin the shark speeds up to a maximum of more than 60 kilometres per hour. When a white shark catches a seal in such a surprise attack, its helpless prey will usually become lethally injured by just one (the first) quick and powerful bite. After a great

Above: White sharks can be attracted by bait towed behind a boat and sometimes come at least partly out of the water to inspect the offer.
Right: The elegant hunters at the top of the oceanic food chain are not at all the ferocious menace of the seas as they are popularly looked upon.

Following pages: However, their unrivalled and extremely impressive 'aerial' hunting technique with leaps and somersaults has only recently been documented on film. The unusual photo series on the following pages shows a 4-m-long female and the target, a simple seal dummy. The shark stays in the air for less than a second!

white has attacked larger mammalian prey such as seals, sea lions, sea elephants or even walrus with one first bite, it patiently waits until its hapless meal has bled to death or drowned, before it returns to start eating. Such shark behaviour minimises the risk of injury or loss of energy when fighting with prey.

In 1980 Peter Klimley, an experienced senior US shark researcher at the University

A large white shark penetrates the smooth surface of the sea suddenly and without any warning. Attracted by its silhouette, the shark grabs the simple seal dummy from below and mangles it between its jaws.

of California, witnessed the first attack of a great white shark on an elephant seal of about 200 kg. In this case the shark completely left the water with the massive elephant seal in its jaws. "It was crazy," he recalls. "After ambushing the elephant seal, the shark attacked its victim several times and removed three or four big pieces from it." This spectacular hunting technique has been observed only very few times and is almost impossible to photograph.

It is winter in South Africa. Robben (Seal) Island swarms with those funny-looking, quick and manoeuvrable animals, it has been named after. The white sharks now present in the area seem to know exactly that the seals will soon give birth to their babies. And the 'silent death' is patiently waiting for its chance to hunt, whenever a

This great white shoots out of the water like a rocket. Such behaviour is not exceptional, but part of a regular hunting technique which only recently has been investigated more thoroughly. All white sharks that catch pinnipeds can do that. Also mako, thresher and some requiem sharks are known for their spectacular leaps into the air, which, however, serve in getting rid of parasites.

careless baby seal strays too far from the safety of the island. But neither did we want to be responsible for the death of a seal, nor did we intend to follow a careless baby seal with our camera, seeking sensations like paparazzi! So we decided to try a trick.

"The seals are so fast and manoeuvrable that a shark hunting them does not have much time for a second, closer look on its prey before it attacks," André explains while lowering a simple, board-like seal dummy into the water. "Probably we are lucky! Maybe the whites will attack this dummy before taking a closer look, if conditions are good for us."

As soon as our seal dummy is floating on the sea surface at some distance from the boat to which it is tied, I focus the camera and put my finger on the trigger. Again, I try to make myself as comfortable as possible in this posture. Although I am mentally prepared to wait for an extended period of time, I am also totally strained after crouching in the boat for an hour. "Will we really be able to fool 'Mr. Teeth'? Will 'he' really fall for our dummy and give us a demonstration of

Great white with seal dummy shortly before 're-entry'. The shark does not really 'jump' clear of the water, but attacks its prey with such overwhelming speed (more than 60 km/h), approaching at a steep angle from below, that it is completely carried out of the water by its own momentum.

a spectacular attack?" I hardly dare to hope for it and continue to cheer myself up: "Don't put the camera down, not now, wait just a little bit longer, it is not really heavy, come on, just a little bit longer."

For more than two hours now I am constantly focusing the seal dummy - in vain. "Probably it was not such a good idea after all," I'm telling myself while the sun is setting and the light level rapidly approaches minimum photographing threshold. Turning to André, I say "It's enough for today," and just want to put down my camera and relax. All of a sudden the unexpected, the incredible happens! Without the slightest warning, just like coming out of nowhere, a pair of wide open jaws breaks through the calm surface of the sea. I can clearly see the dagger-like teeth through the viewfinder of my camera. They seem to be very near, almost too close. I even get afraid and start twitching a little bit before I press the trigger and the camera rapidly clicks away. Still looking through the viewfinder, I witness how our seal dummy gets torn into pieces by the shark while it is thrown high up into the air. The dummy is forced out of the water with the most impressive animal power I ever have seen. This white shark is a female of about four metres in length that makes her surprise attack with such an irresistible power, that she herself is also 'flying' more than four metres high through the air, making a spectacular somersault before diving back into the water 'head over fins'. In less than one second the drama is over. The remnants of our seal dummy float on the water not far behind the boat. The shark's teeth have torn it completely into pieces. For several seconds, André and I stare at each other, left speechless after what we just saw, before we start shouting out loud in joy. Neither of us had seen such dramatic and breath-taking breaching behaviour of a shark before!

Suddenly all the strain of the seemingly endless waiting falls off me. What we just had witnessed was so indescribable, so incredibly dramatic and deeply impressive that I will never ever forget the view of these instantly appearing rows of razor-sharp teeth, coming out of nowhere, and also the unbelievable power of the breaching white shark. Until today, I am reviewing this scene in my mind over and over again.

INDEX: SCIENTIFIC NAMES

BIBLIOGRAPHY

BOOKS AND ARTICLES USED FOR WRITING/CITED IN THE TEXT:

Castro, J.I. (1983) The sharks of North American waters. 180 pp. Texas A&M University Press, College Station, USA.

Clark, E. & D. Doubilet 1975 Into the lairs of "sleeping" sharks. National Geographic 147(4/April):570-584. Washington, D.C., USA.

Clark, E. & D. Doubilet 1981 Sharks, magnificent and misunderstood. National Geographic 160(2/August): 138-187. Washington, D.C., USA.

Clark, E. 1974 The Red Sea's sharkproof fish. National Geographic, November 1974. Washington, D.C., USA.

Clark, E., E. Kristof & D. Lee 1986 Sharks at 2,000 feet. National Geographic 170(5/November): 680-691. Washington, D.C., USA.

Compagno, L.J.V. 1984 FAO Species Catalogue Vol.4 Sharks of the world. Part 1. Hexanchiformes to Lamniformes. pp. 1-249. FAO, Rome, Italy.

Compagno, L.J.V. 1984 FAO Species Catalogue Vol.4 Sharks of the world. Part 2. Carcharhiniformes. pp. 251-655. FAO, Rome, Italy.

Compagno, L.J.V., D.A. Ebert & M.J.Smale 1989 Guide to the sharks and rays of southern Africa. 158 pp. New Holland (Publishers) Ltd., London, England. Also: Struik Publishers, Cape Town, South Africa.

Coupe, S. & R. 1990 Sharks. Weldon Owen Pty Ltd., Sydney, Australia.

Cousteau, J.-M. & M. Richards 1992 Cousteau's great white shark. Harry N. Abrams, Inc., Publishers, New York, NY, USA.

Darom, D. & A. Baranes 1980 The shark. 119 pp. Masada Ltd., Israel. (in Hebrew).

Debelius, H. 1997 Mediterranean and Atlantic Fish Guide. IKAN-UW-ARCHIV, Frankfurt, Germany.

Debelius, H. 1998 Red Sea Reef Guide. IKAN-UW-ARCHIV, Frankfurt, Germany.

Debelius, H. 1999 Indian Ocean Reef Guide. IKAN-UW-ARCHIV, Frankfurt, Germany.

Diverse authors 1998 (2. ed.) Sharks - silent hunters of the deep. Reader's Digest Australia.

Ellis, R. & J.E. McCosker 1991 (2. ed. 1995) Great White Shark. Harper Collins Publishers, London, England & Stanford University Press, USA.

Ellis, R. 1983 The book of sharks. 256 pp. Robert Hale Ltd., London, England.

FAO 1994 Overview of world elasmobranch fisheries. FAO Fisheries technical paper 341. 119 pp. FAO, Rome.

Ferguson, A. & G. Cailliet 1990 Sharks and rays of the Pacific coast. 64 pp. Monterey Bay Aquarium, Monterey, California, USA.

Garman, S. 1913 The Plagiostoma (sharks, skates and rays). Reprint 1997. Benthic Press, Los Angeles, USA.

Gilbert, P.W. 1963 Sharks and survival. D.C. Heath and Co., Boston, Massachusetts, USA.

Grosvenor, M.B. (ed.) 1965 Wondrous world of fishes. 368 pp. National Geographic Society, Washington D.C.

Gruber, S.H. (ed.) 1991 Discovering sharks. American Littoral Society, Highlands, New Jersey, USA.

Healy, C.J. & Caira, J.N. 1998 Identifying cryptic carcharhiniform shark species: tapeworms as diagnostic tools. American Elasmobranch Society 1998 Annual Meeting, Abstracts.

Herald, E.S. 1960 Living Fishes of the World.

Klimley, A.P. & D.G. Ainley 1996 Great white sharks. The biology of *Carcharodon carcharias*. 517 pp. Academic Press, San Diego, California, USA.

Last, P.R. & J.D. Stevens 1994 Sharks and rays of Australia. 513 pp., 84 Tables. CSIRO, Australia.

Long, J.A. 1995 The rise of fishes. 500 million years of evolution. UNSW (University of New South Wales) Press, Sydney, Australia.

McLeish, W.H. 1981 Sharks. Woods Hole Oceanographic Institution, Massachusetts, USA.

Michael, S.W. 1993 Reef sharks & rays of the world. 107 pp. Sea Challengers, Monterey, California, USA.

Moss, S.A. 1984 Sharks. An introduction for the amateur naturalist. 246 pp. Prentice-Hall, New Jersey, USA.

Nelson, D.R. et al. 1991 An acoustic tracking of a megamouth shark, *Megachasma pelagios*. American Elasmobranch Society. 1991 Annual Meeting, Abstracts.

Ross, R.A. & F. Schäfer 2000 Freshwater rays. 192 pp. Aqualog/Verlag A.C.S., Mörfelden-Walldorf, Germany.

Shirai, S. 1986 Ecological encyclopedia of the marine animals of the Indo-Pacific. Vol. 1 (Vertebrata). 352 pp. Shin Nippon Kyoiku Tosho Co., Ltd., Tokio, Japan.

Springer, V.G. & J.P. Gold 1989 Sharks in question: the Smithsonian answer book. 187 pp. Smithsonian Institution Press, Washington D.C., London.

Stafford-Deitsch, J. 1987 Shark - a photographer's story. 200 pp. Headline Book Publishing PLC, London.

Taylor, G. 1994 Whale sharks - the giants of Ningaloo Reef. 176 pp. Angus & Robertson, Sydney, Australia.

Tinker, S.W. & C.J. DeLuca 1976 (1. ed. 1973) Sharks & rays. A handbook of the sharks and rays of Hawaii and the Central Pacific Ocean. 80 pp. Charles E. Tuttle Company, Rutland, Vermont, USA & Tokio, Japan.

Whitehead, P.J.P., M.-L. Bauchot, J.-C. Hureau, J. Nielsen & E. Tortonese (eds.) 1984 Fishes of the North-eastern Atlantic and the Mediterranean. Vol. I. 510 pp. UNESCO, Paris, France.

Yano, K., J. Morrissey, K. Yabumoto & K. Nakaya 1997 Biology of the Megamouth Shark. Tokai University Press, Japan.

CITED WEB SITES (SEE ALSO INTRODUCTION)

http://www.flmnh.ufl.edu/fish/Sharks/InNews
http://www.westpacfisheries.net/actionalert.html
http://www.envirowatch.org